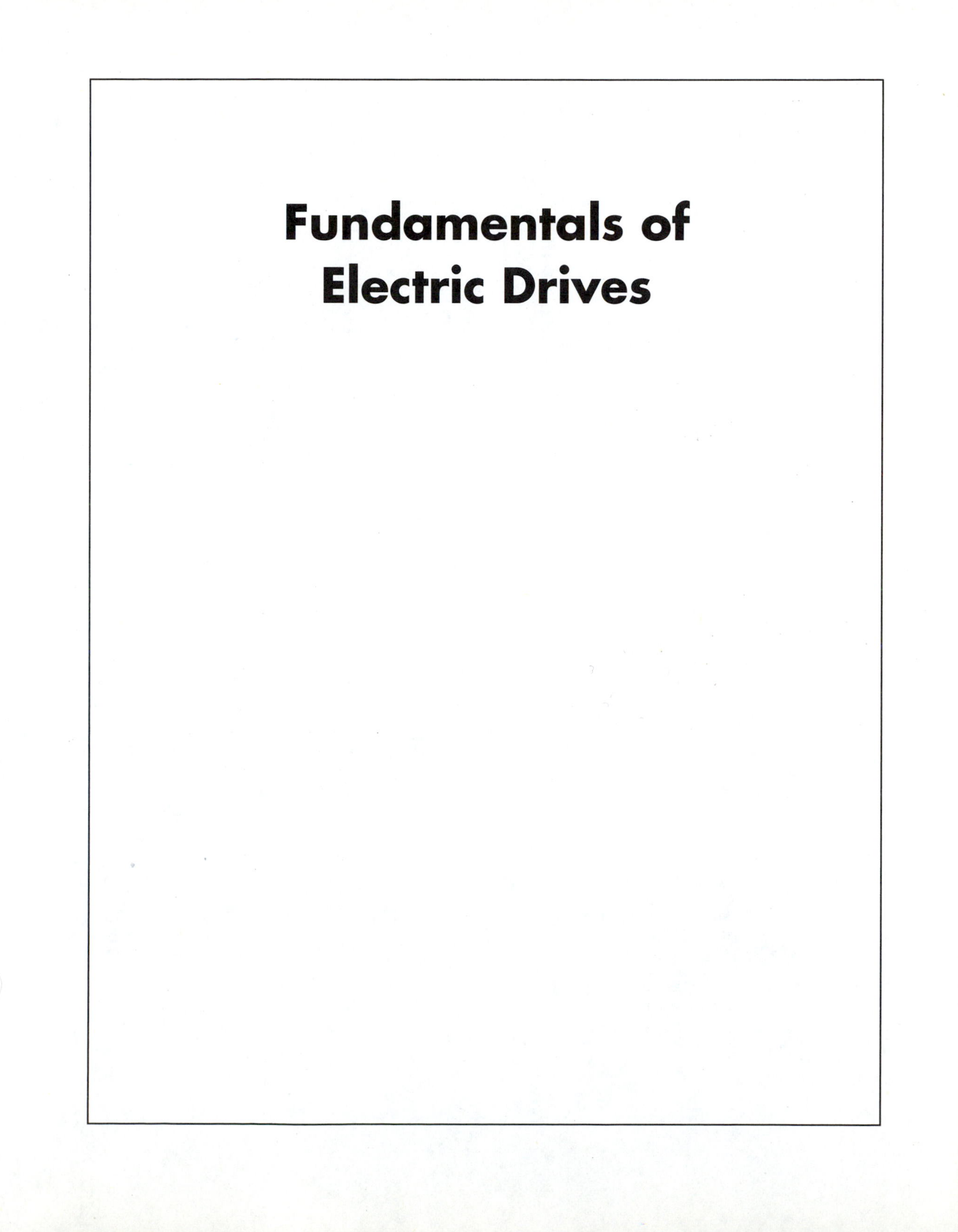

Fundamentals of Electric Drives

Fundamentals of Electric Drives

Mohamed A. El-Sharkawi
University of Washington

Australia • Canada • Mexico • Singapore • Spain • United Kingdom • United States

Sponsoring Editor: *Heather Shelstad*
Marketing: *Nathan Wilbur, Samantha Cabaluna, Christina De Veto*
Editorial Assistant: *Shelley Gesicki*
Production Coordinator: *Laurel Jackson*
Production Service: *Susan Graham*
Manuscript Editor: *Helen Walden*
Permissions Editor: *Mary Kay Hancharick*
Interior Design: *Quinn Design*
Cover Design Coordinator: *Roy Neuhaus*
Cover Design: *Laurie Albrecht*
Cover Photo: *PhotoDisc*
Interior Illustration: *Quinn Design*
Print Buyer: *Vena Dyer*
Typesetting: *Carlisle Communications*
Printing and Binding: *R.R. Donnelley & Sons, Crawfordsville Mfg. Division*

For more information about this or any other Brooks/Cole products, contact:
BROOKS/COLE
511 Forest Lodge Road
Pacific Grove, CA 93950 USA
www.brookscole.com
1-800-423-0563 (Thomson Learning Academic Resource Center)

Printed in the United States of America

10 9 8 7 6 5 4 3 2 1

Library of Congress Cataloging-in-Publication Data

El-Sharkawi, Mohamed A.
Fundamentals of electric drives / Mohamed A. El-Sharkawi.
p. cm.
Includes index.
ISBN 0-534-95222-4
1. Electric driving. I. Title
TK4058 .E39 2000
621.46--dc21 00-022934

Dedicated to my family and my students

Preface

A modern electric drive system consists of a motor, an electric converter, and a controller that are integrated to perform a mechanical maneuver for a given load. Because the torque/volume ratio of modern electric drive systems is continually increasing, hydraulic drives are no longer the only option to use for industrial applications. In addition to their use in industrial automation, modern electric drives have other widespread applications, ranging from robots to automobiles to aircraft. Recent advances in the design of electric drives have resulted in low-cost, lightweight, reliable motors; advances in power electronics have resulted in a level of performance that was not possible a few years ago. For example, induction motors were never used in variable-speed applications until variable frequency and rapid switching were developed. Due to advances in power electronics, several new designs of electric motors are now available.

Modern electric drive systems are used increasingly in such high-performance applications as robotics, guided manipulations, and supervised actuation. In these applications, controlling the rotor speed is only one of several goals; the full range includes controlling the starting, speed, braking, and holding of the electric drive systems. The exploration of these control functions forms the core of this text.

This book is designed to be used as a teaching text for a one-semester course on the fundamentals of electric drives. Readers are expected to be familiar with the basic circuit theories and the fundamentals of electronics, as well as with three-phase analysis and basic electric machinery.

In this book, I cover the basic components of electric drive systems, including mechanical loads, motors, power electronics, converters, and gears and belts. Each component is first discussed separately; then various components are combined in a discussion of the complete drive system. If instructors use this book in the first course on the subject, they will not need to seek and use additional material because this book is self-contained.

The focus of this book is on the fundamentals of electric drive systems. The general types of electric loads and their dependence on speed are explained early in the book, and load characteristics are considered throughout. In addition, I explain and analyze industrial motors from the drive perspective. To help the reader understand why a particular motor is selected for a particular application, I present and highlight the differences and similarities of electric motors.

Power converters are discussed in some detail, with ample mathematical analysis. Early in the book, I present several solid-state switching devices and specific characteristics of each; this comparison of solid-state devices allows the reader to understand their features, characteristics, and limitations.

Converters are divided into several groups: ac/dc, ac/ac, dc/ac, and dc/dc. Several circuits are given for these converters and are analyzed in detail to help readers understand their performance. Detailed analyses of the electrical waveforms of power-converter circuits demonstrate the concepts of power and torque in a harmonic environment.

After readers become familiar with electric machines and power converters, they can comprehend the integration of these two major components that creates an electric drive system. This book includes detailed explanations of the various methods for speed control and braking. Well-known applications appear throughout the book in order to demonstrate the theories and techniques. Discussions of the merits, complexities, and drawbacks of the various drive techniques help readers form opinions from the perspective of a design engineer.

Finally, in each chapter, examples and problems simulate several aspects of drive performance. The problems are designed to address key design and performance issues and are therefore more than mere mathematical exercises.

An Instructor's Solutions Manual (0-534-37167-1) to accompany this book is available from the publisher.

ACKNOWLEDGMENTS

I want to thank the following reviewers for their valuable comments and suggestions: Bimal K. Bose, The University of Tennessee, Knoxville; E. R. (Randy) Collins, Clemson University; Thomas G. Habetler, Georgia Institute of Technology; Martin E. Kaliski, California Polytechic, San Luis Obispo; and Ali Keyhani, Ohio State University.

I would also like to acknowledge my students for their input on this text and, especially, for their critical eyes and minds.

Mohamed A. El-Sharkawi

Contents

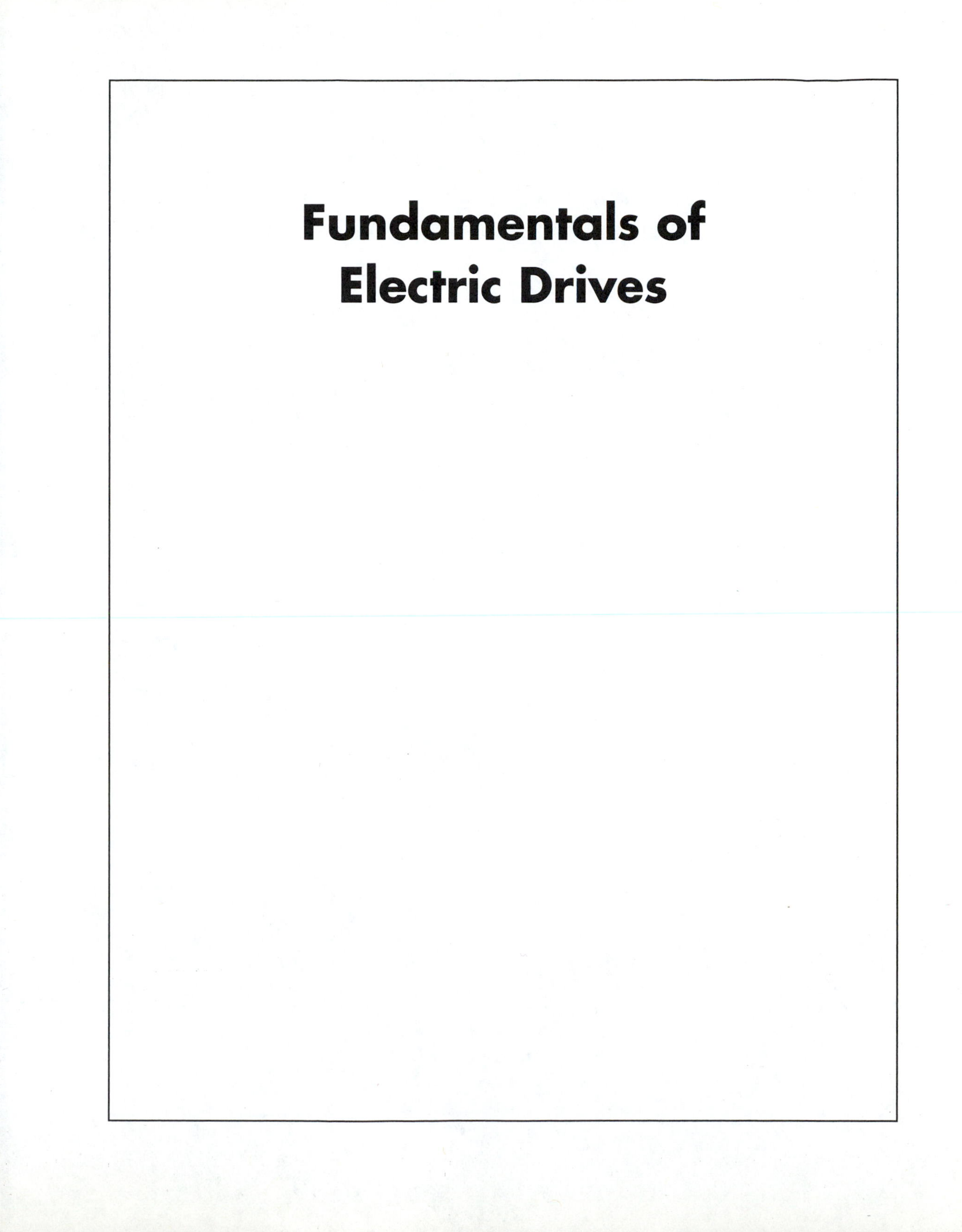

Fundamentals of Electric Drives

1

Elements of Electric Drive Systems

The study of electric drive systems involves controlling electric motors in the steady state and in dynamic operations, taking into account the characteristics of mechanical loads and the behaviors of power electronic converters.

In the not-so-distant past, designing a versatile drive system with broad performance was a difficult task that required bulky, inefficient, and expensive equipment. The speed of an electric motor was controlled by such restrictive methods as resistance insertion, use of autotransformers, or complex multimachine systems. Motor selection for a given application was limited to the available type of power source. For instance, dc motors were used with direct current sources, and induction motors were driven by ac sources.

To alleviate the problem of matching up the motor and the power source and to provide some form of speed control, a common and elaborate scheme such as that shown in Figure 1.1 was commonly used. Because its terminal voltage is relatively easy to adjust, the dc motor was regularly selected for applications requiring speed control. Given the status of the available technology, controlling the speeds of alternating-current machines was much more difficult.

The system in Figure 1.1 consists of three electromechanical machines: an ac motor, a dc generator, and a dc motor. The ac motor (induction), which drives a dc generator, is powered by a single- or multiphase ac source. The speed of the induction

FIGURE 1.1
Multimachine system for speed control

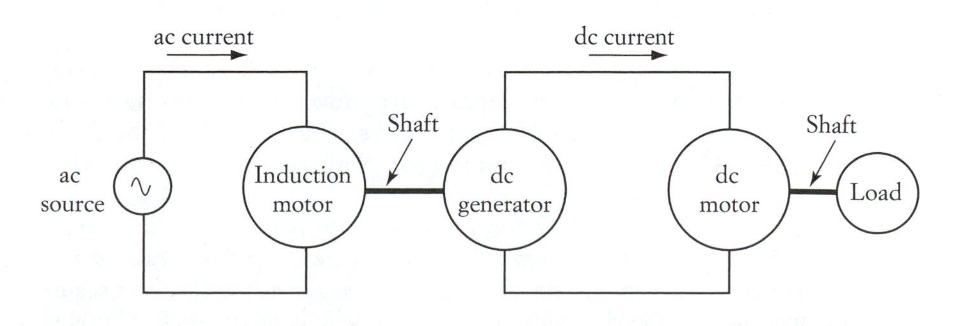

motor is fairly constant. The output of the dc generator is fed to the dc motor. The output voltage of the dc generator is adjusted by controlling its excitation current. Adjusting the field current of the generator controls the terminal voltage of the dc motor. Hence, the speed of the dc motor is controlled accordingly.

The system described here is expensive, inefficient, and complex, and it requires frequent maintenance. However, because of the limited technology available during the first half of the 20th century, this system was the leading option for speed control. In fact, a number of these systems are still in service; for example, old elevators may still use this system today.

During the past few years, however, enormous strides have been made in the areas of power electronics, digital electronics, and microprocessors. With the advances in power electronic devices, cheaper, more efficient, and versatile options for speed control are now available.

Continuous improvements in solid-state technologies are yielding even more reliable and better-performance devices, as well as new types of solid-state switches. Solid-state devices can now handle larger amounts of current and voltage at higher efficiencies and speeds. Additionally, the prices of these devices are continually dropping.

Among the important developments in solid-state power electronics technology is the integrated module. Solid-state switches can now be found in various configurations, such as H-bridge or six-pack modules. Complete driving circuits are now a part of very sophisticated and elegant designs. Most designs now have built-in options for speed control and overcurrent protection. Previously, building such modules took several months.

With the development of power electronic devices and circuits, virtually any type of power source can now be used with any type of electric motor. Speed control can now be achieved by using a single converter. In fact, the older, inefficient drive systems currently being used in some industrial applications are now being replaced with solid-state drives. This retrofitting process is estimated to be a multi-billion-dollar business in the United States alone.

With modern solid-state power technology, motors can be used in more precise applications, such as position control of robots and airplane actuation. Hydraulic and pneumatic systems are now being replaced by electric drives.

1.1 HISTORICAL BACKGROUND

Due to the lack of technology, electric drives historically were designed to provide crude power without consideration of performance. Advances in industrial manufacturing led to a need for more sophisticated drives, which stimulated the development of modern systems. The drive systems have various forms:

1. *Line shaft drives.* This is the oldest form of an electric drive system. (An example is shown in Figure 1.2.) The system consists of a single electric motor that drives equipment through a common line shaft or belt. This system is inflexible because it cannot change the speed of one of the loads alone. It is also inefficient because

the line shaft continuously rotates regardless of the number of pieces of equipment in operation. This system is presented here for historical reasons but is rarely used.

2. *Single-motor, single-load drives.* This is the most common form of electric drive. In this system, a single motor is dedicated to a single load. (Examples are shown in Figure 1.3.) Applications include household equipment and appliances such as electric saws, drills, disk drives, fans, washers, dryers, and blenders. The computer hard disk drive in the figure employs high-performance drives for the rotation of the disk. The head actuation of the hard-disk drive is controlled by a separate system.

FIGURE 1.2
Single-motor, multiple-load drive system

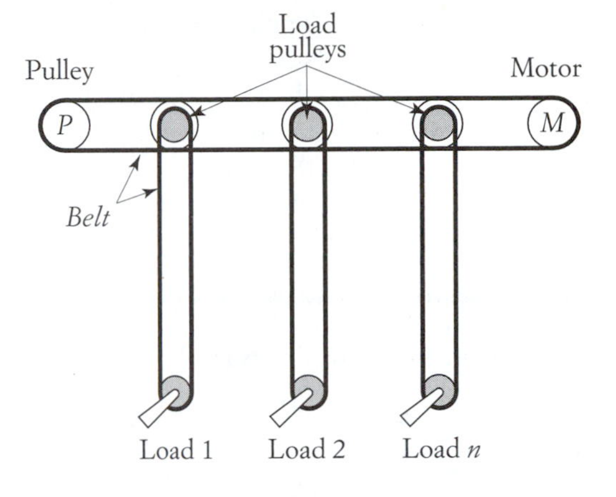

FIGURE 1.3
Single-motor, single-load drive systems

Hard-disk drive

Electric vehicle

Golf cart

Household appliance

The traction of most electric cars is a single-motor drive system. The motor replaces the internal combustion engine of conventional vehicles.

3. *Multimotor drives.* In this type of system, several motors are used to drive a single mechanical load. This form is usually used in complex drive functions such as assembly lines, paper-making machines, and robotics. (Figure 1.4 shows several examples of this type of load.) Airplane actuation is done electrically in most military and several commercial airplane models. Each flap of the airplane is controlled separately by redundant drive systems. Compared to the commonly-used hydraulic actuator, electric actuation (sometimes

FIGURE 1.4
Multiple-motor, single-load drive systems

Mars rover

Airplane actuation system

Robot arm

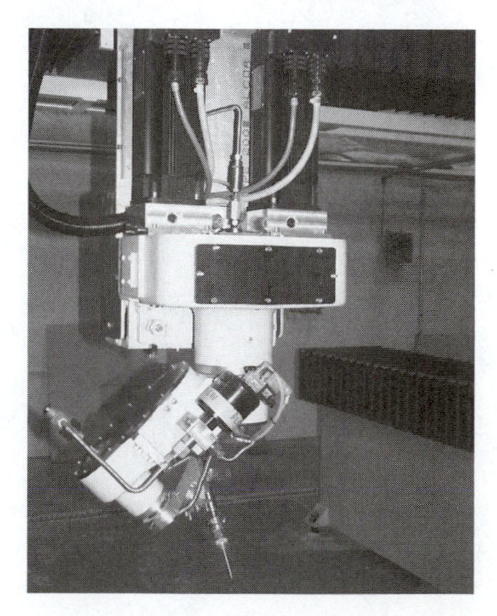

Industrial manipulator

known as fly-by-wire) is much lighter and faster, involves lower maintenance, and does not require the heating of any hydraulic fluid. It is therefore a more popular method in aviation.

The rover used by NASA for Mars exploration is a six-wheeled vehicle of a rocker bogie design to negotiate obstacles. Vehicle navigation is accomplished through control of its drive and steering motors. The energy source of the vehicle is solar—several solar panels capture solar energy and store it in a battery pack.

The industrial manipulator or the robot arm can employ as many motors as the process requires. The manipulator shown in Figure 1.4 is used for waterjet cutting. The end effector of the manipulator or the robot arm must be accurately controlled to achieve the desired precision.

1.2 BASIC COMPONENTS OF AN ELECTRIC DRIVE SYSTEM

A modern electric drive system has five main functional blocks (shown in Figure 1.5): a mechanical load, a motor, a converter, a power source, and a controller. The power source provides the energy the drive system needs. The converter interfaces the motor with the power source and provides the motor with adjustable voltage, current, and/or frequency. The controller supervises the operation of the entire system to enhance overall system performance and stability.

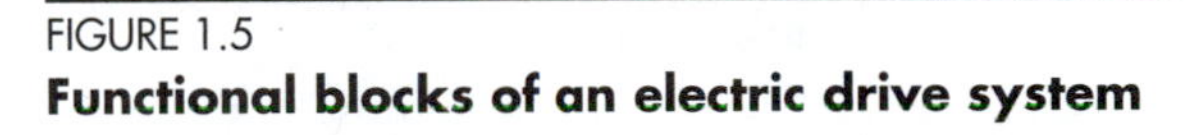

FIGURE 1.5
Functional blocks of an electric drive system

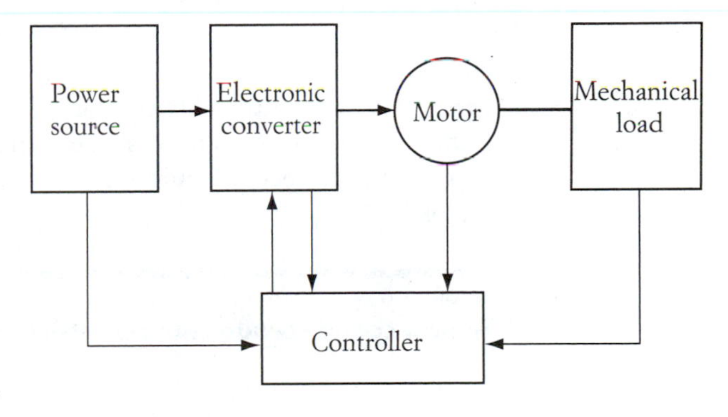

Often, design engineers do not select the mechanical loads or power sources. Rather, the mechanical loads are determined by the nature of the industrial operation, and the power source is determined by what is available at the site. However, designers usually can select the other three components of the drive systems (electric motor, converter, and controller).

The basic criterion in selecting an electric motor for a given drive application is that it meet the power level and performance required by the load during steady-state and dynamic operations. Certain characteristics of the mechanical loads may require a special type of motor. For example, in the applications for which a high starting torque is needed, a dc series motor might be a better choice than an induction motor. In constant-speed applications, synchronous motors might be more suitable than induction or dc motors.

Environmental factors may also determine the motor type. For example, in food processing, chemical industries, and aviation, where the environment must be clean and free from arcs, dc motors cannot be used unless they are encapsulated. This is because of the electric discharge that is generated between the motor's brushes and its commutator segments. In those cases, the squirrel cage induction motor or other brushless machines are probably the better options.

The cost of the electric motor is another important factor. In general, dc motors and newer types of brushless motors are the most expensive machines, whereas squirrel cage induction motors are among the cheapest.

The function of a converter, as its name implies, is to convert the electric waveform of the power source to a waveform that the motor can use. For example, if the power source is an ac type and the motor is a dc machine, the converter transforms the ac waveform to dc. In addition, the converter adjusts the voltage (or current) to desired values. The controller can also be designed to perform a wide range of functions to improve system stability, efficiency, and performance. In addition, it can be used to protect the converter, the motor, or both against excessive current or voltage.

1.2.1 MECHANICAL LOADS

Mechanical loads exhibit wide variations of speed–torque characteristics. Load torques are generally speed dependent and can be represented by an empirical formula such as

$$T = CT_r \left(\frac{n}{n_r} \right)^k \tag{1.1}$$

where C is a proportionality constant, T_r is the load torque at the rated speed n_r, n is the operating speed, and k is an exponential coefficient representing the torque dependency on speed. Figure 1.6 shows typical characteristics of various mechanical loads.

FIGURE 1.6
Typical speed–torque characteristics of mechanical loads

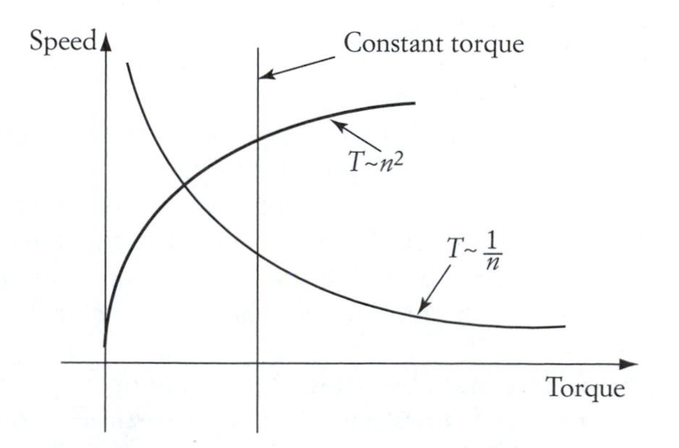

The mechanical power of the load is given by Equation (1.2), where ω is the angular speed in rad/sec, and n is the speed in r/min (or rpm). Figure 1.7 shows the mechanical power characteristics that correspond to the loads shown in Figure 1.6.

$$P = T\omega; \qquad \omega = 2\pi \frac{n}{60} \tag{1.2}$$

Figure 1.8 shows several types of mechanical loads that are commonly used in households. In general, the load characteristics can be grouped into one or more of the following types:

FIGURE 1.7
Typical speed–power characteristics of mechanical loads

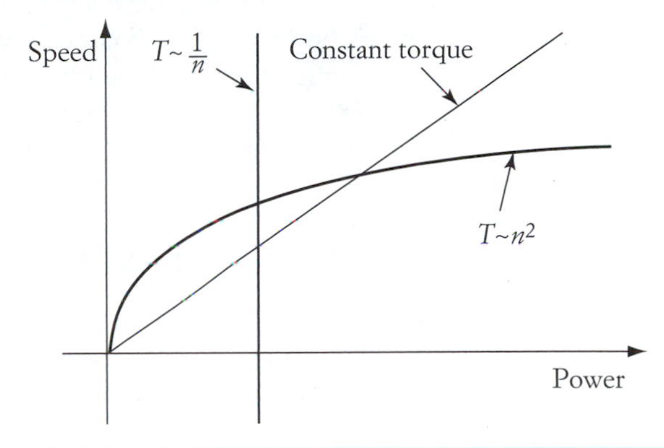

1. *Torque independent of speed.* The characteristics of this type of mechanical load are represented by Equation (1.1) when k is set equal to zero and C equals 1. While torque is independent of speed, the power that the load consumes is linearly dependent on speed. There are many examples of this type of load, such as hoists or the pumping of water or gas against constant pressure.
2. *Torque linearly dependent on speed.* The torque is linearly proportional to speed $k = 1$, and the mechanical power is proportional to the square of the speed. This is an uncommon type of load characteristic and is usually observed in a complex form of load. An example would be a motor driving a dc generator connected to a fixed-resistance load, and the field of the generator is constant.

FIGURE 1.8
Types of common mechanical loads

Fan: $k = 2$

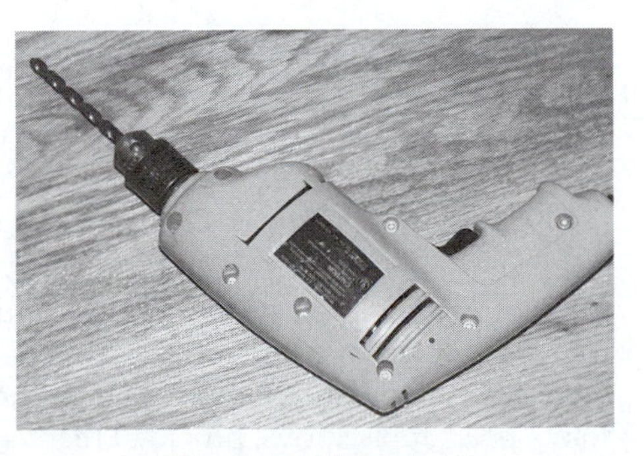

Electric drill: $k = -1$

3. *Torque proportional to the square of speed.* The torque–speed characteristic is parabolic, $k = 2$. Examples of this type of load are fans, centrifugal pumps, and propellers. The load power requirement is proportional to the cube of the speed and may be excessive at high speeds.
4. *Torque inversely proportional to speed.* In this case, $k = -1$. Examples of this type include milling and boring machines. This load usually requires a large torque at starting and at low speeds. The power consumption of such a load is independent of speed. This is why the motor of an electric saw does not always get damaged (due to overcurrent) when the saw disk is blocked.

Some loads may have a combination of the characteristics listed. For example, the friction torque exhibits a complex form of speed–torque that varies according to the operating speed. At low speeds, the friction torque is almost inversely proportional to the speed due to the magnitude of the static and coulomb frictions. At high speeds, it is almost linearly proportional to the speed due to the viscous friction.

1.2.2 ELECTRIC MOTORS

Electric motors exhibit wide variations of speed–torque characteristics, some of which are shown in Figure 1.9. Synchronous or reluctance motors exhibit a constant-speed characteristic similar to that shown by curve I. At steady-state conditions, these motors operate at constant speed regardless of the value of the load torque. Curve II shows a dc shunt or a separately excited motor, where the speed is slightly reduced when the load torque increases. Direct current series motors exhibit the characteristic shown in curve III; the speed is high at light loading conditions and low at heavy loading. Induction motors have a somewhat complex speed characteristic similar to the one given by curve IV; during steady state, they operate at the linear portion of the speed–torque characteristic, which resembles the characteristic of a dc shunt or a separately excited motor. The maximum developed torque of induction motors is limited to T_{max}.

FIGURE 1.9
Speed–torque characteristics of electric motors

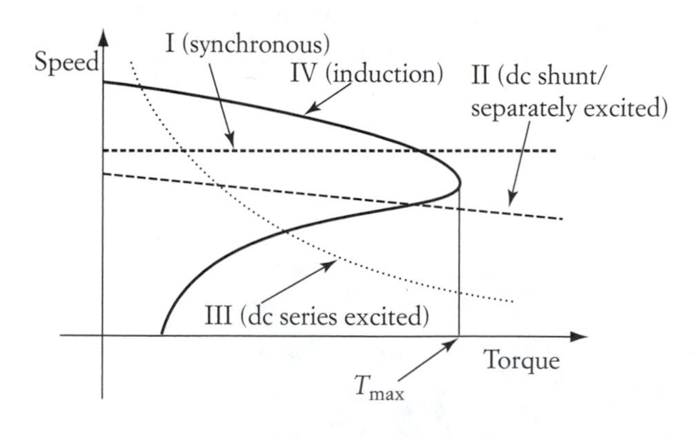

In electric drive applications, electric motors should be selected to match the intended performance of the loads. For example, in constant-speed applications, the synchronous motor is probably the best option. Other motors, such as induction or dc, can also be used in constant-speed applications, provided that feedback circuits are used to compensate for the change in speed when the load torque changes.

1.2.3 POWER SOURCES

Two major types of power sources are used in industrial applications: alternating current (ac) and direct current (dc). Alternating current sources are common in industrial installations and residences. These sources can either be single-phase or multiphase systems. Single-phase power sources are common in residences, where the demand for electric power is limited. Multiphase power sources are used in high power consumption applications. The most common type of power source in the United States is the three-phase, 60-Hz power source. In Europe, most of the Middle East, Africa, and Asia, the frequency is 50 Hz.

Extensive industrial installations usually have more than one type of power source at different voltages and frequencies. Commercial airplanes, for example, may have a 400-Hz ac source in addition to a 270-volt dc source.

1.2.4 CONVERTERS

The main function of a converter is to transform the waveform of a power source to that required by an electric motor in order to achieve the desired performance. Most converters provide adjustable voltage, current, and/or frequency to control the speed, torque, or power of the motor. Figure 1.10 shows the four basic types of converters.

1. *dc to ac.* The dc waveform of the power source is converted to a single- or multiphase ac waveform. The output frequency, current, and/or voltage can be adjusted according to the application. This type of converter is suitable for ac motors, such as induction or synchronous motors.
2. *dc to dc.* This type is also known as a "chopper." The constant-input dc waveform is converted to a dc waveform with variable magnitude. The typical application of this converter is in dc motor drives.
3. *ac to dc.* The ac waveform is converted to dc with adjustable magnitude. The input could be a single- or multiphase source. This type of converter is used in dc drives.
4. *ac to ac.* The input waveform is typically ac with fixed magnitude and frequency. The output is an ac with variable frequency, magnitude, or both. The conversion can be done directly or through a dc link. The dc link system consists of two converters connected in cascade; the first is an ac/dc, and the second is a dc/ac. Typical applications of the dc link converter are ac motors.

In addition to electric drives, dc link converters are also used in such applications as the uninterruptable power supply (UPS). Figure 1.11 shows the basic components of a UPS. The dc link between the two converters has a rechargeable battery. In normal operation, the input current i_{in} is converted to a dc current I_{dc_1}. This current is divided into two parts: one, I_B, charges the battery; the other, I_{dc_2}, is converted to the ac current i_{out} that feeds the load.

FIGURE 1.10
Four types of converters

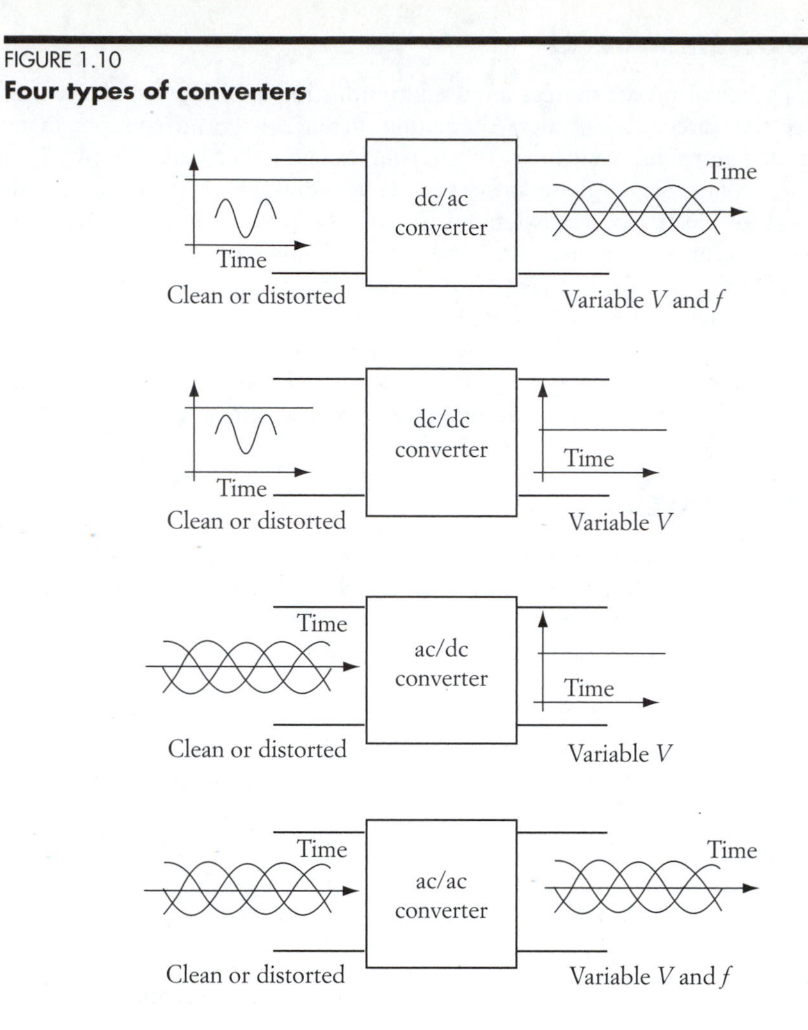

When the source power is lost (such as in a power outage), the input current i_{in} and the dc current I_{dc_1} are each zero. In this case, the energy stored in the battery is used to feed the load. The battery current I_B becomes the input current to the dc/ac converter so that the load power is not interrupted. The capacity of the rechargeable battery and the magnitude of the load current determine the time by which the system can feed to load during an outage.

1.2.5 CONTROLLERS

A well-designed controller has several functions. The most basic function is to monitor system variables, compare them with some desired values, and then readjust the converter output until the system achieves a desired performance. This feature is used in such applications as speed or position control. Some drive systems may lack stability due to limitations in the converter or load characteristics. In such cases, a controller may also be designed to enhance overall stability.

High-performance drives (HPDs)—such as robotics, guided manipulation, and supervised actuation—are good examples of elaborate controllers. In these applications, the system must follow a preselected track at all times. A multirobot system performing a complementary function must move the end effectors about the space of operation according to a preselected, time-tagged trajectory. To achieve this, each motor in the robot arm must follow a specific track so that the aggregate motion of all motors keeps the end effector alongside its trajectory at all times, even when the system loads, inertia, and parameters vary. In addition, the stability of the system must be guaranteed in all operating conditions. The controllers in these applications are complex; their structure and/or parameters must be adaptively tuned to achieve the two basic objectives: (a) to provide the best possible tracking performance without overstressing the hardware and (b) to enhance system stability and robustness.

FIGURE 1.11
A UPS system

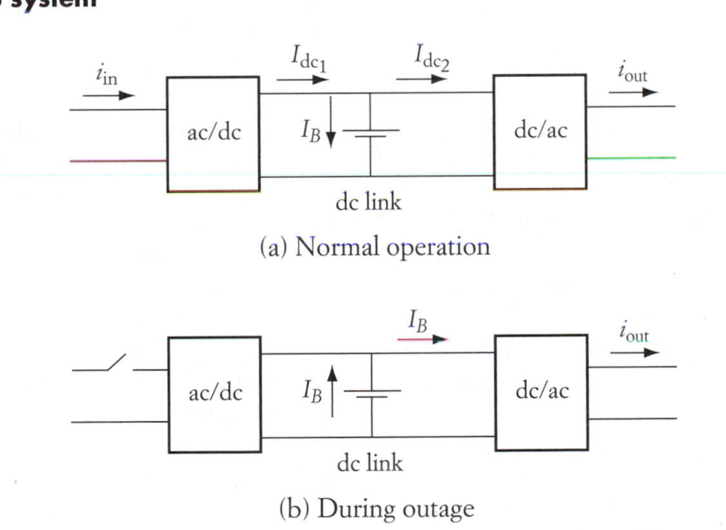

2

Introduction to Solid-State Devices

In electric drive applications, the source waveforms often do not match the waveforms needed by the electric motors to perform the needed functions. For example, if the power source is a dc type (battery) and the motor is an ac type, the motor cannot be powered directly from the source. In this case, a converter composed of electronic devices is used to change the dc waveform into ac with suitable frequency and voltage.

The main building blocks of a converter are the solid-state power electronic switches. They are similar to devices used in analog and digital circuits, such as transistors and diodes. However, power devices are designed to handle high currents and voltages, operate at low junction losses, and withstand high rates of change of voltage and current (*dv/dt* and *di/dt*). The switching speeds of solid-state devices should be as high as possible in order to reduce the size of the circuit magnetic components and to reduce audible noise due to the switching action. Because the human ear may detect sounds between 20 Hz and 20 kHz, the technology of solid-state power switches is continually pushing the switching frequency beyond the audible range.

Most solid-state power electronic devices are used to mimic mechanical switches by connecting and disconnecting electric loads. Ideally, a mechanical switch has the current–voltage characteristics depicted in Figure 2.1. When the switch is open (off), it carries no current, and its terminal voltage is equal to the source voltage. When the switch is closed (on), its terminal voltage is zero, and its current is determined by the load impedance (Ohm's laws).

FIGURE 2.1
Current–voltage characteristics of an ideal switch

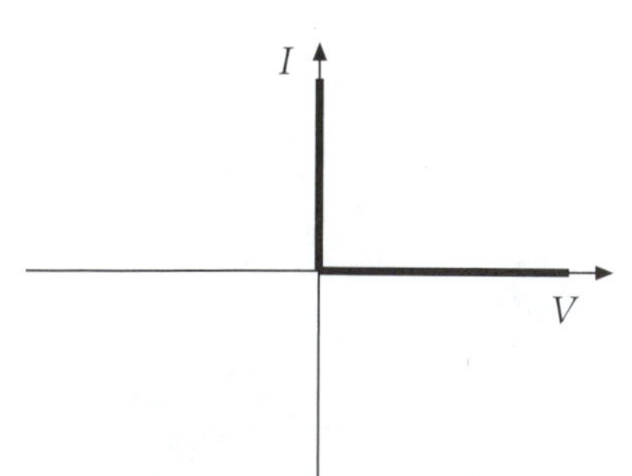

A large part of the ongoing development in power electronics research is devoted to improving the characteristics of solid-state devices and making them comparable to the ideal switch. In this chapter, we will discuss some of the basic operational concepts that make solid-state switches behave more like ideal mechanical switches. Although this chapter covers some of the well-established power electronic devices, keep in mind that the technology in this area is rapidly changing. (For more up-to-date information, consult the manufacturers' catalogs and the technical literature.) The switching devices presented in this chapter are divided into two groups: transistors and thyristors. Each group can be divided into several subgroups with distinct features.

2.1 TRANSISTORS

Transistors can be used as either amplifiers or electronic switches. In power applications, transistors are used mainly as electronic switches. Two types of transistors are commonly used in power applications: the bipolar transistor and the field effect transistor (FET). Other power devices, such as the insulated gate bipolar transistor (IGBT), are hybrids of these two types. Figure 2.2 shows various sizes of bipolar power transistors.

2.1.1 BIPOLAR TRANSISTOR

The bipolar transistor consists of three layers of semiconductor materials in the *n-p-n* or *p-n-p* structure, as shown in Figure 2.3. The ratings of these transistors can be as high as a few hundred amperes. The operation of the bipolar transistor is based on the capability of the *p-n* junction to inject or collect minority carriers. When the emitter is forward-biased, electrons are injected from the *n* (emitter) to the *p* (base) region. If the other *n* layer (collector) is reverse-biased, the electrons in the *p* layer are collected in that *n* layer. Keep in mind that electric current flows in the opposite direction of electrons.

The base layer is very thin compared to the layers of the emitter or collector because it is an obstacle to the flow of current. However, it serves a very useful purpose—it controls the flow of electrons from emitter to collector. If a base current is injected in the *p* junction, more current is allowed to pass from the emitter to the collector. The relationship between the base and collector currents, however, is nonlinear.

FIGURE 2.2
Power transistors

Single devices Integrated components

In Figure 2.3, the symbol of the transistor is shown in the middle, and its positive bias is shown on the right side. The collector is positive-biased with respect to the emitter or base, and the base voltage is positive with respect to the emitter. The base–emitter junction is a simple diode. Hence, the voltage difference between the base and emitter is very small (about 0.6 V).

The basic equations of the bipolar transistor can be written as

$$I_C = \beta I_B + I_{CEO} \tag{2.1}$$

$$I_E = I_B + I_C \tag{2.2}$$

$$V_{CE} = V_{CB} + V_{BE} \tag{2.3}$$

where β is the current gain (ratio of collector to base currents), and I_{CEO} is the leakage current of the collector–emitter junction. The rest of the symbols are explained in Figure 2.3. Because the leakage current is very small compared with βI_B, it is often ignored.

A transistor can be connected in a common-base or common-emitter form. Figure 2.4 shows the characteristic of an *n-p-n* transistor connected in the common emitter. The base characteristic, shown in Figure 2.4(a), is very similar to that of the diode. In the forward direction, the base-emitter voltage is below 0.7 V. A substantial increase in the base current occurs at a slightly higher value of the base-emitter voltage.

FIGURE 2.3
Bipolar transistor

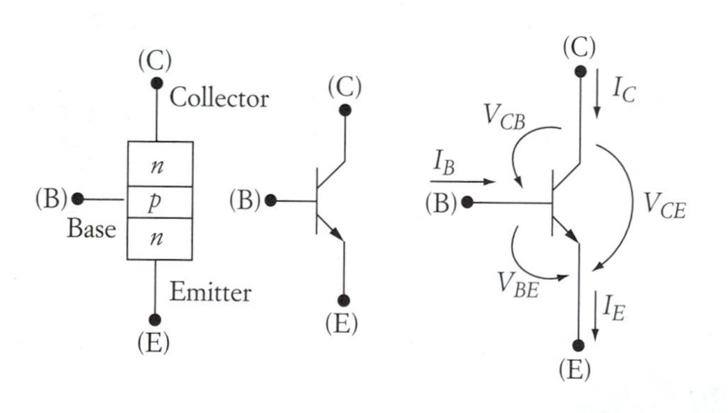

Figure 2.4(b) shows the collector characteristics, which can be divided into three basic regions: the linear region, the cutoff region, and the saturation region. In the linear region, the transistor operates as an amplifier, where β is almost constant and in the order of a few hundreds. Any base current is amplified a few hundred times in the collector circuit. This is the region in which most audio amplifiers operate when using bipolar transistors.

The cutoff region is the area of the characteristic in which the base current is zero. In this case, the collector current is negligibly small regardless of the value of the collector–emitter voltage. In the saturation region, the collector–emitter voltage is very small at high base currents.

The magnitude of the current gain β is dependent on the operating region of the transistor. In the linear region, β is in the range of hundreds. In the saturation region, it is often less than 30. When a transistor is used as a switch, it operates in two regions: cutoff and saturation. On a continuous basis, power transistors do not operate in the linear region because the current of the power transistor is high, and in the linear region, the voltage V_{CE} is also large. Therefore, the losses of the transistor (the current multiplied by V_{CE}) are excessive and lead to thermal damage of the transistor.

In the cutoff region, the transistor acts as an open switch, where the collector current is almost zero regardless of V_{CE}. In the saturation region, the transistor operates as a closed switch because the voltage across the switch is very small, and the external circuit determines the magnitude of the collector current. Compare these features with the mechanical switch characteristics in Figure 2.1.

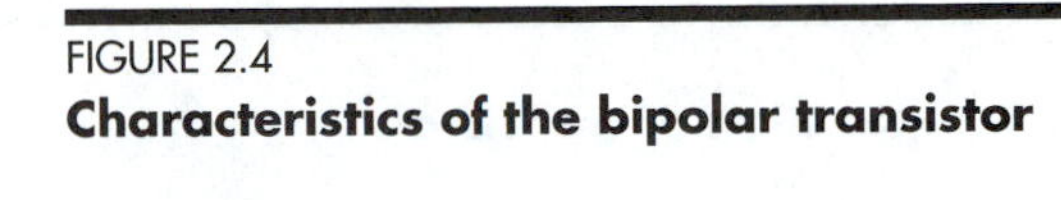
FIGURE 2.4
Characteristics of the bipolar transistor

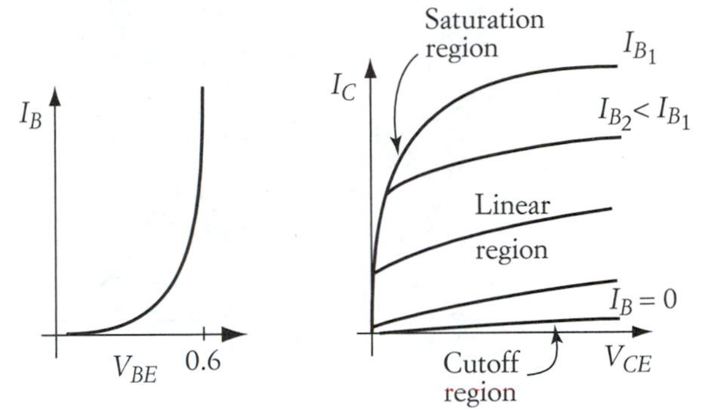

(a) Base characteristics (b) Collector characteristics

The circuit in Figure 2.5 explains the operation of a transistor in a switching circuit. The transistor is connected to an external circuit that consists of a dc source V_{CC} and a load resistance R_L. The base circuit of the transistor is connected to a current source to produce the base current of the transistor.

The loop equation of the collector circuit is represented by

$$V_{CC} = V_{CE} + R_L I_C \tag{2.4}$$

This equation, which demonstrates a linear relationship between I_C and V_{CE}, is shown in the characteristics of Figure 2.5 and is known as the load-line equation. The equation has a negative slope and intersects the V_{CE} axis at a value equal to V_{CC} and the I_C axis at a value equal to V_{CC}/R_L.

If the base current is set equal to zero, the operating point of the circuit is in the cutoff region point 1. The collector current in this case is very small and can be ignored. The collector emitter voltage of the transistor is almost equal to the source voltage V_{CC}. This operation resembles an open mechanical switch.

Now, assume that the base current is set to the maximum value. The transistor operates in the saturation region at point 2. The voltage drop across the collector–emitter terminals of the transistor is small and can be ignored. The collector (or load) current is almost equal to V_{CC}/R_L. In this case, the transistor is equivalent to a closed mechanical switch.

The bipolar transistor is a current-driven device. To open the transistor, the base current should be set to zero. To close the transistor, the base current should be set as high as the ratings permit. Keep in mind that the base current must exist

FIGURE 2.5
Switching operation of transistors

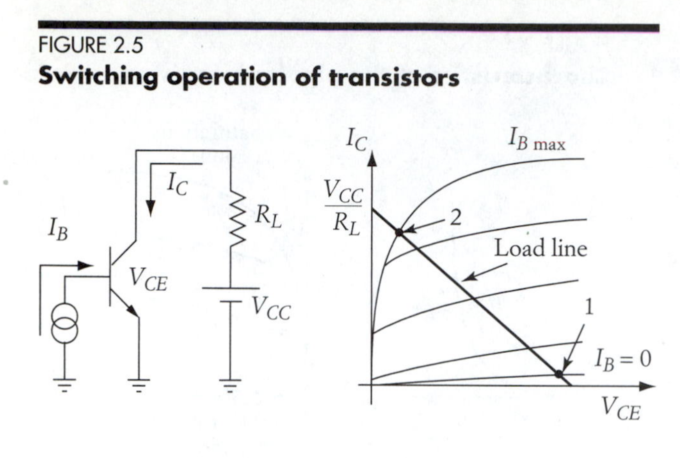

for as long as the transistor is closed. Because β is small in the saturation region and the collector current is high in power applications, the base current is also high in magnitude. This situation creates two major problems: the first is that there are relatively high losses in the base circuit. The second is that the driving circuit must be capable of producing a large base current for as long as the transistor is closed. Such a circuit is large, of low efficiency, and complex to build.

EXAMPLE 2.1

A transistor has $\beta_1 = 200$ in the linear region and $\beta_2 = 10$ in the saturation region. Calculate the base current when the collector current is equal to 10 A, assuming that the transistor operates in the linear region. Repeat the calculation for the saturation region.

SOLUTION
In the linear region,

$$I_B = \frac{I_C}{\beta_1} = \frac{10}{200} = 50 \text{ mA}$$

In the saturation region,

$$I_B = \frac{I_C}{\beta_2} = \frac{10}{10} = 1 \text{ A}$$

Note that the base current in the saturation region is 20 times that in the linear region. This ratio is the same as β_1/β_2.

2.1.2 FIELD EFFECT TRANSISTOR (FET)

Field effect transistors (FETs) are widely used as electronic switches in computer and logic circuits. There are several subspecies of FETs. The most common are the junction gate FET (JFET), the metal-oxide-semiconductor FET (MOSFET), and the insulated gate FET (IGFET).

The operation of FETs is based on the principle that the current near the surface of a semiconductor material can be changed when an electric field is applied at the surface. An example is shown in Figure 2.6. Two *n*-junctions (source and drain) are embedded in a *p* material. The gate, which is metal, is connected to the positive side of a dc supply. The source and drain are connected to another dc supply, with the drain on the positive side and the source on the negative side. The voltage difference between the drain and source creates a current flowing in the channel. The magnitude of the current is affected by the strength of the electric field from the gate. Thus, the gate voltage controls the drain current. Remember that the base current of the bipolar transistor controls the collector current, which makes the FET much easier to control than the bipolar transistor.

FIGURE 2.6
N-channel IGFET

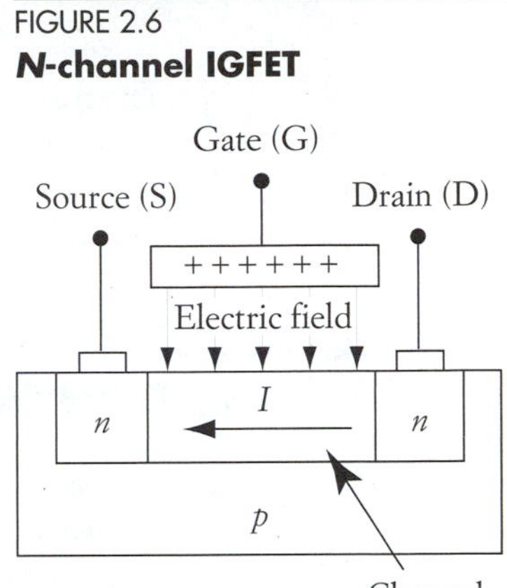

Several characteristics can be obtained from the many FET subspecies. Figures 2.7 and 2.8 show the characteristics of a MOSFET in enhanced mode and in enhanced/depletion mode. The difference between the two is that a MOSFET designed for enhanced/depletion mode has a narrow-doped conducting layer diffused into the channel. The

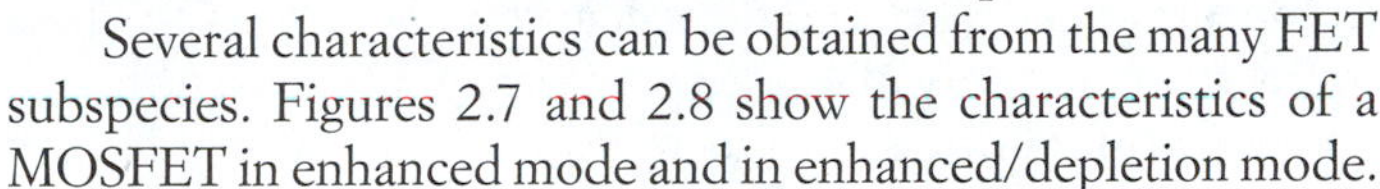

FIGURE 2.7
Enhanced-mode MOSFET

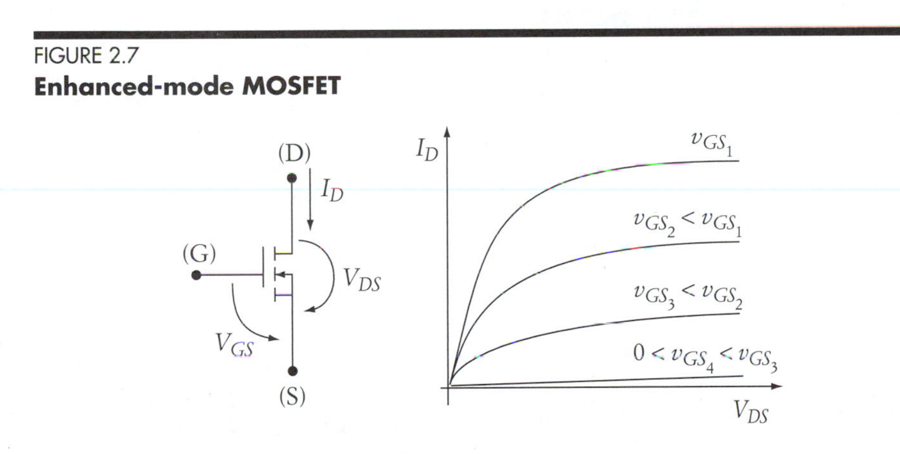

FIGURE 2.8
Enhanced/depletion mode MOSFET

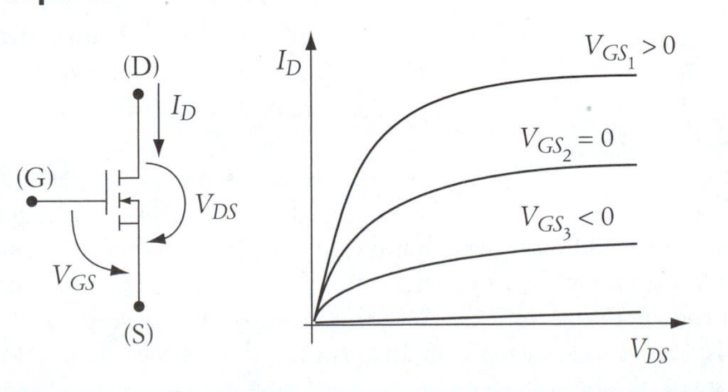

presence of this layer results in current I_D flowing inside the channel even if the gate-to-source voltage V_{GS} is negative.

The main advantage of FETs over bipolar transistors is in the way the current in the switching circuit is controlled. The FETs are voltage-driven devices, unlike the current-driven bipolar transistors. The gate voltage controls the drain current I_D of a FET, which is relatively easy to implement.

2.2 THYRISTORS

Thyristor is a name given to a family of devices that include the silicon-controlled rectifier (SCR), the bidirectional switch (triac), and the gate turnoff SCR (GTO). These devices can handle large currents and are widely used in power applications. Although not commonly used, other thyristor devices are also available for low-current control circuits, such as the silicon unilateral switch (SUS) and the bilateral diode (diac).

2.2.1 FOUR-LAYER DIODE

The four-layer diode is the basic form of thyristor. Its structure is shown on the left in Figure 2.9. The four-layer diode consists of four layers of semiconductor materials constructed in the *p-n-p-n* order. To understand its operation, reconstruct the thyristor in the form shown in the middle of Figure 2.9. The four-layer diode can be modeled as two three-layer devices (Q_1 and Q_2) with a common *n-p* junction. These two devices are in fact transistors, where Q_1 is a *p-n-p* transistor and Q_2 is an *n-p-n* transistor. The circuit representing these transistors is shown on the right side of Figure 2.9. The symbols A and K are for anode and cathode, which represent the emitters of Q_1 and Q_2, respectively.

FIGURE 2.9
Four-layer diode

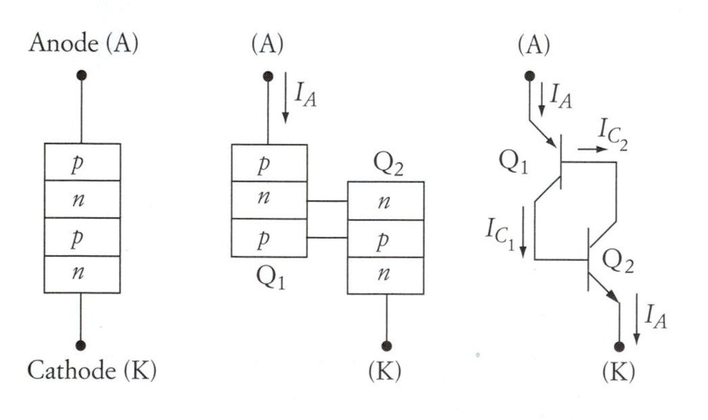

When the anode-to-cathode voltage is negative, the base-emitter junctions of the two transistors are reverse bias. Only leakage current is flowing between the anode and cathode. The leakage current is very small and the device is actually open. If the magnitude of the applied voltage increases to a breakdown limit (called reverse breakdown V_{RB}), the device is destroyed and a permanent short occurs. This phenomenon is known as thermal runaway.

When the anode-to-cathode voltage is positive, the base-emitter junctions of both transistors are in forward bias. In this case, the following equations apply:

$$I_{C_1} = \alpha_1 I_A + I_{CBO_1} \tag{2.5}$$

$$I_{C_2} = \alpha_2 I_A - I_{CBO_2} \tag{2.6}$$

$$I_A = I_{C_1} + I_{C_2} \tag{2.7}$$

$$I_A = \frac{I_{CBO_1} - I_{CBO_2}}{1 - (\alpha_1 + \alpha_2)} \tag{2.8}$$

where α is the current gain (the ratio of collector current to emitter current), and I_{CBO} is the leakage current in the collector–base junction. The current gain varies in magnitude depending on the collector-to-emitter voltage. If the voltage is small, the quantity $\alpha_1 + \alpha_2$ is much less than one, and the anode current is very small. The device in this case is considered open. When the voltage increases, the current gain also increases. When the value of $\alpha_1 + \alpha_2$ approaches unity, a breakover occurs, and the anode current tends to increase without limits. However, the impedance of the external circuit will limit the current according to Ohm's laws. The voltage at breakover is called the breakover voltage or V_{BO}.

The basic current–voltage characteristic of the four-layer diode is shown in Figure 2.10. In the forward direction (quadrant 1), the anode current is negligibly small when the anode-to-cathode voltage is less than the breakover voltage V_{BO}. At the breakover voltage, the device starts to conduct and the external circuit determines the magnitude of the forward current. During the conduction period, the voltage across the device is very small ($\approx$ 1 V). This voltage drop results in power losses inside the device; these losses produce heat that must be dissipated to protect the device from thermal damage.

FIGURE 2.10
Basic characteristics of the four-layer diode

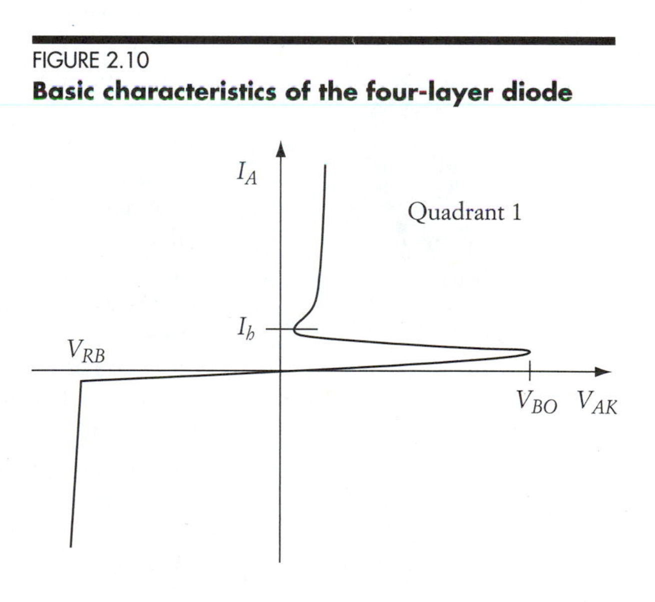

If the anode current is reduced below a certain value called *holding current* I_h, the device opens, and the current drops to zero. The voltage across the device is now equal to the source voltage. This process is called commutation.

2.2.2 SILICON-CONTROLLED RECTIFIER

Figure 2.11 shows various silicon-controlled rectifiers (SCRs) used in power applications. SCRs are frequently used in power electronic circuits. However, this may change in the future, as other devices such as the IGBT are getting larger ratings and are easier to control. Nevertheless, the popularity of the SCR is due to several factors, including the following:

- SCRs are cheaper to manufacture than other types of solid-state switches, such as the bipolar transistors and FETs.
- A single pulse, instead of the continuous signal needed by a bipolar transistor, can turn on an SCR. Hence, losses are reduced.
- In ac circuits, the SCR is self-commutated and may not need an external circuit to turn it off.
- An SCR can have much larger current and voltage ratings than the transistor.

The construction of an SCR is almost identical to that of the four-layer diode. The only difference is that the SCR has a third terminal, called the gate, which is connected to the third layer as shown in Figure 2.12. When the anode-to-cathode voltage is positive (but less than V_{BO}), the SCR is open, and no anode current flows. The characteristic of the SCR, in this case, is identical to that of the four-layer diode. Now assume that a gate current signal I_G is applied when the anode–cathode voltage is less than V_{BO}. The base current of transistor Q_2 momentarily equals the gate current, and Q_2 is turned on. Consequently, the collector current of Q_2 provides the base current for Q_1. Then Q_1 is also turned on, and the entire SCR is closed. If the gate signal is removed after the SCR is turned on, the SCR remains in the on state because the base current of Q_2 is provided by the collector current of Q_1. The SCR is only turned off when the external circuit forces the anode current to fall below the holding value I_h.

FIGURE 2.11
Various sizes of SCRs

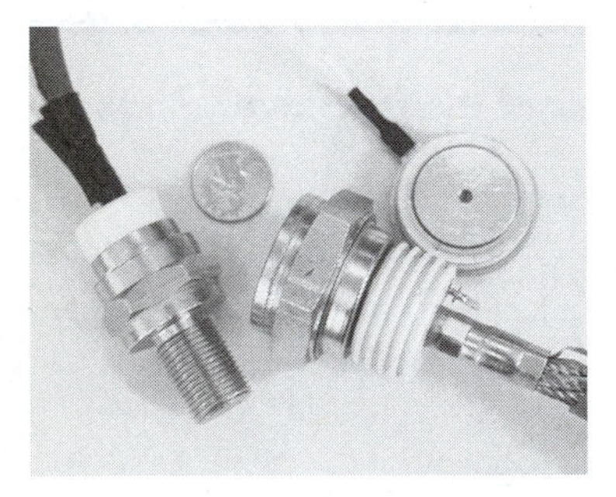

The SCR symbol and the current–voltage characteristics are shown in Figure 2.13. The characteristics are similar to those for the four-layer diode, except that the SCR can be turned on at a lower voltage than V_{BO}.

The turn-on voltage V_{TO} is dependent on the magnitude of the gate current—the higher the gate current, the lower the turn-on voltage. When the gate pulse is as high as the rating permits, the SCR can be turned on at a very low anode-to-cathode voltage.

A general expression relating the turn-on voltage V_{TO} to the equivalent dc gate current I_G can be written as

$$V_{TO} = V_{BO}e^{-(I_G K)} \tag{2.9}$$

where K is a constant whose value is dependent on the device characteristics. Keep in mind that the formula of Equation (2.9) is empirical. For more accurate information about SCR triggering characteristics, you should consult with the specification sheet of the particular device.

FIGURE 2.12
Silicon-controlled rectifier

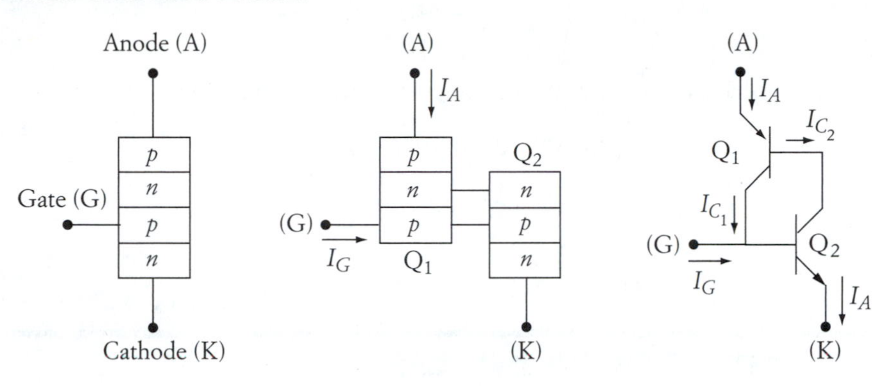

FIGURE 2.13
SCR symbol and characteristics

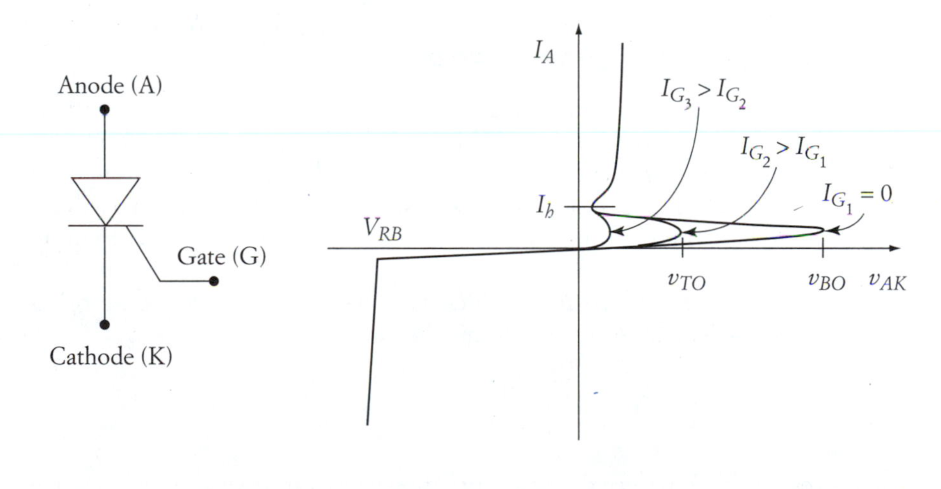

EXAMPLE 2.2

A SCR is connected in series with an ac voltage source of 120 V (rms value) and a load resistance. The breakover voltage of the SCR $V_{BO} = 200$ V, and $K = 0.2\ \text{mA}^{-1}$. Calculate the approximate value of the dc gate current required to trigger the SCR at 30°.

SOLUTION

The source voltage can be written as

$$V_s = \sqrt{2}(120)\sin(\omega t)$$

When the SCR is open, the voltage across the SCR is equal to the source voltage. For a 30° triggering angle, the voltage across the SCR is

$$V_{TO} = \sqrt{2}(120)\sin(30)$$

The dc triggering current can then be calculated using Equation (2.9):

$$V_{TO} = \sqrt{2}(120)\sin(30) = 200e^{-(0.2I_G)}$$

$$I_G = 4.29 \text{ mA}$$

2.3 OTHER POWER DEVICES

Various newer hybrid designs of power electronic devices are available that offer new characteristics or substantial modifications to the existing characteristics of the single devices. In this section, a few of these hybrid devices are described. However, be aware that changes in this area are very rapid and that newer devices are continually emerging.

2.3.1 DARLINGTON TRANSISTOR

When a bipolar transistor operates as a switch, only the cutoff and saturation regions are used. In the saturation region, the current gain β is very small. Hence, when the transistor is closed, a large base current is needed. This base current must be maintained for as long as the transistor is closed. The continuous large base current results in high transistor losses and demands an extensive control circuit to provide it.

Assume that the load current (which is equal to the emitter current) is 100 A, and β is 4. The base current in this case must be

$$\frac{100}{1 + \beta} = 20 \text{ A}$$

which is very large. To reduce the base current, two transistors can be connected in Darlington fashion as shown in Figure 2.14. The emitter current of Q_1 is $(1 + \beta_1)I_{B_1}$. This emitter current is also equal to the base current of transistor Q_2. The emitter current of transistor Q_2 is

$$I_{E_2} = (1 + \beta_2)I_{B_2} = (1 + \beta_2)(1 + \beta_1)I_{B_1} \tag{2.10}$$

Hence, the ratio of the emitter current of Q_2 (which is the load current) and the base current of Q_1 (which is the triggering current) is

$$\frac{I_{E_2}}{I_{B_1}} = (1 + \beta_1)(1 + \beta_2) \tag{2.11}$$

Assume that Q_1 and Q_2 are identical transistors with $\beta_1 = \beta_2 = 4$. Also assume that the load current is 100 A. In this case, the base current for the Darlington transistor is 4 A, which is one-fifth of the base current computed for the single transistor in the preceding case.

2.3.2 INSULATED GATE BIPOLAR TRANSISTOR (IGBT)

Bipolar transistors are devices with relatively low losses in the power circuit (collector circuit) during the conduction period, due to their relatively low forward drop V_{CE} when closed. Bipolar transistors are also more suitable for high switching frequencies than SCRs. These are very desirable features for power applications. However, bipolar transistors have very low current gains at the saturation region (when closed). Thus, the base currents are relatively high, which makes the triggering circuits bulky, expensive, and of low efficiency.

On the other hand, MOSFETs are voltage-controlled devices that require very small input current. Consequently, the triggering circuit is much simpler and less expensive to build. In addition, the forward voltage drop V_{DS} of a MOSFET is small for low-voltage devices (<200 V). At this voltage level, the MOSFET is a fast-switching power device. Because of these features, MOSFETs replace bipolar transistors in low-voltage applications (<200 V).

In high-voltage applications (>200 V), both the bipolar transistor and the MOSFET have desirable features and drawbacks. Combining the two in one circuit, as shown in Figure 2.15, enhances the desirable features and diminishes the drawbacks. The MOSFET is placed in the input circuit and the bipolar transistor in the output (power) circuit. The MOSFET is triggered by a voltage signal with a very low gate current. Then the source current of the MOSFET triggers (closes) the bipolar transistor. The losses of the output power circuit are relatively low even for high-voltage applications. Furthermore, because the output circuit is a bipolar transistor, it can be used in high-frequency switching applications. These two devices can now be included on the same wafer; the new device is called the insulated gate bipolar

FIGURE 2.14
Darlington transistor

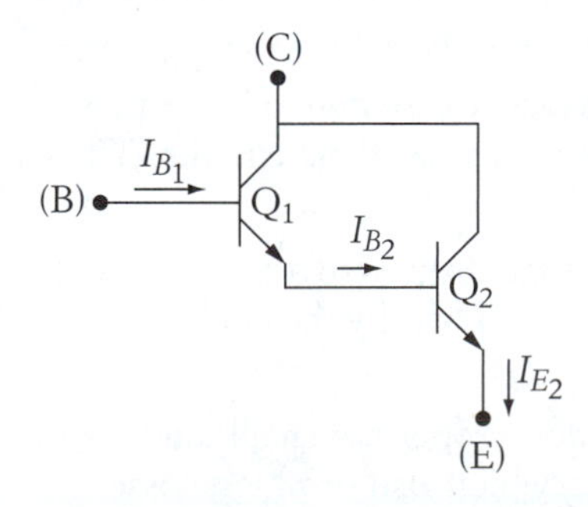

FIGURE 2.15
MOSFET and bipolar circuit

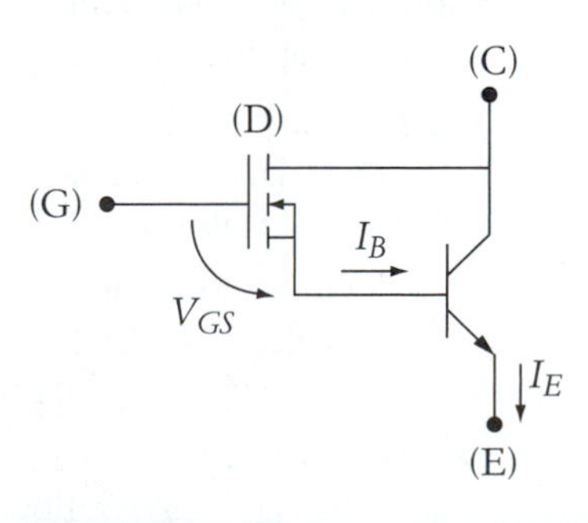

FIGURE 2.16
Characteristics of the IGBT

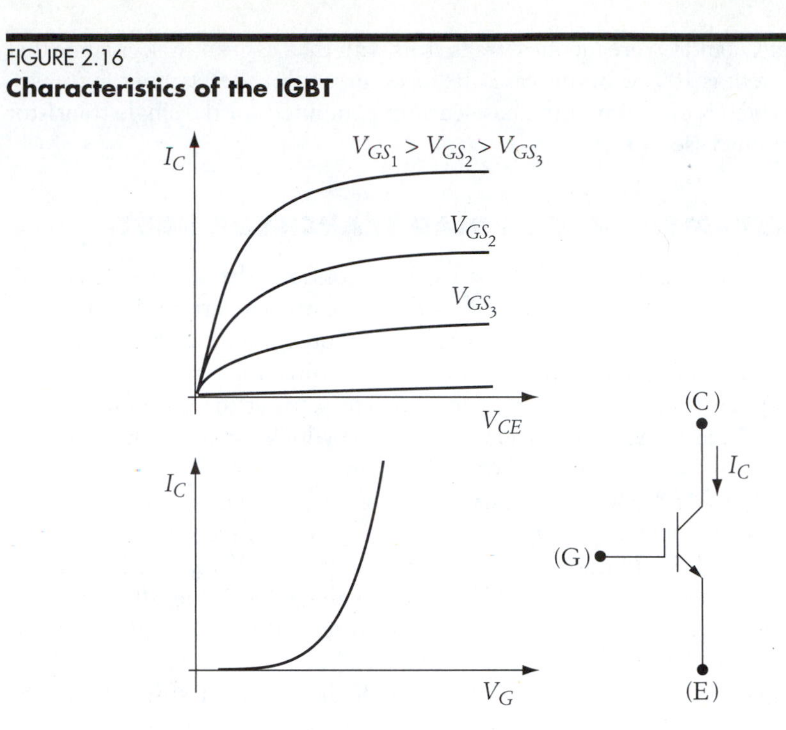

transistor or IGBT. The symbol (for *N*-channel type) and characteristics of an IGBT are shown in Figure 2.16.

2.4 RATINGS OF POWER ELECTRONIC DEVICES

Most power electronic devices are robust with a long operational life, providing they are adequately protected against excessive currents and voltages. A quick summary of various aspects of power device protection follows.

1. *Steady-state circuit ratings.* The steady-state current and voltage of the circuit should always be less than the device ratings. A good rule of thumb is to select a device that has ratings at least double the circuit requirements.
2. *Junction temperature.* Losses inside solid-state devices are due to impurities of their material as well as to the operating conditions of their circuits. These losses are as follows:
   a. During the conduction period, the voltage drop across the solid-state device is about 1 V. This voltage drop multiplied by the current inside the device produces losses.
   b. When the device is in the blocking mode (open), a small amount of leakage current flows inside the device, which also produces losses.

c. The gate circuits of the SCRs and FETs and the base circuits of the transistors produce losses due to their triggering signals.

d. Every time the solid-state device is turned on or off, switching losses are produced. These losses are usually higher for faster devices and for devices operating in high-frequency modes.

All these losses produce heat inside the solid-state junctions. The heat can be tolerated up to a critical level; 125°C is a typical value. For excessive heat beyond the critical value, the device may suffer from temporary failure or permanent damage.

To help dissipate the junction temperature, a cooling method must be used. The most common cooling device is the heat sink made of a metal sheet with fins. Examples of heat sinks are shown in Figure 2.17. The solid-state device is mounted directly on the heat sink. Any junction heat is transferred to the heat sink, which in turn dissipates it in the surrounding air. Liquid and gas cooling are also common methods used for high-power devices.

3. *Surge current.* It is the absolute maximum of the nonrepetitive impulse current.
4. *Switching time.* The turn-on time is the interval between applying the triggering signal and the turn-on of the device. The turn-off time is the interval elapsed while the device is switching from the on state to the off state. The longer the switching time is, the lower is the operating frequency of the circuit.
5. *Critical rate of rise of current (or maximum di/dt).* A solid-state device can be damaged if the *di/dt* of the circuit exceeds the maximum allowable value of the device, and *di/dt* damage can occur even if the current is below the surge limit of the device. To protect the device from this damage, a *snubbing* circuit for *di/dt* must be used.
6. *Critical rate of rise of voltage (or maximum dv/dt).* When *dv/dt* across a device exceeds its allowable limit, the device is forced to close. This is a form

FIGURE 2.17
Heat sinks

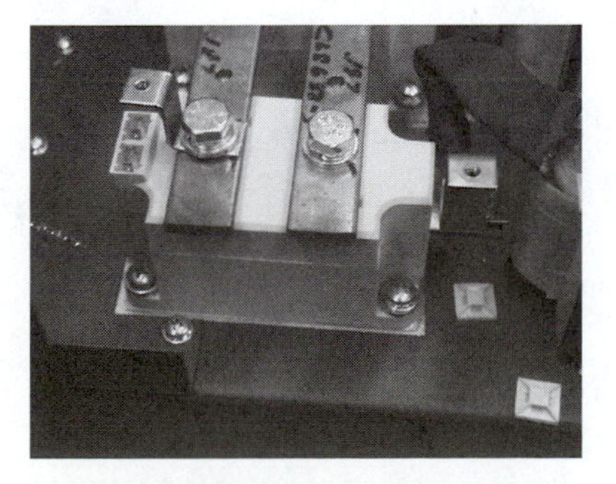

Power electronic module mounted on heat sink

Power transistors mounted on heat sinks

of false triggering that may lead to excessive current or excessive *di/dt*. To protect the device against excessive *dv/dt*, a snubbing circuit for *dv/dt* must be used.

2.5 *di/dt* and *dv/dt* PROTECTION

To protect a power electronic device against excessive *di/dt* and *dv/dt,* a snubbing circuit must be used. The function of this circuit is to limit the current and voltage transients. A simple snubbing circuit for an SCR is shown in Figure 2.18. The circuit is composed of a source voltage, a load, and an SCR. The circuit has a snubbing inductor L_s to limit the *di/dt* in the current path. It also contains an *RC* circuit to limit the *dv/dt* across the SCR.

Let us first assume that the load has the following impedance:

$$Z_L = R_L + j\omega L_L + \frac{1}{j\omega C_L} \tag{2.12}$$

Now examine the path of the current i_1, which can be written in Laplace form as

$$I_1(S) = \frac{V(S)}{\left(R + SL + \dfrac{1}{CS}\right)} \tag{2.13}$$

where $R = R_s + R_L$, $L = L_s + L_L$, and $C = C_S\,C_L\,/\,(C_S + C_L)$. Assume that the source voltage is step input. Hence, $V(S) = V/S$. Equation (2.13) can be written as

$$I_1(S) = \frac{CV\omega_n^2}{S^2 + 2\xi\,\omega_n S + \omega_n^2} \tag{2.14}$$

$$\omega_n = \sqrt{\frac{1}{CL}}$$

$$\xi = \frac{R}{2}\sqrt{\frac{C}{L}}$$

where ξ is called the damping coefficient, and ω_n is the natural frequency of oscillation. The time domain solution of Equation (2.14) is

$$i_1(t) = \frac{CV\omega_n}{\sqrt{(1 - \xi^2)}}\, e^{-\xi\omega_n t} \sin\left(\omega_n \sqrt{(1 - \xi^2)}\,t\right) \tag{2.15}$$

The *di/dt* of i_1 can be computed using Equation (2.15):

$$\frac{di_1}{dt} = \frac{-CV\omega_n^2\xi}{\sqrt{(1 - \xi^2)}}\, e^{-\xi\omega_n t}\sin\,[\omega_n\sqrt{(1 - \xi^2)}\,t] + CV\omega_n^2 e^{-\xi\omega_n t}\cos\,[\omega_n\sqrt{(1 - \xi^2)}\,t] \tag{2.16}$$

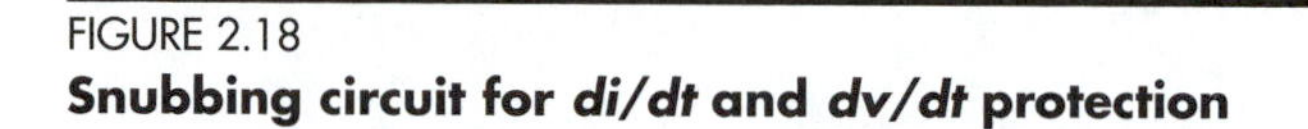

FIGURE 2.18
Snubbing circuit for *di/dt* and *dv/dt* protection

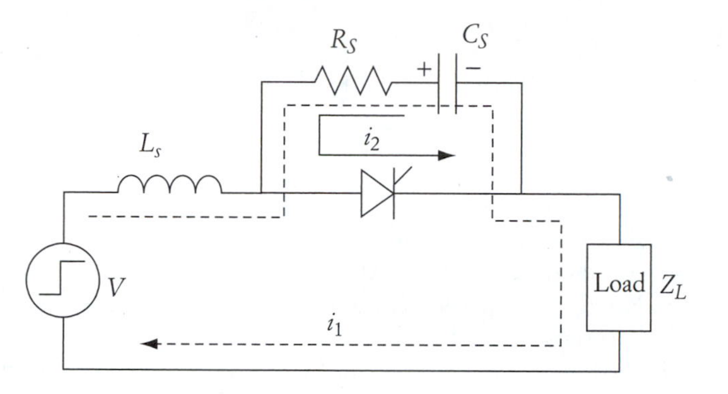

Let us assume that the capacitors are initially uncharged. Furthermore, the charge on the capacitors cannot instantly change. The maximum *di/dt* then occurs at the initial time ($t = 0$). Hence,

$$\left[\frac{di_1}{dt}\right]_{\max} = \frac{V}{L} \tag{2.17}$$

The snubbing inductor is then calculated as

$$L_s = \frac{V}{\left(\dfrac{di_1}{dt}\right)_{\max}} - L_L \tag{2.18}$$

For adequate protection, L_s should be selected so that V is substituted by the maximum nonrepetitive forward blocking voltage V_{BO}, and *di/dt* by, say, half the maximum *di/dt* rating of the SCR.

$$L_s = \frac{V_{BO}}{0.5\left(\dfrac{di}{dt}\right)_{\text{rating}}} - L_L \tag{2.19}$$

Equation (2.19) should be adequate to protect the SCR from excessive *di/dt* due to supply surges. This is only one of three types of transients that a SCR should be able to withstand without damage. The other two transients are the *dv/dt* and the *di/dt* created by the RC snubbing circuit itself when the SCR is turned on.

The RC circuit (R_s and C_s) can protect the SCR from the other two transients. Let us assume that the charge on the capacitor C_s is zero when the voltage V is applied. With this assumption, the voltage across the SCR V_{SCR} at the initial time is

$$V_{\text{SCR}} = R_s\, i_1 \tag{2.20}$$

Then

$$\frac{dV_{\text{SCR}}}{dt} = R_s \frac{di_1}{dt} \tag{2.21}$$

Substituting di_1/dt of Equation (2.17) into (2.21) yields

$$\frac{dV_{\text{SCR}}}{dt} = R_s \frac{V}{L} \tag{2.22}$$

Equation (2.22) shows that the smaller the resistance R_s is, the smaller is the dv/dt across the SCR. In fact, Equation (2.22) shows that there will be no dv/dt if R_s is set equal to zero. However, as we will see next, R_s is needed to limit the di/dt created by the snubbing capacitor.

Let us assume that the user triggers the SCR. The current going through the SCR has two components: one is i_1 from the source, and the other is i_2 from the snubbing capacitor, as shown in Figure 2.18. We already discussed the di/dt attributed to i_1. The current i_2 will also cause a di/dt and must also be limited to a tolerable value.

Let us first write the equation for i_2.

$$i_2 = \frac{V_o}{R_s} e^{-t/(R_s C_s)} \tag{2.23}$$

where V_o is the capacitor voltage due to its initial charge before the SCR is triggered. You may assume the worst value for V_o, which is equal to the V_{BO}. The di/dt of this circuit is then

$$\frac{di_2}{dt} = \frac{-V_o}{R_s^2 C_s} e^{-t/(R_s C_s)} \tag{2.24}$$

The maximum di_2/dt occurs at $t = 0$. Hence,

$$\left[\frac{di_2}{dt}\right]_{\text{max}} = \frac{-V_o}{R_s^2 C_s} \tag{2.25}$$

As in the previous discussion, di_2/dt should also be limited to, say, half the SCR rating.

Now examine Equations (2.22) and (2.25). Equation (2.22) shows that small R_s leads to small dv/dt, but Equation (2.25) shows that if R_s is made small, the di_2/dt will be large. This is a situation in which a compromise is made.

EXAMPLE 2.3

An SCR is connected between an ac source of 120 V (rms) and a resistive load. The maximum di/dt of the SCR is 100 A/μsec, and the maximum nonrepetitive forward blocking voltage V_{BO} is equal to 300 V. Calculate the minimum value of the snubbing inductance.

SOLUTION

Direct substitution in Equation (2.19) yields

$$L = \frac{300}{0.5(100)} = 6\ \mu\text{h}$$

To reduce the size of the inductor, iron-core material could be used. The problem with iron-core inductors, however, is the core saturation, which reduces the value of the inductance at high current values. Air-core inductors do not suffer from saturation but are bulky. Nevertheless, air-core inductors are normally used for snubbing circuits.

EXAMPLE 2.4

Design a snubbing circuit to protect a SCR from excessive *dv/dt*. The SCR has a snubbing inductor of 8 μh. The snubbing circuit must not allow the SCR to exceed the following:

$$V_{BO} = 4000\ \text{V}$$

$$\frac{di}{dt} = 200\ \text{A}/\mu\text{sec}$$

$$\frac{dv}{dt} = 1500\ \text{V}/\mu\text{sec}$$

SOLUTION

Equation (2.22) can be used to select R_s. However, any value selected for R_s should ensure that the di_2/dt of Equation (2.25) is not excessive. The easiest method is to pick a reasonable value for the capacitance C_s, then compute the R_s that limits the maximum di_2/dt to half the SCR rating. Then we must check the value of the dv/dt across the SCR.

Let us select C_s to be 10 μf. Then, from Equation (2.25),

$$R_s = \sqrt{\frac{V_{BO}}{0.5\,\frac{di}{dt}\,C_s}} = \sqrt{\frac{4000}{0.5 \times 200 \times 10}} = 2\ \Omega$$

Substituting R_s into Equation (2.22) yields

$$\frac{dV_{\text{scr}}}{dt} = R_s\,\frac{V_{BO}}{L} = 2\,\frac{4000}{8} = 1\ \text{kV}/\mu s$$

The dv/dt is below the rating of the SCR.

Another factor that should be kept in mind is the losses of the snubbing circuit due to the presence of R_s. When the SCR is not triggering, the current i_1 causes losses in the snubbing circuit that can be expressed by

$$P = i_1^2 R_s$$

A bypass diode can be used in parallel with R_s to reduce the losses. The diode can also reduce the dv/dt in Equation (2.22).

CHAPTER 2 PROBLEMS

2.1 A bipolar transistor is connected to a resistive load as shown in Figure 2.19. The source voltage V_{CC} is 60 V and R_L is 5 Ω.

FIGURE 2.19

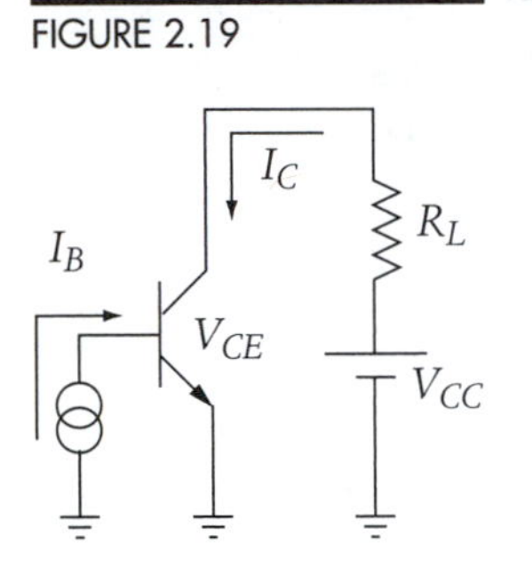

In the saturation region, the collector–emitter voltage V_{CE} is 5V and β is 6. While the transistor is in the saturation region, calculate the following:

a. Load current
b. Load power
c. Losses in the collector circuit
d. Losses in the base circuit
e. Efficiency of the circuit

2.2 For the transistor in Problem 2.1, compute the load current, load power, and efficiency of the circuit when the transistor is in the cutoff region. Assume that the collector current is 100 mA in the cutoff region.

2.3 Design a snubbing circuit for a power bipolar transistor that operates in a circuit between a resistive load of 20 Ω and an ac source of 120 V (rms) at 60 Hz. The available ratings of the transistor are

$$\left(\frac{dv_{ce}}{dt}\right)_{\max} = 300 \text{ V}/\mu\text{sec} \qquad \left(\frac{di_c}{dt}\right)_{\max} = 20 \text{ A}/\mu\text{sec} \qquad (V_{ce})_{\max} = 600 \text{ V}$$

2.4 Design a snubbing circuit for a power transistor that operates in a circuit consisting of an inductive load and an ac source. The resistive component of the load is 20 Ω and the inductive component is 10 Ω. The ac source is 120 V (rms) at 60 Hz. The transistor has the same ratings as the one in Problem 2.3.

3

Introduction to Solid-State Switching Circuits

Reduction of system losses is one of the major achievements resulting from the use of solid-state power devices. As an example, the induction motor is known for its low efficiency at light loading conditions. To reduce its losses, the terminal voltage of the motor should be reduced during no-load or light loading conditions. This can be achieved by using an autotransformer equipped with control mechanisms for voltage adjustment. This is an expensive option that also requires much maintenance. An alternative method is to use a power electronic circuit designed to control the motor voltage; this option is often much cheaper and more efficient.

In addition to improving system efficiency, power electronic devices can greatly enhance system operation and can provide features that may not otherwise be achieved. For example, when an induction motor is operating in a variable-speed environment, the frequency of the supply must vary. (Speed control of the induction motor is explained in a later chapter.) Without a power electronic circuit, the frequency can only be varied by using complex electromechanical systems employing several electric machines.

In this chapter, we will introduce the concept of solid-state switching circuits applied to static loads (resistive, inductive, and capacitive). Solid-state switching for dynamic loads, such as electric motors, is treated in later chapters. In the following analyses, we will assume that the solid-state devices are ideally switched without losses or transients.

3.1 SINGLE-PHASE, HALF-WAVE, ac/dc CONVERSION FOR RESISTIVE LOADS

Figure 3.1 shows a simple circuit consisting of an alternating current source of potential v_s, a load resistance, and an SCR connected between the source and the load. An external circuit, not shown in the figure, triggers the SCR. The load voltage is labeled v_t. Assume that the SCR is an ideal device with no voltage drop during conduction. When the SCR is open, the current of the circuit is zero and the load voltage is also zero. When the SCR is closed, the voltage across the resistance is equal to the source voltage.

FIGURE 3.1
Half-wave SCR with resistive load

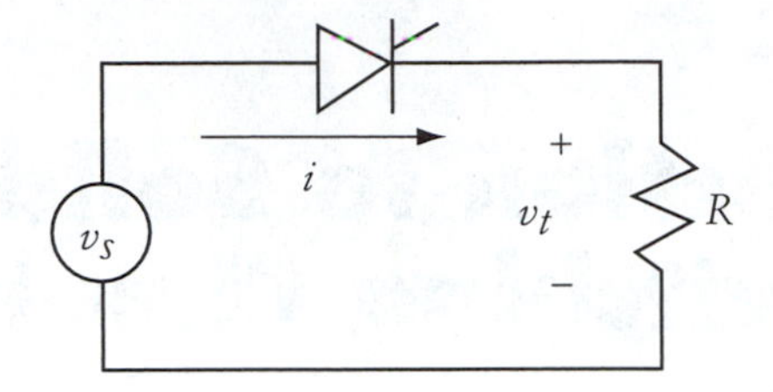

Let us assume that the source voltage is sinusoidal, expressed by the equation

$$v_s = V_{\max} \sin(\omega t) \tag{3.1}$$

At an angle $\omega t = \alpha$, the SCR is triggered (closed). No current existed in the circuit before the SCR was triggered. For the discontinuous current operation of this switching circuit, the current waveform can be expressed by

$$i = \frac{v_s}{R}(u_\alpha - u_\beta) \tag{3.2}$$

where u_α and u_β are step functions defined as

$$u_\alpha = u(\omega t - \alpha), \qquad u_\alpha = 0 \quad \text{for } \omega t < \alpha, \quad \text{otherwise } u_\alpha = 1$$

$$u_\beta = u(\omega t - \pi), \qquad u_\beta = 0 \quad \text{for } \omega t < 180°, \quad \text{otherwise } u_\beta = 1$$

The waveforms of the current and voltages are shown in Figure 3.2. Before α, the current and the load voltage are both zero. After α, the SCR is closed and the load voltage is equal to the source voltage. The current, as expressed in Equation (3.2), is equal to zero except between α and π. At π, the current reaches zero and the SCR is turned off. (Remember that the SCR is turned off when its current falls below its holding value.) The voltage across the SCR is zero while closed. When the SCR is open, the voltage across its terminals is equal to the source voltage. The voltage across the load can be expressed by

$$v_t = iR = v_s(u_\alpha - u_\beta) \tag{3.3}$$

The average voltage across the resistance (V_{ave}) is a function of the area under the waveform of v_t.

$$V_{ave} = \frac{1}{2\pi}\int_0^{2\pi} v_s(u_\alpha - u_\beta)\, d\omega t = \frac{1}{2\pi}\int_\alpha^{\pi} v_s\, d\omega t = \frac{1}{2\pi}\int_\alpha^{\pi} V_{\max}\sin(\omega t)\, d\omega t$$

$$V_{ave} = \frac{V_{\max}}{2\pi}(1 + \cos\alpha) \tag{3.4}$$

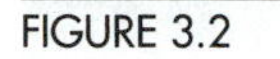

FIGURE 3.2
Waveforms of the circuit in Figure 3.1

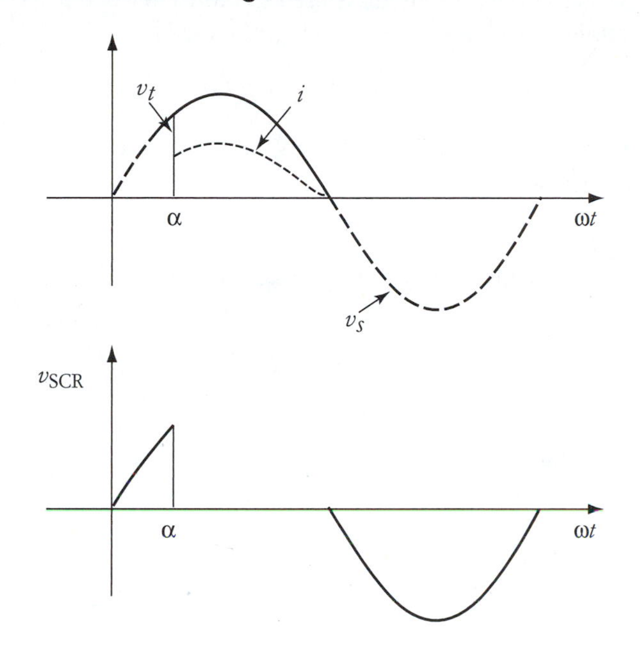

FIGURE 3.3
Average voltage across the load

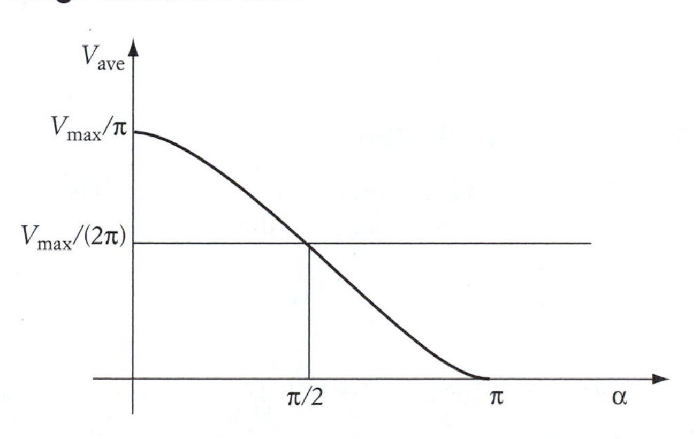

Figure 3.3 shows the load average voltage, which is expressed by Equation (3.4). The maximum average voltage is V_{max}/π, which is obtained when the triggering angle is zero. The minimum average voltage is zero at $\alpha = \pi$. The average current of the circuit is linearly proportional to the load average voltage

$$I_{ave} = \frac{V_{ave}}{R} \tag{3.5}$$

EXAMPLE 3.1

A single-phase, half-wave SCR circuit is used to reduce the average voltage across a nonlinear resistance with a resistive value represented by

$$R = 0.2\,V_{ave}^2 + 5\,\Omega$$

The source voltage of the circuit is 110 V(rms). Calculate the average current when the triggering angle is 90°.

SOLUTION

The first step is to calculate the average voltage across the load as given in Equation (3.4):

$$V_{ave} = \frac{V_{max}}{2\pi}(1 + \cos\alpha) = \frac{\sqrt{2}\,110}{2\pi}[1 + \cos(90)] = 24.75 \text{ V}$$

The resistance of the load is a nonlinear function of the average voltage. At 24.75 V, the load resistance is

$$R = 0.2\,V_{ave}^2 + 5 = 0.2\,(24.75)^2 + 5 = 127.6\,\Omega$$

The average current of the load can be obtained by dividing the average voltage over the load resistance:

$$I_{ave} = \frac{V_{ave}}{R} = \frac{24.75}{127.6} \approx 0.2 \text{ A}$$

3.1.1 ROOT-MEAN-SQUARES

Electric quantities (current, voltage, and power) are often expressed by their root-mean-square (rms) values. In a purely sinusoidal system, the rms value is

$$V_{rms} = \frac{V_{max}}{\sqrt{2}}$$

where V_{max} is the maximum (peak) of the voltage waveform and V_{rms} is the root-mean-square.

In nonsinusoidal systems, the above equation is not valid. The rms then must be computed using the basic definition for the rms values. To do that, you first compute the square of the waveform, find its average value, and then compute its square root. These three steps are shown in Equation (3.6).

$$V_{rms} = \sqrt{\frac{1}{2\pi}\int_0^{2\pi}[v(t)]^2\,d\omega t} = \sqrt{\frac{1}{2\pi}\int_0^{2\pi}[V_{max}\sin(\omega t)]^2\,d\omega t} \tag{3.6}$$

Since the load voltage exists only during the conduction period (between α and π), the integration limits of Equation (3.6) are also the limits of the conduction period.

$$V_{rms} = \sqrt{\frac{V_{max}^2}{2\pi} \int_{\alpha}^{\pi} [\sin(\omega t)]^2 \, d\omega t} = \sqrt{\frac{V_{max}^2}{4\pi} \int_{\alpha}^{\pi} [1 - \cos(2\omega t)] \, d\omega t}$$

$$V_{rms} = \frac{V_{max}}{2} \sqrt{\left[1 - \frac{\alpha}{\pi} + \frac{\sin(2\alpha)}{2\pi}\right]} \tag{3.7}$$

$$V_{rms} = \frac{V_{s\,rms}}{\sqrt{2}} \sqrt{\left(1 - \frac{\alpha}{\pi} + \frac{\sin(2\alpha)}{2\pi}\right)} \tag{3.8}$$

where $V_{s\,rms}$ is the rms voltage of the sinusoidal source

$$V_{s\,rms} = \frac{V_{max}}{\sqrt{2}}$$

Figure 3.4 and Equation (3.8) show that the rms voltage across the load V_{rms} is a function of the triggering angle α. When $\alpha = 0$,

$$V_{rms} = \frac{V_{s\,rms}}{\sqrt{2}}$$

The rms current can also be calculated similarly to the way given in Equation (3.6). A simpler method is to use the expression

$$I_{rms} = \frac{V_{rms}}{R}$$

FIGURE 3.4
rms voltage of the load

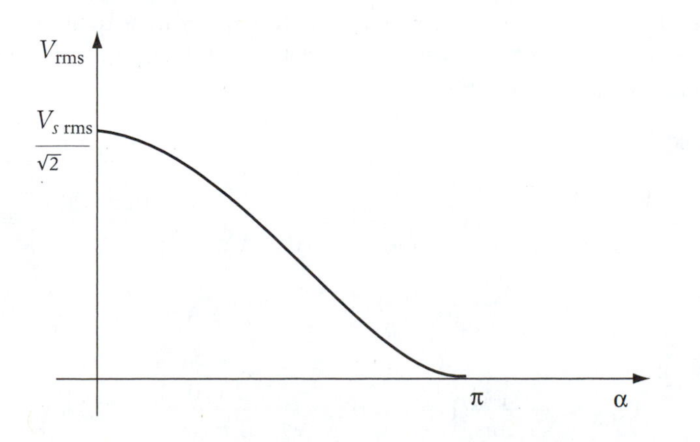

EXAMPLE 3.2

An ac source of 110 V (rms) is connected to a resistive element of 2 Ω through a single SCR. For $\alpha = 45°$ and 90°, calculate the following:

a. rms voltage across the load resistance
b. rms current of the resistance
c. Average voltage drop across the SCR

SOLUTION

For $\alpha = 45°$

a.
$$V_{rms} = \frac{V_{s\,rms}}{\sqrt{2}}\sqrt{\left[1 - \frac{\alpha}{\pi} + \frac{\sin(2\alpha)}{2\pi}\right]}$$

$$= \frac{110}{\sqrt{2}}\sqrt{\left[1 - \frac{45(\pi/180)}{\pi} + \frac{\sin(90)}{2\pi}\right]} = 74.13 \text{ V}$$

b.
$$I_{rms} = \frac{V_{rms}}{R} = \frac{74.13}{2} = 37.07 \text{ A}$$

c. As shown in Figure 3.2, the voltage across the SCR is equal to the source voltage when the SCR is not conducting. When the SCR is closed, the voltage across its terminals is zero. Hence, the average voltage across the SCR can be computed as

$$V_{SCR} = \frac{1}{2\pi}\left[\int_0^{\alpha} v_s \, d\omega t + \int_{\pi}^{2\pi} v_s \, d\omega t\right] = -\frac{V_{max}}{2\pi}(1 + \cos\alpha)$$

$$V_{SCR} = -\frac{\sqrt{2}\,110}{2\pi}[1 + \cos(45)] = -42.27 \text{ V}$$

Note that the average voltage across the SCR is a negative quantity. This is expected, because most of the waveform across the SCR is negative when the SCR is not conducting, as shown in Figure 3.2.

For $\alpha = 90°$

a.
$$V_{rms} = \frac{V_{s\,rms}}{\sqrt{2}}\sqrt{\left[1 - \frac{\alpha}{\pi} + \frac{\sin(2\alpha)}{2\pi}\right]} = \frac{110}{\sqrt{2}}\sqrt{\left[1 - \frac{90(\pi/180)}{\pi}\right]} = 55 \text{ V}$$

b.
$$I_{rms} = \frac{V_{rms}}{R} = \frac{55}{2} = 27.5 \text{ A}$$

c.
$$V_{SCR} = -\frac{V_{max}}{2\pi}(1 + \cos\alpha) = -\frac{\sqrt{2}\,110}{2\pi} = -24.76 \text{ V}$$

It is expected that the rms voltage and rms current are lower when the triggering angle increases. However, it may not be obvious why the average voltage across the SCR is also lower. It is lower because more of the positive ac voltage waveform will be across the SCR when it is not conducting. As shown in Figure 3.2, when the positive section of the SCR voltage increases, the magnitude of the SCR's average voltage is reduced.

3.1.2 ELECTRIC POWER

Electric power is the average quantity of the instantaneous power, which is the direct multiplication of the instantaneous current and instantaneous voltage. The integration of power is known as energy. The power computation is very essential in power electronic circuits because it provides information on the flow of energy in the circuit as well as on circuit performance and efficiency. In circuits with purely sinusoidal waveforms, the computation of power is very straightforward. However, in circuits with discontinuities, the power must be computed by special methods. Three of these methods are discussed here.

3.1.2.1 THE RMS METHOD

This is the most common method for calculating electric power. It is based on the rms values of current or voltage. For a resistive load, electric power can be computed as

$$P = \frac{V_{rms}^2}{R} = I_{rms}^2 R \tag{3.9}$$

Substituting the value of V_{rms} of Equation (3.7) into Equation (3.9) yields

$$P = \frac{V_{max}^2}{8\pi R}[2(\pi - \alpha) + \sin(2\alpha)] \tag{3.10}$$

3.1.2.2 THE INSTANTANEOUS POWER METHOD

Electric power is also the average value of the instantaneous power $p(t)$.

$$p(t) = i(t)\, v(t) \tag{3.11}$$

where $i(t)$ is the instantaneous current and $v(t)$ is the instantaneous voltage. For the particular case of a resistive load, the instantaneous power can also be represented by

$$p(t) = \frac{v(t)^2}{R} \tag{3.12}$$

The power P consumed by the load is the average value of Equation (3.11) or (3.12):

$$P = \frac{1}{2\pi}\int_0^{2\pi} p(t)\, d\omega t = \frac{1}{2\pi}\int_0^{2\pi} \frac{v(t)^2}{R}\, d\omega t \tag{3.13}$$

FIGURE 3.5
Electric power of the load

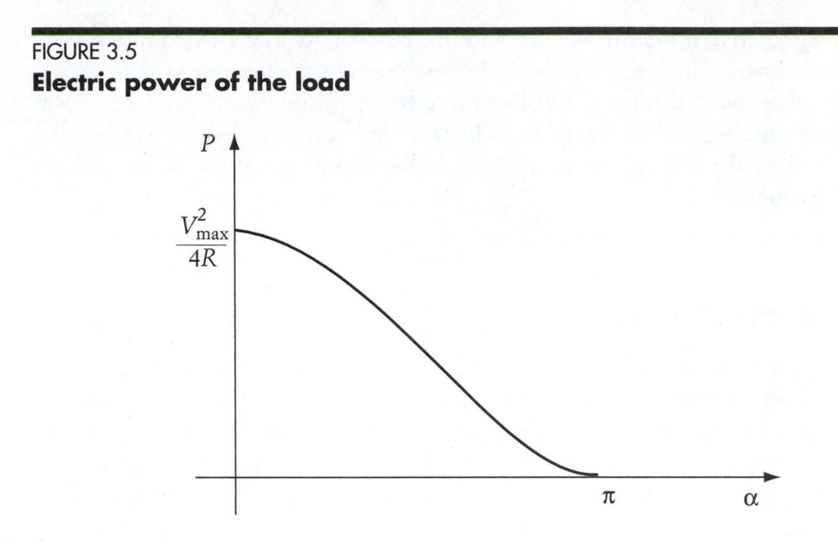

For the conduction period between α and π, the integration limits of Equation (3.13) can be changed to the limits of the conduction period.

$$P = \frac{1}{2\pi}\int_{\alpha}^{\pi} \frac{v(t)^2}{R}\, d\omega t = \frac{1}{2\pi}\int_{\alpha}^{\pi} \frac{[V_{max}\sin(\omega t)]^2}{R}\, d\omega t$$

$$P = \frac{V^2_{max}}{2\pi R}\int_{\alpha}^{\pi} [\sin(\omega t)]^2\, d\omega t = \frac{V^2_{max}}{4\pi R}\int_{\alpha}^{\pi} [1 - \cos(2\omega t)]\, d\omega t$$

$$P = \frac{V^2_{max}}{8\pi R}[2(\pi - \alpha) + \sin(2\alpha)] \tag{3.14}$$

Note that Equation (3.14) is identical to Equation (3.10).

Figure 3.5 shows the electric power as a function of the triggering angle α. The average power consumed by the load is determined by the triggering angle α (assuming the source voltage and load resistance are both constant). The maximum value of the average power occurred at $\alpha = 0$. At $\alpha = \pi$, the load power is zero.

3.1.2.3 THE HARMONICS METHOD

Based on the law of conservation, load power can also be calculated by using the equations at the source side. The process is more involved since the waveform of the source voltage is sinusoidal and that of the current is discontinuous. Let us start by defining the instantaneous power at the source side as

$$p(t) = v_s(t)\, i(t) \tag{3.15}$$

where $v_s(t)$ is the source voltage and $i(t)$ is the current of the circuit. Since the current is discontinuous, it can be represented by the Fourier series

$$i(t) = i_{dc} + i_1(t) + i_2(t) + i_3(t) + \cdots \tag{3.16}$$

where i_{dc} is a dc component of $i(t)$, $i_1(t)$ is the fundamental frequency component, $i_2(t)$ is the second harmonic current, and so on.

Only voltages and currents of the same frequency produce power. If the frequencies of the current and voltage are not identical, the average value of Equation (3.15) is zero. Assume that the source voltage is purely sinusoidal. Then the average power can only be produced by the fundamental component of $i(t)$.

$$P = V_{s\,\text{rms}} I_{1\,\text{rms}} \cos \phi_1 \tag{3.17}$$

where ϕ_1 is the phase shift between the fundamental frequency current and the source voltage. $I_{1\,\text{rms}}$ is the rms value of $i_1(t)$. $V_{s\,\text{rms}}$ is the rms value of the source voltage. Both ϕ_1 and $I_{1\,\text{rms}}$ can be computed by using the Fourier formula for fundamental components

$$i_1(t) = c_1 \sin(\omega t + \phi_1) \tag{3.18}$$

where

$$c_1 = \sqrt{a_1^2 + b_1^2} \tag{3.19}$$

$$\phi_1 = \tan^{-1}\left(\frac{a_1}{b_1}\right) \tag{3.20}$$

$$a_1 = \frac{1}{\pi} \int_0^{2\pi} i(\omega t) \cos(\omega t)\, d\omega t$$

$$b_1 = \frac{1}{\pi} \int_0^{2\pi} i(\omega t) \sin(\omega t)\, d\omega t$$

where $i(\omega t)$ is the total current of Equation (3.16), including all harmonics. This is the source voltage divided by the load resistance during the conduction period. Outside the conduction range, the current is equal to zero. Hence, a_1 and b_1 can be computed as follows:

$$a_1 = \frac{1}{\pi} \int_\alpha^{\pi} I_{\max} \sin(\omega t) \cos(\omega t)\, d\omega t = \frac{I_{\max}}{2\pi} [\cos(2\alpha) - 1] \tag{3.21}$$

$$b_1 = \frac{1}{\pi} \int_\alpha^{\pi} I_{\max} [\sin(\omega t)]^2\, d\omega t = \frac{I_{\max}}{4\pi} [\sin(2\alpha) + 2(\pi - \alpha)] \tag{3.22}$$

where

$$I_{\max} = \frac{V_{s\,\max}}{R}$$

The phase shift angle of the fundamental current is

$$\phi_1 = \tan^{-1}\left(\frac{a_1}{b_1}\right) = \tan^{-1}\left[\frac{2\ (\cos(2\alpha) - 1)}{\sin(2\alpha) + 2(\pi - \alpha)}\right] \tag{3.23}$$

The angle ϕ_1 is also called the displacement angle. The cosine of this angle is called the displacement power factor, or *DPF.*

$$DPF = \cos\phi_1 = \frac{b_1}{c_1} \tag{3.24}$$

Equation (3.23) shows that when α is equal to zero, the current of the fundamental frequency and the source voltage are in phase, and the displacement angle is equal to zero.

Let us go back to the calculation of the power. We rewrite Equation (3.17) as

$$P = V_{s\,\text{rms}} I_{1\,\text{rms}} \cos\phi_1 = \frac{V_{s\,\max} I_{1\,\max}}{2} \cos\phi_1 = \frac{V_{s\,\text{rms}}\, c_1}{2} \cos\phi_1$$

Replacing c_1 by the formula in Equation (3.19) and substituting for $\cos\phi_1$ yields

$$P = \frac{V_{s\,\max}\sqrt{a_1^2 + b_1^2}}{2} \frac{b_1}{\sqrt{a_1^2 + b_1^2}} = V_{s\,\max}\frac{b_1}{2}$$

$$P = \frac{V_{s\,\max} I_{\max}}{8\pi}\left[2(\pi - \alpha) + \sin\,(2\alpha)\right] = \frac{V_{s\,\max}^2}{8\pi R}\left[2(\pi - \alpha) + \sin\,(2\alpha)\right] \tag{3.25}$$

Note that the calculations of power using Equation (3.10), (3.14), or (3.25) give identical results.

3.1.3 dc POWER

Another widely used term for electric power is known as dc power, P_{dc}, which is defined as

$$P_{dc} = V_{ave} I_{ave} \tag{3.26}$$

Keep in mind that the dc power is not equal to the power computed in Equation (3.17); it is mainly used as a simple and approximate number.

EXAMPLE 3.3

An ac/dc, single-phase SCR converter is connected to a 10 Ω resistive load. The voltage on the ac side is 110 V (rms). The triggering angle of the converter is 60°. Compute the power dissipated in the load resistance using the instantaneous power and the harmonics method.

SOLUTION

The power computed by the instantaneous power method is given in Equation (3.14):

$$P = \frac{V_{\max}^2}{8\pi R}\left[2(\pi - \alpha) + \sin(2\alpha)\right] = \frac{2(110)^2}{8\pi(10)}\left[2\left(\pi - 60\frac{\pi}{180}\right) + \sin(120)\right] \approx 486\ \text{W}$$

Now we use the harmonics method:

$$a_1 = \frac{I_{max}}{2\pi}[\cos(2\alpha) - 1] = \frac{V_{max}}{2\pi R}[\cos(120) - 1] = -3.71$$

$$b_1 = \frac{I_{max}}{4\pi}[\sin(2\alpha) + 2(\pi - \alpha)] = 6.24$$

$$c_1 = \sqrt{a_1^2 + b_1^2} = 7.26$$

$$\phi_1 = \tan^{-1}\left(\frac{a_1}{b_1}\right) = 30.73° \text{ lagging}$$

$$P = V_{s\,rms}\, I_{1\,rms} \cos(\phi_1) = \frac{V_{s\,rms}\, c_1}{2}\cos(\phi_1) = \frac{\sqrt{2}(110)(7.26)}{2}\cos(30.73) \approx 486 \text{ W}$$

3.1.4 POWER FACTOR

As described in Equation (3.11), instantaneous power is defined as the multiplication of the instantaneous current by the instantaneous voltage. In a purely sinusoidal circuit, the voltage and current can be expressed as

$$v(t) = V_{max} \sin(\omega t)$$

$$i(t) = I_{max} \sin(\omega t - \phi)$$

where ϕ is the phase shift between the voltage and current waveforms. In a purely resistive load, the phase shift is zero. In a purely capacitive and inductive load, $\phi = 90°$ and $\phi = -90°$, respectively. The general expression of instantaneous power is

$$p(t) = v(t)\, i(t) = V_{max}\, I_{max} \sin(\omega t) \sin(\omega t - \phi)$$

$$p(t) = \frac{V_{max}\, I_{max}}{2}[\cos(\phi) - \cos(2\omega t - \phi)]$$

Recalling that the rms voltage and current are $V_{max}/\sqrt{2}$ and $I_{max}/\sqrt{2}$, we can write the equation of instantaneous power as

$$p(t) = V_{rms}\, I_{rms} \cos\phi - V_{rms}\, I_{rms} \cos(2\omega t - \phi)] \tag{3.27}$$

The first term of Equation (3.27) is the real power, and the second term is the reactive power. The power that produces work is the average value of $p(t)$. When we compute the average power (P) by using Equation (3.27), only the first term will be nonzero.

$$P = \frac{1}{2\pi}\int_0^{2\pi} p(t)\, d\omega t = V_{rms}\, I_{rms} \cos\phi \tag{3.28}$$

where $\cos\phi$ is called the power factor, or *pf.* The apparent power S is defined as

$$S = V_{\text{rms}} I_{\text{rms}}$$

Hence,

$$pf = \cos(\phi) = \frac{P}{S} \tag{3.29}$$

The *pf* is a good indicator of how much of the apparent power is converted into work. For given values of voltage and current, the maximum value of P is for a resistive load when $\phi = 0$.

In circuits with current harmonics, the power factor is defined in a similar way. As you recall from Equation (3.17), a voltage and a current of the same frequency produce power. Assume that the ac supply is a sinusoidal voltage source; then

$$P = V_{\text{rms}} I_{1\ \text{rms}} \cos\phi_1$$

where $I_{1\ \text{rms}}$ is the rms value of the current component at the frequency of the supply voltage. This is known as the fundamental component or first harmonic. The phase angle ϕ_1 is defined in Equation (3.20); this is the phase shift between the voltage waveform and the fundamental component of the current waveform. Since ϕ_1 is the phase shift of the fundamental component alone, it is called the displacement power factor angle. As given in Equation (3.24), $\cos\phi_1$ is called the displacement power factor, or *DPF.*

The *pf* of a system with harmonics is defined the same way as given in Equation (3.29):

$$pf = \frac{P}{S} = \frac{V_{\text{rms}} I_{1\ \text{rms}} \cos\phi_1}{V_{\text{rms}} I_{\text{rms}}} = \frac{I_{1\ \text{rms}}}{I_{\text{rms}}} \cos\phi_1 = \frac{I_{1\ \text{rms}}}{I_{\text{rms}}} DPF \tag{3.30}$$

Note that when the harmonics are severe, the ratio of the fundamental current to the total current is small. This results in a low power factor even for a resistive load. A poor power factor is a very undesirable aspect of power electronics, since it indicates that only a small portion of apparent power S is converted to work.

EXAMPLE 3.4

For the circuit given in Example 3.3, compute the power factor at the ac side.

SOLUTION

In Example 3.3, we computed the following parameters:

$$a_1 = -3.71$$

$$b_1 = 6.24$$

$$c_1 = \sqrt{a_1^2 + b_1^2} = 7.26$$

$$\phi_1 = \tan^{-1}\left(\frac{a_1}{b_1}\right) = 30.73° \text{ lagging}$$

Now let us compute the rms value of the fundamental current:

$$I_{1\ \mathrm{rms}} = \frac{c_1}{\sqrt{2}} = \frac{7.26}{\sqrt{2}} = 5.13 \text{ A}$$

The total rms current can be computed by using the load rms voltage as given in Equation (3.8):

$$I_{\mathrm{rms}} = \frac{V_{\mathrm{rms}}}{R} = \frac{V_{s\ \mathrm{rms}}}{R}\sqrt{\left[1 - \frac{\alpha}{\pi} + \frac{\sin(2\alpha)}{2\pi}\right]} = 9.86 \text{ A}$$

The power factor can now be computed as

$$pf = \frac{I_{1\ \mathrm{rms}}}{I_{\mathrm{rms}}}\cos(\phi_1) = \frac{5.13}{9.86}\cos(30.73) = 0.447$$

Note that the power factor is poor, even for this resistive load. If the circuit has no switching device, the harmonics will not be present, and the power factor will be equal to one.

3.2 SINGLE-PHASE, FULL-WAVE, ac/dc CONVERSION FOR RESISTIVE LOADS

Figure 3.6 shows an example of a full-wave SCR circuit, which is known as a full-wave bridge. The circuit consists of four solid-state switches (SCRs) and a resistive load. $S_1$and S_2 are triggered when point A of v_s is positive, whereas S_3 and S_4 are triggered when it is negative. The waveforms of the full wave circuit are shown in Figure 3.7. When S_1 and S_2 are triggered, the current i_1 flows in the direction of the solid arrows. The current will cease conduction when the potential of point A reaches zero at 180°. The waveform of the load voltage during this period is identical to that of v_s. During the negative half of the sine wave, point B is positive. When S_3 and S_4 are triggered, the current i_2 flows in the direction of the dashed arrows. Due to this bridge arrangement, i_1 and i_2 flow in the same direction in the load branch. This makes point C always positive with respect to point D. Hence, the load current and the load voltage do not change direction in either part of the ac cycle of v_s.

The analysis of the full-wave circuit is almost identical to that of the half-wave circuit. Since we are using both halves of the sine wave, the average voltage across the load is double that given in Equation (3.4):

$$V_{\mathrm{ave}} = \frac{1}{\pi}\int_{\alpha}^{\pi} v_s \, d\omega t = \frac{1}{\pi}\int_{\alpha}^{\pi} V_{\max}\sin(\omega t)\, d\omega t = \frac{V_{\max}}{\pi}(1 + \cos\alpha) \quad (3.31)$$

FIGURE 3.6
Full-wave SCR circuit

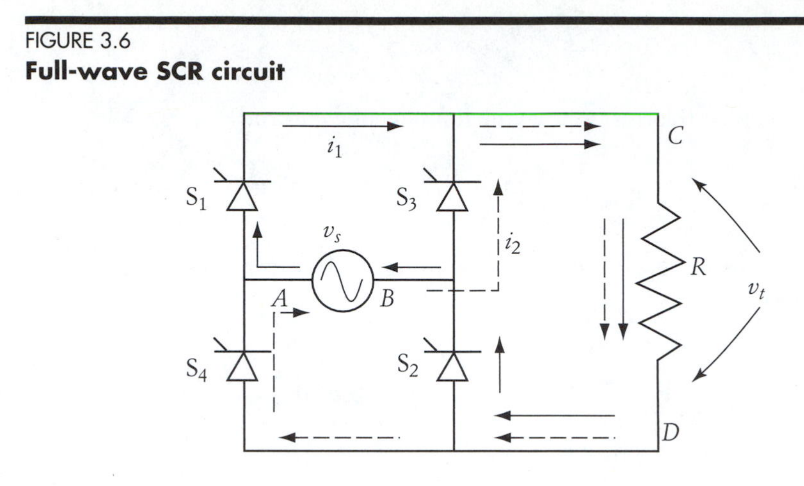

FIGURE 3.7
Waveforms of full-wave SCR circuit

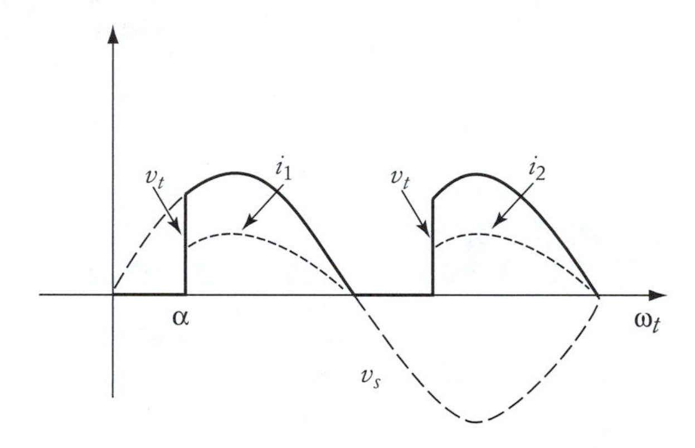

Note that the rms voltage across the load in full-wave circuits *is not* double the value in the half-wave circuit. To explain this, let us modify Equations (3.6) through (3.8) to accommodate the full-wave analysis.

$$V_{\text{rms}} = \sqrt{\frac{1}{2\pi}\int_0^{2\pi} v(t)^2 \, d\omega t} = \sqrt{\frac{1}{2\pi}\int_0^{2\pi} [V_{\max}\sin(\omega t)]^2 \, d\omega t} \tag{3.32}$$

If we assume that the triggering of the SCRs is symmetrical in both halves of the ac cycle, Equation (3.32) can be written as

$$V_{\text{rms}} = \sqrt{\frac{V_{\max}^2}{\pi}\int_\alpha^{\pi} \sin(\omega t)^2 \, d\omega t} = \sqrt{\frac{V_{\max}^2}{2\pi}\int_\alpha^{\pi} [1 - \cos(2\omega t)] \, d\omega t} \tag{3.33}$$

$$V_{rms} = \frac{V_{max}}{\sqrt{2}}\sqrt{\left[1 - \frac{\alpha}{\pi} + \frac{\sin(2\alpha)}{2\pi}\right]}$$

The term $\frac{V_{max}}{\sqrt{2}}$ is the rms voltage of the sinusoidal source, $V_{s\,rms}$:

$$V_{rms} = V_{s\,rms}\sqrt{\left[1 - \frac{\alpha}{\pi} + \frac{\sin(2\alpha)}{2\pi}\right]} \quad (3.34)$$

Comparing Equations (3.8) and (3.34), we can conclude that the load rms voltage of the full-wave circuit is $\sqrt{2}$ larger than that for the half-wave circuit (not double!).

The electric power of the load P in a full-wave circuit can also be calculated by the method described in Equations (3.9) and (3.10).

$$P = \frac{V_{rms}^2}{R} = \frac{V_{max}^2}{4\pi R}[2(\pi - \alpha) + \sin(2\alpha)] \quad (3.35)$$

Note that the load power in the full-wave circuit is double that for the half-wave circuit.

EXAMPLE 3.5

A full-wave, ac/dc converter is connected to a resistive load of 5 Ω. The voltage of the ac source is 110 V(rms). It is required that the rms voltage across the load be 55 V. Calculate the triggering angle and the load power.

SOLUTION

The rms voltage across the load is given as

$$V_{rms} = V_{s\,rms}\sqrt{\left[1 - \frac{\alpha}{\pi} + \frac{\sin(2\alpha)}{2\pi}\right]}$$

$$55 = 110\sqrt{\left[1 - \frac{\alpha}{\pi} + \frac{\sin(2\alpha)}{2\pi}\right]}$$

$$2.25 = \alpha\frac{\pi}{180} - \frac{\sin(2\alpha)}{2}$$

By iteration, we can solve for α:

$$\alpha \approx 112.5^\circ$$

The load power can be computed using the rms voltage across the load:

$$P = \frac{V_{rms}^2}{R} = \frac{(55)^2}{5} = 605 \text{ W}$$

3.3 SINGLE-PHASE, HALF-WAVE, ac/dc CONVERSION FOR INDUCTIVE LOADS WITHOUT FREEWHEELING DIODE

Inductors do not consume energy but rather exchange energy with the source. In any half cycle, an inductor can temporarily store energy during one-fourth of the cycle and then return the energy back to the source during the next one-fourth. Inductors cannot permanently store energy, and hence the net energy during the period of current conduction must be zero.

Assume that a mechanical switch is used to interrupt the current of an inductor. When the switchblades open while the inductor energy is not fully returned to the source, the voltage across the switchblades increases rapidly, creating an arc between the blades. The arc keeps the current flowing until the inductor energy is totally dissipated. This explains the arc between an ac outlet and a load plug when we disconnect highly inductive loads such as electric saws or home appliances.

In the steady-state operation, the current of a purely inductive load lags the voltage by 90°. Hence,

$$v(t) = V_{\max} \sin(\omega t)$$

$$i(t) = -I_{\max} \cos(\omega t)$$

The instantaneous power consumed by a given load is the multiplication of the instantaneous voltage across the load times the instantaneous load current. For a purely inductive load, the instantaneous power is

$$p(t) = v(t)\, i(t) = -\frac{V_{\max} I_{\max}}{2} \sin(2\omega t) \tag{3.36}$$

Two basic characteristics can be obtained from Equation (3.36). The first is that the average power (the one that produces energy) is zero. The second is that the frequency of the instantaneous power $p(t)$ is double the frequency of the voltage or current.

Figure 3.8 shows the waveforms associated with a sinusoidal circuit with purely inductive load. The average power of an inductor is equal to zero for any half of the cycle. Note that the power is averaged to zero at all the zero crossings of the current. Hence, if the switch is opened at exactly the zero crossings of the current, the inductor energy is zero and no arc is produced.

Let us now consider the solid-state switching circuit of Figure 3.9. It consists of an inductance and a resistance and is powered from a single-phase, half-wave, ac-to-dc converter. In purely sinusoidal circuits, the voltage across the resistance v_R is in phase with the current, and the voltage across the inductor v_L is leading the current by 90°.

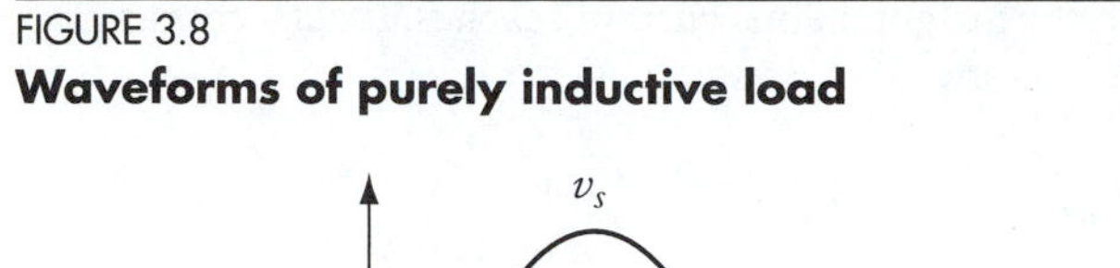

FIGURE 3.8
Waveforms of purely inductive load

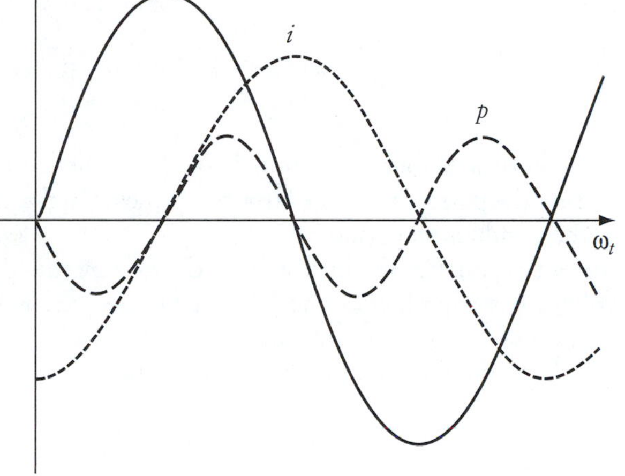

FIGURE 3.9
Half-wave inductive circuit

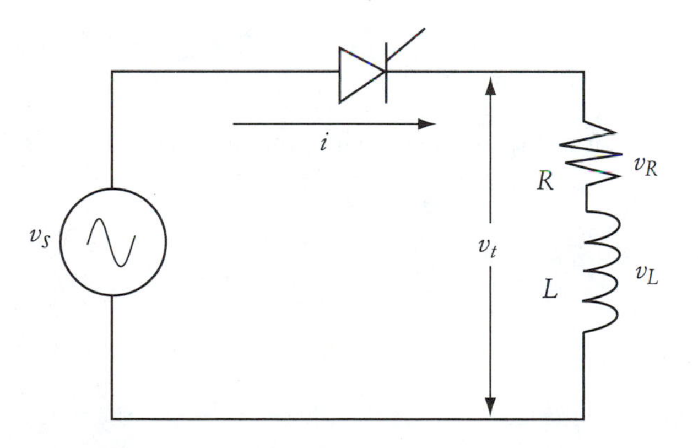

Let us assume that the voltage source of the circuit in Figure 3.9 is sinusoidal. We also assume that the SCR is triggered at an angle α. Because of the SCR switching, we would expect the current to be discontinuous. To calculate the load current, let us start by defining the terminal voltage across the entire load v_t as

$$v_t = v_s(u_\alpha - u_\beta) = [V_{\max} \sin(\omega t)](u_\alpha - u_\beta) \qquad (3.37)$$

where β is the angle at which the instantaneous current reaches its zero crossing. u_α and u_β are step functions defined by

$$u_\alpha = u(\omega t - \alpha) \qquad \begin{cases} u_\alpha = 0 & \text{for } \omega t < \alpha \\ u_\alpha = 1 & \text{for } \omega t \geq \alpha \end{cases}$$

$$u_\beta = u(\omega t - \beta) \qquad \begin{cases} u_\beta = 0 & \text{for } \omega t < \beta \\ u_\beta = 1 & \text{for } \omega t \geq \beta \end{cases}$$

Equation (3.37) indicates that the terminal voltage of the load is equal to the source voltage during the period when the current is flowing in the circuit. This period is known as the conduction period.

The load current is equal to the load voltage divided by the load impedance. Since the load voltage has step functions and the load impedance is a complex variable, the computation of the load current can be simplified by using Laplace transformations. Let $V(S)$ denote the Laplace transformation of v_t.

$$V(S) = \mathcal{L}v_t = V_{\max}\left[\frac{S\sin\alpha + \omega\cos\alpha}{S^2 + \omega^2} e^{(-\alpha S/\omega)} - \frac{S\sin\beta + \omega\cos\beta}{S^2 + \omega^2} e^{(-\beta S/\omega)}\right] \tag{3.38}$$

The load current is the load voltage divided by the load impedance

$$I(S) = \frac{V(S)}{Z(S)} = \frac{V(S)}{R + SL}$$

$$I(S) = \frac{V_{\max}}{R + SL}\left[\frac{S\sin\alpha + \omega\cos\alpha}{S^2 + \omega^2} e^{(-\alpha S/\omega)} - \frac{S\sin\beta + \omega\cos\beta}{S^2 + \omega^2} e^{(-\beta S/\omega)}\right] \tag{3.39}$$

Inverting the current of Equation (3.39) into the time domain yields

$$i(t) = \mathcal{L}^{-1} I(S)$$

$$i(t) = \frac{V_{\max}}{z}[(u_\alpha - u_\beta)\sin(\omega t - \phi) +$$

$$u_\alpha[\sin(\phi - \alpha)]\, e^{-[(\omega t - \alpha)/\omega\tau]} - u_\beta \sin(\phi - \beta) e^{-[(\omega t - \beta)/\omega\tau]}] \tag{3.40}$$

where z, ϕ, and τ are the load impedance, phase angle, and the time constant of the load, respectively.

$$z = \sqrt{R^2 + (\omega L)^2}$$

$$\phi = \tan^{-1}\left(\frac{\omega L}{R}\right)$$

$$\tau = \frac{L}{R}$$

Upon a close examination of Equation (3.40), one would note that the load current has three components. The first is sinusoidal, shifted by the load power factor angle ϕ. This component is only present during the conduction period (between α and β). The second and third components are exponentially decaying functions. The second component is present after the triggering angle α, whereas the third component is active after β. One may simplify Equation (3.40) during the conduction period by setting $u_\alpha = 1$ and $u_\beta = 0$.

$$i(t) = \frac{V_{\max}}{z}\left[\sin(\omega t - \phi) + [\sin(\phi - \alpha)]\, e^{-[(\omega t - \alpha)/\omega\tau]}\right] \tag{3.41}$$

Figure 3.10 shows the waveforms of the load current and voltage. Note that the current is a deformed half-wave. To calculate the angle at which the current is at maximum, we set the derivative of Equation (3.41) to zero:

$$\frac{\partial\, i(t)}{\partial\, \omega t} = \frac{V_{\max}}{z}\left[\cos(\omega t - \phi) - \frac{1}{\omega\tau}[\sin(\phi - \alpha)]\, e^{-[(\omega t - \alpha)/\omega\tau]}\right] = 0$$

or

$$\omega\tau \cos(\omega t - \phi) = [\sin(\phi - \alpha)]\, e^{-[(\omega t - \alpha)/\omega\tau]}$$

The computation of ωt is numerical. A reasonable approximation is to assume that the exponential term is fast decaying. In this case, the peak of the current occurs at $\omega t = 90 + \phi$.

FIGURE 3.10
Current waveform of the circuit in Figure 3.9

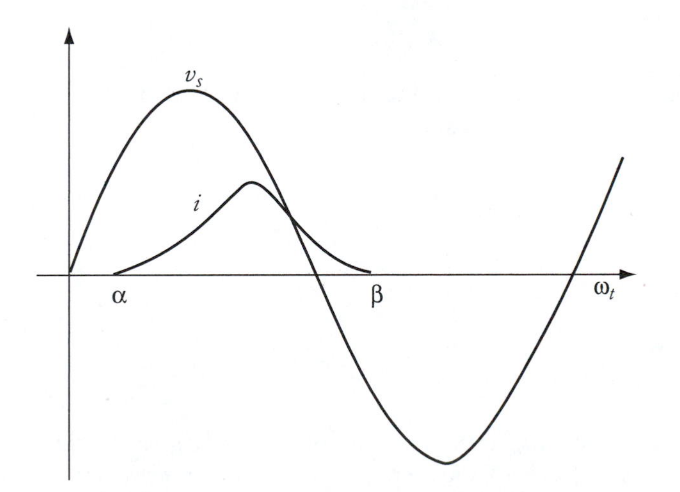

The instantaneous current of Equations (3.40) and (3.41) is equal to zero at the boundary conditions (at α and β). The conduction period $\gamma = \beta - \alpha$ can be computed by setting ωt of Equation (3.40) or (3.41) equal to β and the instantaneous current equal to zero.

$$i(\beta) = \frac{V_{max}}{z}\left[\sin(\beta - \phi) + [\sin(\phi - \alpha)]\, e^{-[(\beta-\alpha)/\omega\tau]}\right] = 0 \qquad (3.42)$$

If the load impedance is known and the system voltage is given, β of Equation (3.42) can be computed using an iterative method.

To compute the average current, you can integrate the expression of Equation (3.41), which is a rather laborious process. Another simpler method is first to compute the average load voltage, and then calculate the average load current. As we explained earlier, the load voltage is equal to the source voltage during the conduction period. Hence, the average voltage of the load V_{ave} can be obtained by the following integration:

$$V_{ave} = \frac{1}{2\pi}\int_0^{2\pi} v_s(u_\alpha - u_\beta)\, d\omega t = \frac{1}{2\pi}\int_\alpha^\beta v_s\, d\omega t = \frac{1}{2\pi}\int_\alpha^\beta V_{max}\sin(\omega t)\, d\omega t$$

$$= \frac{V_{max}}{2\pi}[\cos\alpha - \cos\beta)] \qquad (3.43)$$

3.3.1 AVERAGE VOLTAGE ACROSS INDUCTANCE AND RESISTANCE

The average voltage across the load V_{ave} is equal to the average voltage across the resistance $V_{R\,ave}$ plus the average voltage across the inductance $V_{L\,ave}$.

$$V_{ave} = V_{R\,ave} + V_{L\,ave}$$

The average voltage across the inductance is zero and can be shown by integrating the instantaneous voltage across the inductance v_L.

$$v_L = L\frac{di}{dt}$$

Hence,

$$V_{L\,ave} = \frac{1}{T}\int_{t_\alpha}^{t_\beta} v_L\, dt = \frac{L}{T}\int_{t_\alpha}^{t_\beta} di = \frac{L}{T}[i(t_\beta) - i(t_\alpha)]$$

where T is the period, $i(t_\alpha)$ is the load current at α, and $i(t_\beta)$ is the current at β. Since the current at α or β is zero, the average voltage across the inductive element of the

FIGURE 3.11
Waveforms of inductive load in a switching circuit

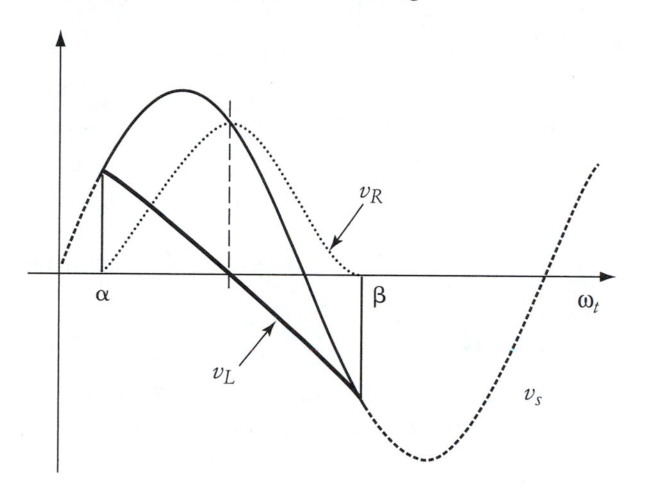

load is zero. Hence, the average voltage across the entire load must equal the average voltage across its resistive element alone.

$$V_{\text{ave}} = V_{R\,\text{ave}} = RI_{\text{ave}} \tag{3.44}$$

Figure 3.11 shows four waveforms: the source voltage, the voltage across the load, the voltage across the resistive component, and the voltage across the inductive component. The conduction started at α and stopped at β. The voltage waveform across the load is identical to the waveform of the source voltage during the conduction period. Note that the load voltage becomes negative because, due to the presence of the inductance, the current is in conduction beyond 180°. The voltage across the resistive component of the load must have the same exact shape as the current. Hence, it is also unidirectional. The voltage across the inductive component of the load is the difference between the total voltage across the load and the voltage across the resistive element. The inductive voltage is bidirectional (has positive and negative values) where the positive area must equal the negative area by the end of the conduction period $\omega t = \beta$.

3.3.2 AVERAGE POWER OF INDUCTANCE

As shown in Figure 3.8, the inductance cannot store energy and the average power of the inductance P_L is zero. This is evident when you integrate the instantaneous power of the inductive load.

$$P_L = \frac{1}{2\pi}\int_{\alpha}^{\beta} p_L \, d\omega t = 0$$

where p_L is the instantaneous power of the inductive load. This can be proven by the following analysis:

$$p_L = v_L i = L \frac{di}{dt} i$$

Hence, the average power of the inductive load is

$$P_L = \frac{1}{T} \int_{t_\alpha}^{t_\beta} Li\, di = \frac{L}{2T} [i^2(t_\beta) - i^2(t_\alpha)]$$

Since the current at α or β is zero, the average power of the inductance is zero.

3.3.3 RMS VOLTAGE

To calculate the rms voltage across the load, repeat the process described in Equation (3.6). Notice that the integration limits should be between α and β.

$$V_{rms} = \sqrt{\frac{1}{2\pi} \int_0^{2\pi} v_t(\omega t)^2\, d\omega t} = \sqrt{\frac{1}{2\pi} \int_\alpha^\beta [V_{max} \sin(\omega t)]^2\, d\omega t}$$

$$V_{rms} = \sqrt{\frac{V_{max}^2}{4\pi} \int_\alpha^\beta [1 - \cos(2\omega t)]\, d\omega t} = \frac{V_{max}}{2\sqrt{\pi}} \sqrt{\left[\gamma - \frac{\sin(2\beta) - \sin(2\alpha)}{2}\right]} \quad (3.45)$$

where the conduction period $\gamma = \beta - \alpha$.

In the case of inductive or capacitive load, the rms current cannot be assumed equal to the rms voltage divided by the load impedance at the steady-state frequency, because the current is composed of various harmonics and the impedance is frequency-dependent. In this case, the rms current should be computed using the instantaneous current in Equation (3.41). The rms power is still equal to the square of the rms current multiplied by the load resistance.

EXAMPLE 3.6

A half-wave SCR ac/dc converter is powering an inductive load. The resistance of the load is 10 Ω, and the inductance is 30 mH. The ac source is 100 V(rms) at 60 Hz. The SCR is triggered at 60°. Calculate the conduction period.

SOLUTION

To compute the conduction period, we need to compute β using equation (3.42).

$$i(\beta) = \frac{V_{max}}{z} [\sin(\beta - \phi) + [\sin(\phi - \alpha)]\, e^{-[(\beta-\alpha)/\omega\tau]}] = 0$$

$$\sin(\phi - \beta) = [\sin(\phi - \alpha)]\, e^{-[(\beta-\alpha)/\omega\tau]}$$

where

$$\omega = 2\pi \times 60 = 377 \text{ rad/sec}$$

$$\tau = \frac{L}{R} = \frac{0.03}{10} = 0.003 \text{ sec}$$

$$\phi = \tan^{-1}\left(\frac{\omega L}{R}\right) = 48.52°$$

Thus,

$$\sin(48.52 - \beta) = [\sin(48.52 - 60)]\, e^{-[(\beta-60)\,(\pi/180)/(377\times 0.003)]}$$

This equation can be solved using an iterative technique,

$$\beta \approx 230°$$

The conduction period γ is

$$\gamma = \beta - \alpha = 230 - 60 = 170°$$

Note that the conduction period is less than 180°, which is an indication that the current is discontinuous.

3.4 SINGLE-PHASE, HALF-WAVE, ac/dc CONVERSION FOR INDUCTIVE LOADS WITH FREEWHEELING DIODE

The freewheeling diode is a rectifier connected across the inductive load in opposite polarities to the SCR as shown in Figure 3.12. When the SCR is forward biased and triggered, the current i_s flows from the source to the load impedance; i_s does not go through the freewheeling diode. Since the load is inductive, the current must continue to flow beyond the zero crossing of the source voltage. However, when the terminal voltage of the load tends to reverse its polarity after 180° as shown in Figure 3.11, the diode becomes forward biased and starts conducting. Thus, the diode prevents the terminal voltage of the load from becoming negative (reversed). Once the diode starts conducting, the current i_s falls to zero, and the SCR is commutated (turned off). Beyond 180°, the current of the load is i_d, which flows in the diode–load loop until the inductor energy is totally dissipated.

Because of the presence of the diode, the voltage across the load cannot be negative. Any stored energy inside the inductor during the SCR conduction will be dissipated in the resistive component of the load when the diode conducts.

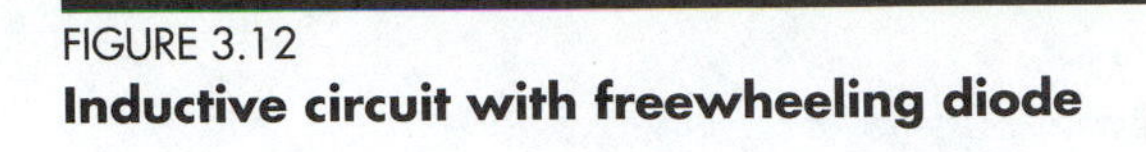
FIGURE 3.12
Inductive circuit with freewheeling diode

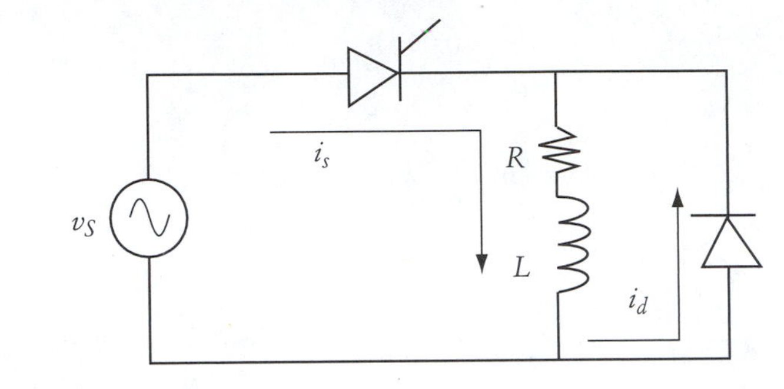

FIGURE 3.13
Analysis of the circuit in Figure 3.12

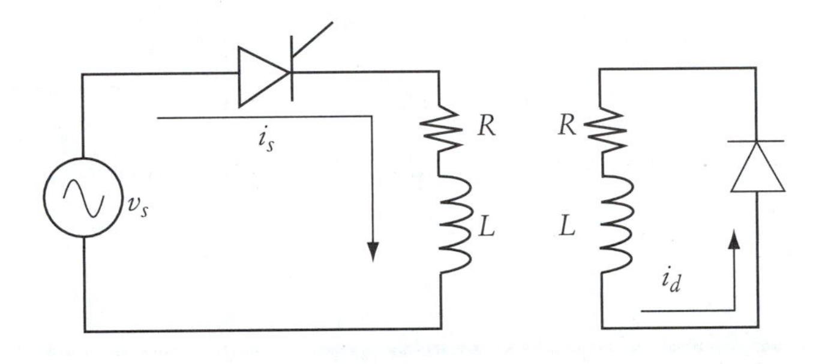

The analysis of this freewheeling circuit can be divided into two steps, as shown in Figure 3.13. In the first, the source current i_s for the period from α to π is computed. In the second, the current of the freewheeling diode during the period from π to β is calculated. The waveforms are shown in Figure 3.14.

The instantaneous value of i_s is calculated by Equation (3.41) up to π. The value of the current at π can be computed from Equation (3.41) by replacing ωt by π. The general form should include the initial condition of the current at the triggering angle α.

$$i_s(\pi) = \frac{V_{\max}}{\sqrt{R^2 + (\omega L)^2}} [\sin(\phi) + \sin(\phi - \alpha)e^{-(\pi - \alpha)/\omega\tau}] + I(\alpha) \qquad (3.46)$$

The current $i(\pi)$ is the load current when the SCR is about to commutate. This value is also the initial condition of the freewheeling diode current i_d. $I(\alpha)$ is the current at α, which is the initial condition when the SCR starts conducting. If the current is discontinuous, $I(\alpha)$ is zero. The diode current i_d flows in an R-L circuit. Thus, its instantaneous values must be exponentially decaying.

FIGURE 3.14
Waveforms of the circuit in Figure 3.13

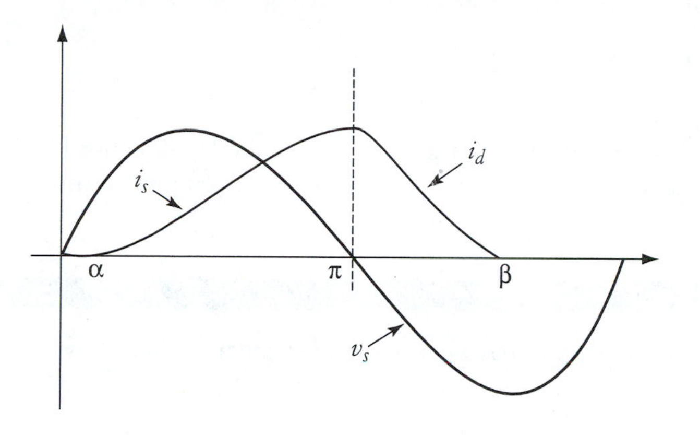

$$i_d = i(\pi)e^{-[(\omega t - \pi)/\omega\tau]}\,u(\omega t - \pi) \tag{3.47}$$

where $u(\omega t - \pi)$ is a unit step function activated at π. The time constant of the circuit τ is

$$\tau = \frac{L}{R}$$

The average current of the load can be calculated by adding the averages of i_s and i_d.

$$I_{ave} = \frac{V_{max}}{2\pi\sqrt{R^2 + (\omega L)^2}} \int_{\alpha}^{\pi} [\sin(\omega t - \phi) + \sin(\phi - \alpha)e^{-[(\omega t - \alpha)/\omega\tau]}]\, d\omega t + \frac{\beta - \alpha}{2\pi} I(\alpha) +$$

$$\frac{V_{max}}{2\pi\sqrt{R^2 + (\omega L)^2}} [\sin(\phi) + \sin(\phi - \alpha)e^{-[(\pi - \alpha)/\omega\tau]}] \int_{\pi}^{\beta} e^{-[(\omega t - \pi)/\omega\tau]}\, d\omega t \tag{3.48}$$

The solution of Equation (3.48) is very involved and requires the value of β. However, because the average voltage across the inductor is always equal to zero for any complete cycle, the computation of the average current is a relatively simple task. Start by calculating the average voltage across the entire load.

$$V_{ave} = \frac{1}{2\pi} \int_{\alpha}^{\pi} V_{max} \sin(\omega t)\, d\omega t = \frac{V_{max}}{2\pi}(1 + \cos\alpha) \tag{3.49}$$

Note that Equation (3.49) does not include the voltage beyond π, because the diode is conducting after π and the voltage across the entire load is equal to zero.

Since the average voltage across the inductor is zero, Equation (3.49) is also valid for the average voltage across the resistive component of the load. Hence, the average current can be calculated as given in Equation (3.50),

$$I_{ave} = \frac{V_{ave}}{R} \tag{3.50}$$

The power can be calculated by the method described in Equation (3.17). The computation requires the rms value of the fundamental component of the current.

EXAMPLE 3.7

An inductive circuit with freewheeling diode similar to that shown in Figure 3.12 has the following data:

$$Vs = 110\text{ V} \qquad L = 20\text{ mH} \qquad R = 10\ \Omega$$

The triggering angle of the SCR is adjusted to 60°. Calculate the following:

a. Conduction period
b. Maximum diode current
c. Average current of the diode
d. Average load current
e. Average current of the SCR

SOLUTION

a. To compute the conduction period, we must find β. To do this, we need to make a simple assumption that when the diode current i_d reaches 5% of its maximum value, the diode circuit is practically open. From Equation (3.47), we can write

$$\frac{i_d}{i(\pi)} = e^{-[(\omega t - \pi)/\omega \tau]}$$

$$\ln(0.05) = -\frac{\beta - \pi}{\omega \tau}$$

$$\beta = \pi - \omega\tau \ln(0.05) = \pi - 377\left(\frac{0.02}{10}\right)\ln(0.05) = 309°$$

The conduction period γ is

$$\gamma = \beta - \alpha = 309 - 60 = 249°$$

Note that the conduction period is more than 180° due to the presence of the diode.

b. The maximum current of the diode occurs at the initial time of the diode conduction, when $\omega t = \pi$. This value is the same as that for the SCR current at π. Using Equation (3.46), we get

$$i_s(\pi) = \frac{V_{\max}}{\sqrt{R^2 + (\omega L)^2}}\left[\sin(\phi) + \sin(\phi - \alpha)e^{-[(\pi-\alpha)/\omega\tau]}\right]$$

$I(\alpha)$ in Equation (3.46) is equal to zero since the conduction period is less than 360°. Now let us compute the parameters of the previous equation.

$$Z = \sqrt{R^2 + (\omega L)^2} = \sqrt{10^2 + (7.54)^2} = 12.52\ \Omega$$

$$\omega\tau = \frac{\omega L}{R} = 0.754$$

$$\phi = \tan^{-1}\left(\frac{\omega L}{R}\right) = 37°$$

Substituting these parameters in the current equation yields

$$i_d(\pi) = i_s(\pi) = \frac{\sqrt{2}\ 110}{12.52}\left[\sin(37) + \sin(37 - 60)e^{-[(\pi - 60(\pi/180))/0.754]}\right] = 7.18\text{ A}$$

If you compute the contribution of the exponential term, you will find it very small due to its fast decaying effect. The first term alone is 7.48 A.

c. The average current of the diode can be computed from Equation (3.47).

$$i_d = i(\pi)e^{-[(\omega t - \pi)/\omega\tau]}\, u(\omega t - \pi)$$

$$I_{d\text{ ave}} = \frac{i(\pi)}{2\pi}\int_{\pi}^{\beta} e^{-[(\omega t - \pi)/\omega\tau]}\, d\omega t = \frac{i(\pi)}{2\pi}(-\omega\tau)\left[e^{-[(\beta-\pi)/\omega\tau]} - 1\right]$$

$$I_{d\text{ ave}} = \frac{7.18}{2\pi}(-0.754)\left[e^{-[(309(\pi/180)-\pi)/0.754]} - 1\right] = 0.81\text{ A}$$

d. The average load current can be computed using Equation (3.50), but first let us calculate the average voltage of the load as given in Equation (3.49).

$$V_{\text{ave}} = \frac{V_{\max}}{2\pi}(1 + \cos\alpha) = \frac{\sqrt{2}\ 110}{2\pi}(1 + \cos 60) = 37.14\text{ V}$$

Then

$$I_{\text{ave}} = \frac{V_{\text{ave}}}{R} = \frac{37.14}{10} = 3.714\text{ A}$$

e. The average current of the SCR is the average current of the load minus the average current of the diode.

$$I_{SCR} = I_{ave} - I_{d\,ave} = 3.714 - 0.81 = 2.904 \text{ A}$$

3.5 THREE-PHASE, HALF-WAVE, ac/dc CONVERSION FOR RESISTIVE LOADS

A single load can be energized by a three-phase system through power electronic switches. Such an arrangement is common for loads with high power demands. Figure 3.15 shows a simple three-phase, half-wave, ac/dc converter. The converter consists of a three-phase supply, an SCR for each phase, and a resistive load. The anode of each SCR is connected to one of the phases of the ac source. The cathodes are commonly connected to the load. The second terminal of the load is connected to the neutral point of the three-phase source.

FIGURE 3.15
Three-phase, half-wave, ac/dc converter

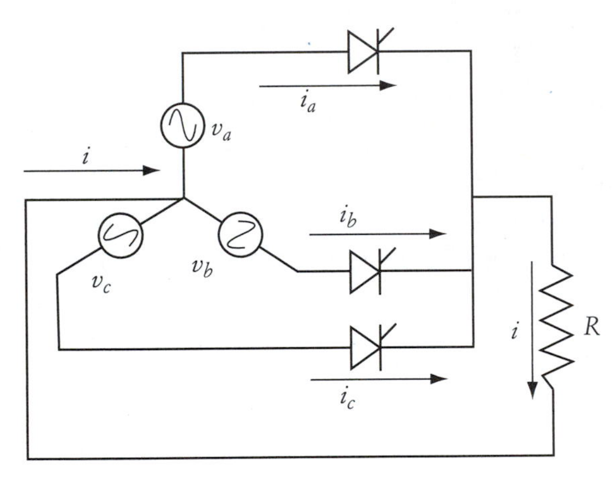

The operation of the circuit can be explained by the waveforms in Figure 3.16. The figure shows the three-phase voltages (v_a, v_b, and v_c) and the load current. As explained earlier, the SCR conducts when the voltage across its terminals (anode to cathode) is positive and the SCR receives a triggering pulse. The SCR commutates when the current falls below its holding value.

Let us start by assuming that the circuit starts its operation when the voltage of phase a is in the positive cycle. Assume that the triggering signal of the SCR of phase a is applied at α_a. The SCR closes and the current i_a flows to the load for the remainder of this half-cycle. The SCR of phase a commutates when the voltage v_a reaches zero. The voltage of phase b is now positive. If the SCR of phase b is triggered at α_b, the current i_b flows into the load until the voltage of phase b is zero. Then the voltage of phase c becomes positive. If the SCR of phase c is triggered at α_c, the current i_c flows into the load. This process is repeated every cycle.

In this case, the triggering angles of the three phases are separated by 120°. In summary, the current of each phase flows starting at its corresponding α and continues for the remainder of the positive part of the phase voltage. The SCR of the corresponding phase will be commutated naturally when its current falls below its holding value.

FIGURE 3.16
Waveforms of the circuit in Figure 3.15

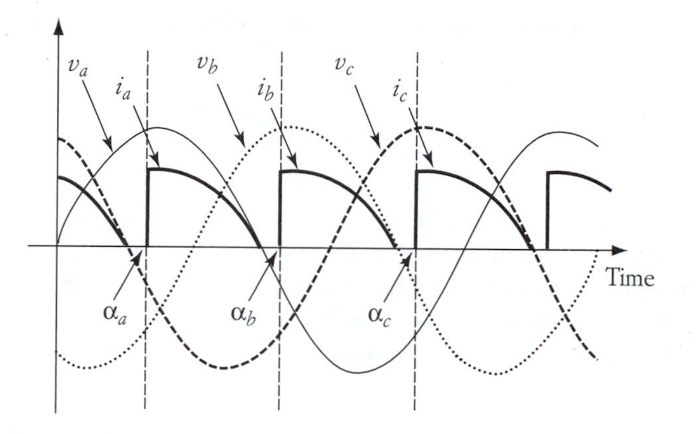

FIGURE 3.17
Waveforms for advanced triggering

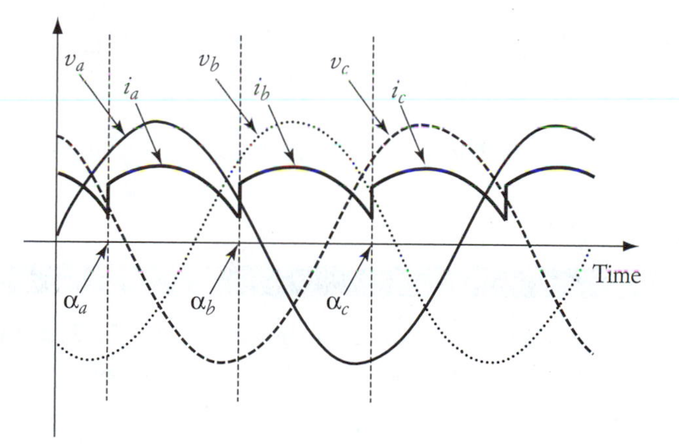

The load current in Figure 3.16 is discontinuous because of the delayed triggering angle. If the triggering is advanced, the current tends to increase in magnitude and duration. If the current duration is $\geq 120°$, the load current is continuous, as shown in Figure 3.17. The condition for continuous current is $\alpha_a \leq 60°$. Keep in mind that when the load is inductive, the current tends to be continuous even for delayed triggering.

To compute the average voltage across the load, we need to integrate the source voltage during the conduction period. The method is similar to the one described in Equation (3.4) for single-phase systems. The difference here is that we must take into account the three-phase quantities. Each phase contributes to the total current only once per cycle—the load current is composed of three, equal-current pulses. Therefore, the total average voltage across the load is the sum of the average voltages contributed by each phase. Nevertheless, since we are assuming that the three-phase system is balanced, we can compute the average voltage of one phase only and multiply it by 3.

$$V_{ave} = \frac{3}{2\pi}\int_{\alpha}^{\beta} v_s \, d\omega t = \frac{3}{2\pi}\int_{\alpha}^{\beta} V_{max}\sin(\omega t)\, d\omega t = \frac{3V_{max}}{2\pi}(\cos\alpha - \cos\beta) \quad (3.51)$$

where V_{max} is the peak value of the phase voltage (phase-to-neutral). If we are using SCRs as switches and the current is continuous, $\beta = 120° + \alpha$. If these switches are transistors, β is determined by the time the base current is turned off. The average current in this circuit can still be computed by using the average voltage across the load resistance and the resistance itself.

$$I_{ave} = \frac{V_{ave}}{R}$$

The average power can also be calculated similar to the single-phase circuits, but here we must take into account the three-phase quantities.

$$P = \frac{3}{2\pi}\int_{\alpha}^{\beta} \frac{v(t)^2}{R}\, d\omega t = \frac{3}{2\pi}\int_{\alpha}^{\beta} \frac{[V_{max}\sin(\omega t)]^2}{R}\, d\omega t \quad (3.52)$$

$$P = \frac{3V_{max}^2}{4\pi R}\int_{\alpha}^{\beta}[1 - \cos(2\omega t)]\, d\omega t = \frac{3V_{max}^2}{8\pi R}[2(\beta - \alpha) + \sin(2\alpha) - \sin(2\beta)]$$

EXAMPLE 3.8

The balanced, three-phase, ac/dc converter shown in Figure 3.15 has the following parameters:

$$V_{ab} = 208 \text{ V} \qquad R = 10\ \Omega$$

Calculate the power delivered to the load when the triggering angle is 80° and 30°.

SOLUTION

At $\alpha_a = 80°$, the current is discontinuous. Thus, $\beta_a = 180°$.

$$P = \frac{3V_{max}^2}{8\pi R}[2(\beta - \alpha) + \sin(2\alpha) - \sin(2\beta)]$$

$$P = \frac{3(\sqrt{2}\,208/\sqrt{3})^2}{8\pi(10)}\left[200\frac{\pi}{180} + \sin(160) - \sin(360)\right] = 1.32 \text{ kW}$$

At $\alpha_a = 30°$ the current is continuous, and $\beta = \alpha_a + 120$.

$$P = \frac{3V_{max}^2}{8\pi R}[2(\beta - \alpha) + \sin(2\alpha) - \sin(2\beta)]$$

$$P = \frac{3(\sqrt{2}\,208/\sqrt{3})^2}{8\pi(10)}\left[240\frac{\pi}{180} + \sin(60) - \sin(300)\right] = 2.042 \text{ kW}$$

3.6 THREE-PHASE, HALF-WAVE, ac/dc CONVERSION FOR INDUCTIVE LOADS

We have explained earlier that inductive loads make currents flow beyond the zero crossing of their corresponding phase voltage. In a three-phase switching circuit, similar to that shown in Figure 3.18, the current of the load could be continuous or pulsating (discontinuous), depending on the triggering angle and the size of the inductive element. If an inductive load causes the current to flow for less than 120° in each phase, the load current is pulsating; that is, the conduction period of each phase γ is

$$\gamma = \beta - \alpha < 120°$$

The current expression is similar to that given in Equation (3.41), but is modified for three-phase circuits with discontinuous currents.

$$i(\omega t) = \frac{V_{max}}{\sqrt{R^2 + (\omega L)^2}}\{[\sin(\omega t - \phi) + \sin(\phi - \alpha)e^{-[(\omega t - \alpha)/\omega\tau]}](u_\alpha - u_\beta) +$$

$$[\sin(\omega t - 120 - \phi) + \sin(\phi - \alpha)e^{-[(\omega\tau - 120 - \alpha)/\omega\tau]}] + (u_{\alpha+120} - u_{\beta+120}) +$$

$$[\sin(\omega t - 240 - \phi) + \sin(\phi - \alpha)e^{-[(\omega\tau - 240 - \alpha)/\omega\tau]}](u_{\alpha+240} - u_{\beta+240})\} \quad (3.53)$$

where $u_{\alpha+120}$ is a step function defined as $u[(\omega t - (\alpha + 120)]$.

Figure 3.19 shows the waveforms of the inductive circuit in Figure 3.18. In Figure 3.19, the conduction period is assumed to be $< 120°$. Unlike the current of a resistive load, the current in an inductive circuit does not swiftly change at the triggering angle. When the conduction period of each phase is 120°, the current is continuous, as shown in Figure 3.20.

FIGURE 3.18
Three-phase, half-wave, ac/dc converter for inductive load

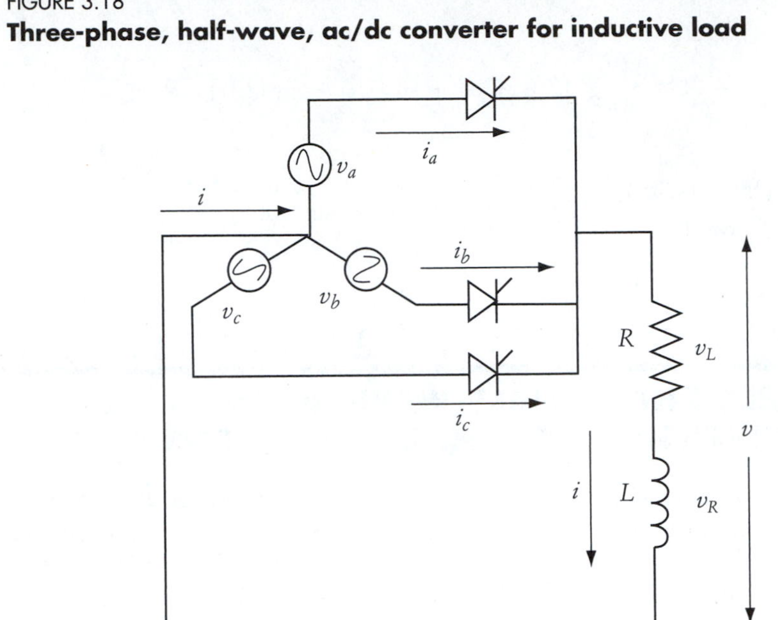

FIGURE 3.19
Pulsating load current

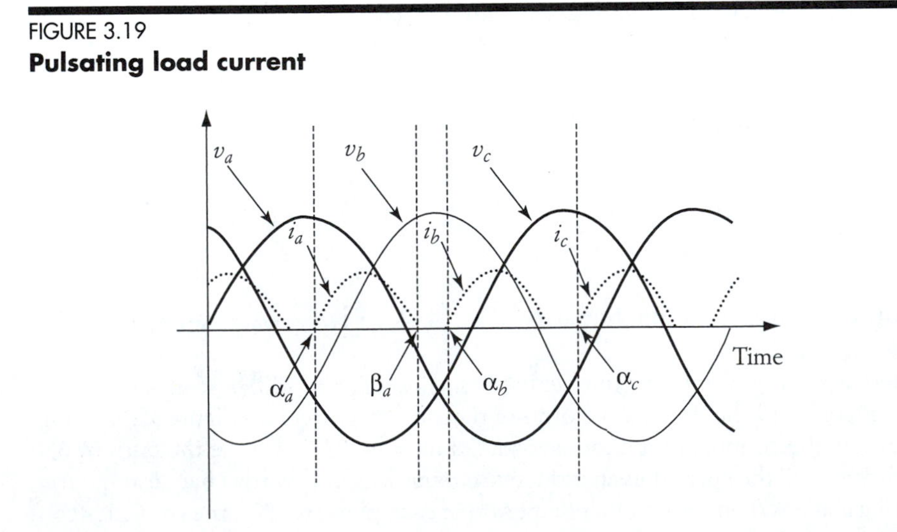

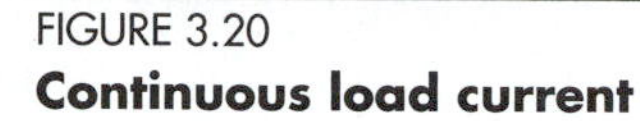

FIGURE 3.20
Continuous load current

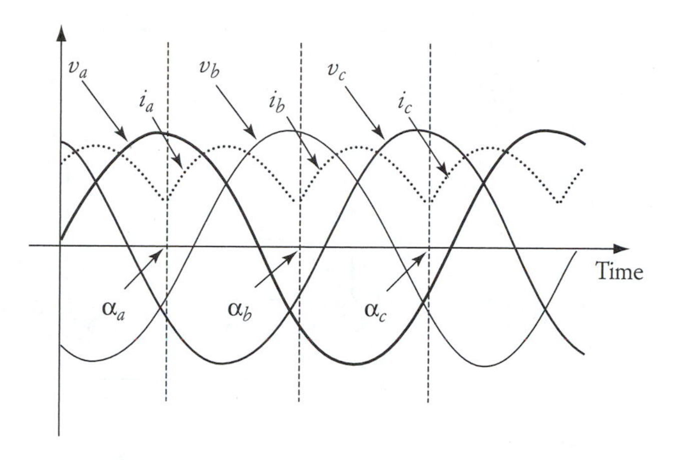

3.7 THREE-PHASE, FULL-WAVE, ac/dc CONVERSION

The three-phase, full-wave converter shown in Figure 3.21 is very popular in heavy-load applications. With this circuit, the three-phase ac is converted to a dc using six switches (S_1 through S_6). Each ac line is connected to the middle of one IGBT leg. On one side, the load is connected to the emitters of the IGBTs, and the other side is connected to the collectors.

The conductions of the IGBTs occur when their forward voltage drops are positive and the triggering signals are present. For example, when $v_{ab} = v_{an} - v_{bn} > 0$, S_1 and S_6 are ready to be triggered. However, when $v_{ab} < 0$, S_3 and S_4 are ready to be triggered. This rule also applies to the other two line-to-line voltages (v_{bc} and v_{ca}).

The triggering of the IGBTs is synchronized with the source voltage, as shown in Figure 3.22. The conduction period of each IGBT is shown at the top section of the figure. One complete cycle is divided into six segments; each is 60° long. Thus, the conduction period of each IGBT is 120°.

Let us assume that we initiate the triggering of S_1, S_3, and S_5 when v_{an}, v_{bn}, and v_{cn}, respectively, are at their peaks. In addition, the triggering of S_2, S_4, and S_6 are initiated when $v_{ab} = 0$, $v_{bc} = 0$, and $v_{ca} = 0$, respectively. Note that at any moment only one switch from the top IGBTs and one from the bottom are closed; no two switches on the same leg are closed.

Now let us examine each interval. In the one starting at the peak of v_{an}, v_{ab} is positive. Thus, S_1 and S_6 are triggered, and the load voltage is v_{ab}. At the beginning of the next interval ($v_{ac} > 0$), S_2 is triggered, and S_1 remains closed. After S_2

FIGURE 3.21
Three-phase, full-wave, ac/dc converter

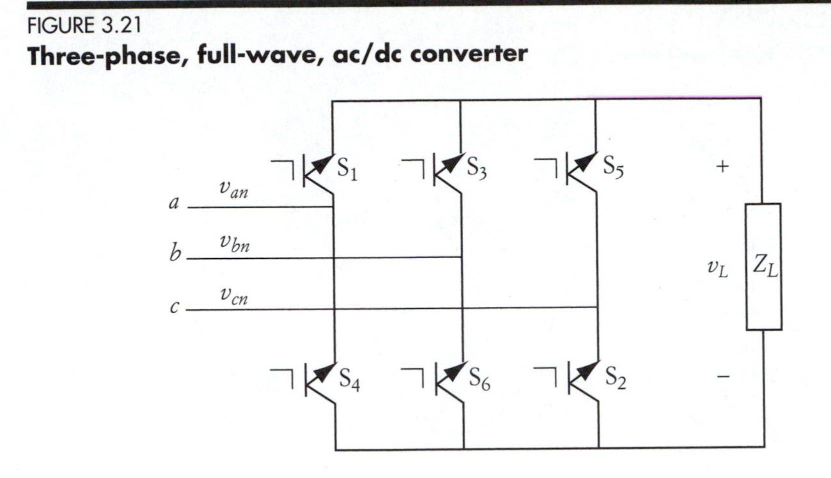

is triggered, S_6 is commutated instantly since $v_{ab} < 0$. Repeating this triggering logic for all other intervals, we can generate the waveforms of the load voltage shown at the bottom of Figure 3.22.

Assume that the phase voltages are expressed by

$$v_{an} = V_{\max} \sin \omega t$$

$$v_{bn} = V_{\max} \sin(\omega t - 120)$$

$$v_{cn} = V_{\max} \sin(\omega t + 120)$$

The average voltage of the load V_{ave} is

$$V_{\text{ave}} = 6\, V_{\text{seg}}$$

where V_{seg} is the average voltage of any 60° segment. The average voltage of the segment when S_1 and S_6 are closed is

$$V_{\text{seg}} = \frac{1}{2\pi} \int_{90}^{90+60} v_{ab}\, d\omega t = \frac{1}{2\pi} \int_{90}^{150} V_{\max} [\sin(\omega t) - \sin(\omega t - 120)]\, d\omega t$$

$$V_{\text{seg}} = \frac{\sqrt{3}\, V_{\max}}{4\pi}$$

$$V_{\text{ave}} = 6\, V_{\text{seg}} = \frac{3\sqrt{3}\, V_{\max}}{2\pi}$$

We can write the equation in a more general form by assuming that each IGBT is triggered at an angle α. This angle is measured from the time the phase voltage is equal

FIGURE 3.22
Waveforms of the converter in Figure 3.21

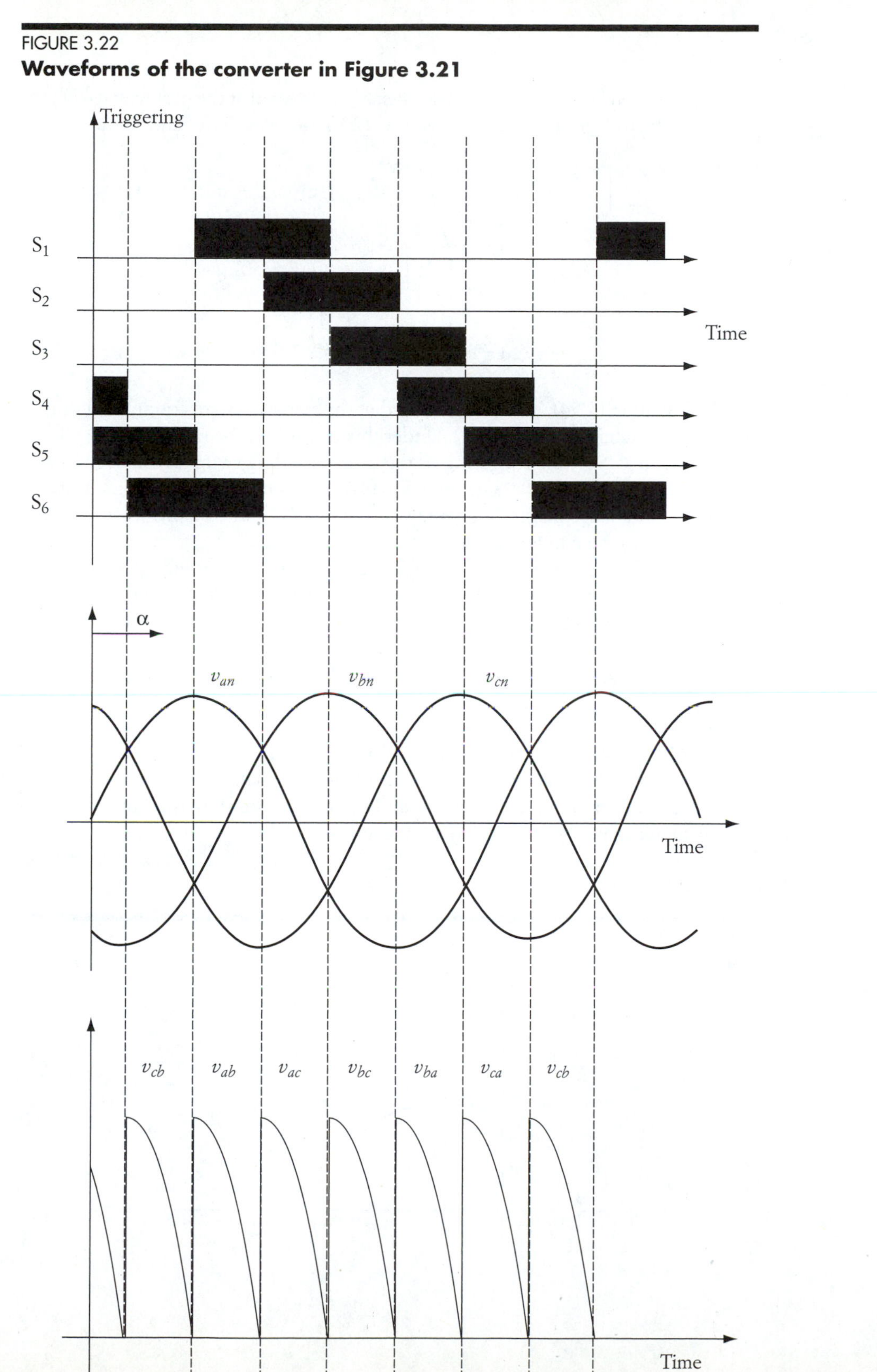

to zero, as shown in Figure 3.22. Thus, when S_1 is triggered at the peak voltage of v_{an}, $\alpha = 90°$. The triggering of S_3 occurs at $\alpha + 120°$, S_5 at $\alpha + 240°$, and so on.

$$V_{ave} = \frac{6}{2\pi} \int_{\alpha}^{\alpha+60} v_{ab}\, d\omega t = \frac{3}{\pi} \int_{\alpha}^{\alpha+60} V_{max} [\sin(\omega t) - \sin(\omega t - 120)]\, d\omega t$$

$$V_{ave} = \frac{3\sqrt{3}\, V_{max}}{\pi} \int_{\alpha}^{\alpha+60} \sin(\omega t + 30)\, d\omega t \tag{3.54}$$

$$V_{ave} = \frac{3\sqrt{3}\, V_{max}}{\pi} \cos(\alpha - 30)$$

In Equation (3.54), we are assuming that the transistors are conducting during the entire switching segment (60°). Under this condition, the range of α is $-30° \leq \alpha \leq 90°$. If $\alpha > 90°$, the triggering of the corresponding switches (such as S_1 and S_6) is not for a complete 60°, and Equation (3.54) does not apply. Keep in mind that when $\alpha > 150°$, the line-to-line voltage v_{ab} across S_1 and S_6 is negative, and the switches cannot be conducting current.

Figure 3.23 shows the average voltage across the load for any given α. From Equation (3.54) and Figure 3.23, the maximum average voltage occurs when the triggering angle is 30°.

$$V_{ave\ max} = \frac{3\sqrt{3}\, V_{max}}{\pi} \tag{3.55}$$

Figure 3.24 shows the waveforms for $\alpha = 30°$. Compare the waveform of the load voltage in this case to the one in Figure 3.22 (for $\alpha = 90°$). Note that the load voltage in Figure 3.24 does not reach zero and the ripples are smoothed out. The harmonic contents at $\alpha = 30°$ are much reduced.

FIGURE 3.23
Load average voltage

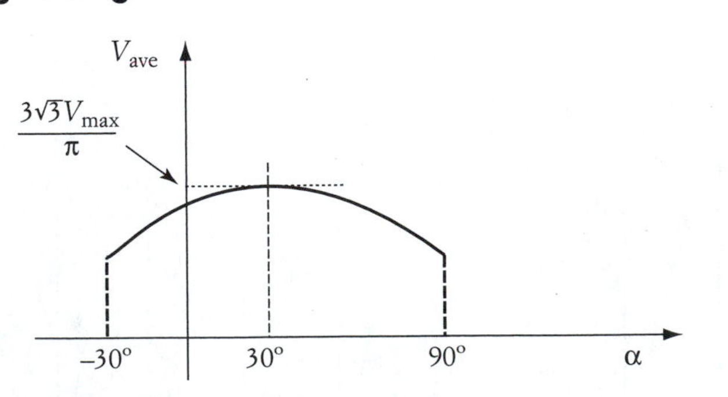

FIGURE 3.24
Waveforms of the converter in Figure 3.21 for α = 60°

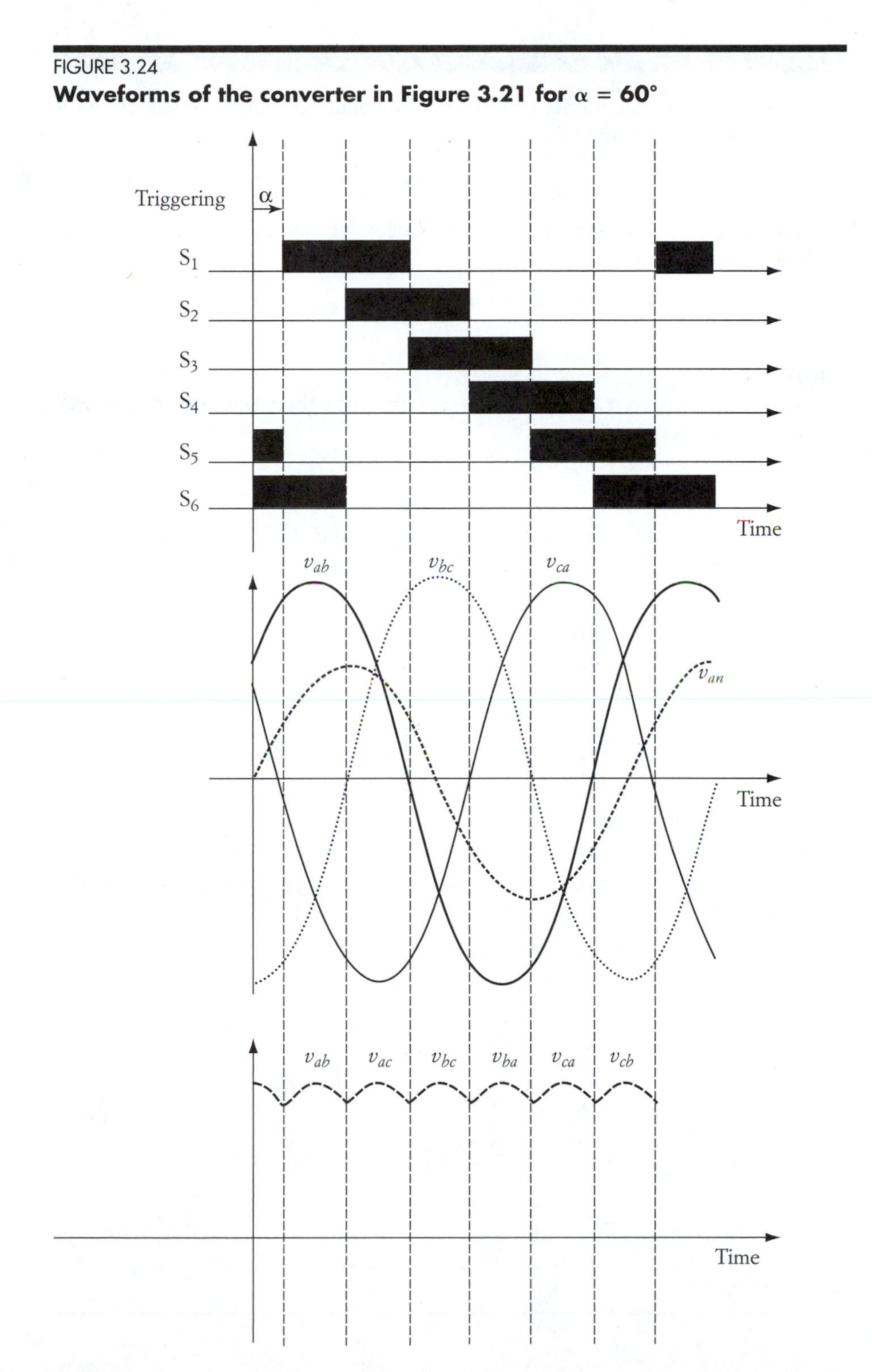

EXAMPLE 3.9

A three-phase, full wave, ac/dc converter has a balanced source voltage of 208 V (line-to-line). Compute the following:

a. Maximum average voltage across the load
b. Triggering angle at which the average voltage of the load equals the peak phase voltage of the source
c. Load voltage when the triggering angle is $-30°$.

SOLUTION

a. The maximum average load voltage occurs when the triggering angle is 30°. Using Equation (3.55),

$$V_{\text{ave max}} = \frac{3\sqrt{3}\,V_{\max}}{\pi} = \frac{3\sqrt{3}\,[\sqrt{2} \times 208/\sqrt{3}\,]}{\pi} = 281\ V$$

b. From Equation (3.54), the average voltage of the load can be made equal to the peak of the phase voltage of the source if

$$\frac{3\sqrt{3}}{\pi}[\cos(\alpha + 30) + \sin\alpha] = 1$$

$$\alpha \approx 82°$$

Note that the average voltage of the load, for any value of α, cannot exceed the peak of the line-to-line voltage of the source.

c. At $\alpha = -30°$, the average voltage is

$$V_{\text{ave}} = \frac{3\sqrt{3}\,V_{\max}}{2\pi}$$

$$V_{\text{ave}} = \frac{3\sqrt{3}\,[\sqrt{2} \times 208/\sqrt{3}\,]}{2\pi} = 140.5\ V$$

Since the magnitude of the average voltage is a sinusoidal waveform as shown in Figure 3.23, the average voltage at $\alpha = -30°$ is $V_{\text{ave max}}/2$.

3.8 dc/dc CONVERSION

Direct-current-to-direct-current converters are normally designed to provide output dc waveforms at adjustable voltage levels. These converters, also known as choppers, can be designed to produce fixed output voltage for variable input voltage or variable output voltage for fixed input voltage. Generally, there are three basic types of dc/dc converters:

1. *Step-down (Buck) converter,* where the output voltage of the converter is lower than the input voltage
2. *Step-up (Boost) converter,* where the output voltage is higher than the input voltage
3. *Step-down/step-up (Buck–Boost) converter,* where the output voltage can be made either lower or higher than the input voltage

In most electric drive applications, the dc/dc converter is a step-down type. The other types are normally used in applications such as power supplies and uninterruptible power supplies.

To understand the fundamentals of the dc/dc buck converter, examine the simple circuit shown in Figure 3.25. The figure shows a bipolar transistor whose emitter is connected to a load, and its collector is connected to the positive side of a dc source.

FIGURE 3.25
Simple chopper circuit

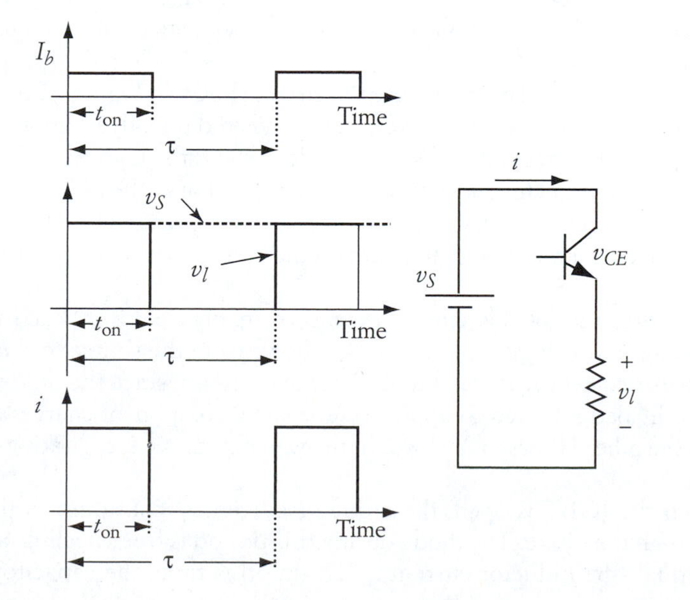

The potential of the voltage source is fixed. The figure also shows the corresponding waveforms. The top waveform is for the base current I_b. Since the base current is present for the period t_{on}, the transistor is in conduction. While conducting, the load voltage v_l is equal to the source voltage V_s. When the base current is not present, the transistor is open and the load voltage is zero. The switching period is labeled τ. If the switching action is repeated in a fixed timing pattern (fixed τ), the average voltage of the load and the load power can be controlled by adjusting t_{on}.

Assume that the transistor is an ideal device. The average voltage across the load V_{ave} is the integration of the source voltage over the period τ.

$$V_{ave} = \frac{1}{\tau} \int_0^{t_{on}} V_s \, dt = \frac{t_{on}}{\tau} V_s = KV_s \tag{3.56}$$

where $K = t_{on}/\tau$ is called the duty ratio. The maximum value of K is equal to 1 when the on time t_{on} is equal to the period τ. Hence, the maximum output voltage of this converter is equal to the source voltage, which is why it is called a step-down converter.

The output voltage of the converter can be controlled by using one of two methods:

1. By fixing the period τ and adjusting the on time t_{on}. This method is known as pulse-width modulation or PWM. Since τ is constant, the switching frequency is constant.
2. By fixing the on time t_{on} and adjusting the period τ. This is called frequency modulation (FM).

Note that the load voltage fluctuates between zero and V_s in any cycle. The current also fluctuates between zero and the maximum value. These fluctuations may not be acceptable in drive applications because they may result in damaging pulsating torque.

To solve this problem, we can use the circuit shown in Figure 3.26. In this figure, the switch is in series with a reversed connected diode. In addition, the circuit has a low-pass filter between the switching device and the load resistance. The filter consists of an inductor and a capacitor. The function of the inductor is to maintain the current fairly constant between the switching segments. The capacitor is normally selected large enough to maintain the voltage reasonably constant across the load.

The waveforms of this circuit are shown in Figure 3.27. When the IGBT switch is closed (during t_{on}), the current i_s flows from the source to the inductor. The inductor current i_L is equal to the source current i_s since the diode is reverse bias. i_L is divided into two components: one small component charges the capacitor, and the other i flows to the load. In the waveforms, we are ignoring the charging current.

When the IGBT is open, the inductor current i_L continues to flow to the load through the diode. The diode during this period is freewheeling and its current i_d equals the inductor current i_L. During this time, the capacitor also dis-

FIGURE 3.26
Chopper circuit with filter

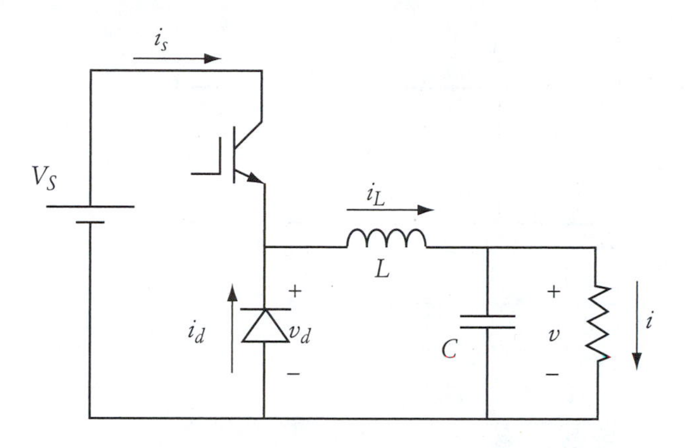

charges into the load. The waveforms in the figure ignore the discharging current. If we maintain high enough switching frequency, the ripples in the inductor current (and load current) as well as the ripples in the load voltage are minimized.

EXAMPLE 3.10

The switching frequency of the chopper shown in Figure 3.26 is 2 KHz. The source voltage is 80 V, and the duty ratio is 30%. The load resistance is 4 Ω. Assume that the inductor and capacitor are ideal and large enough to sustain the load current and load voltage with little ripple. Calculate the following:

a. On time and switching period
b. Average voltage across the load
c. Average voltage across the diode
d. Average current of the load
e. Load power

SOLUTION

a. The period can be computed using the switching frequency.

$$\tau = \frac{1}{f} = 0.5 \text{ msec}$$

$$t_{\text{on}} = K\tau = 0.15 \text{ msec}$$

FIGURE 3.27
Waveforms of chopper circuit

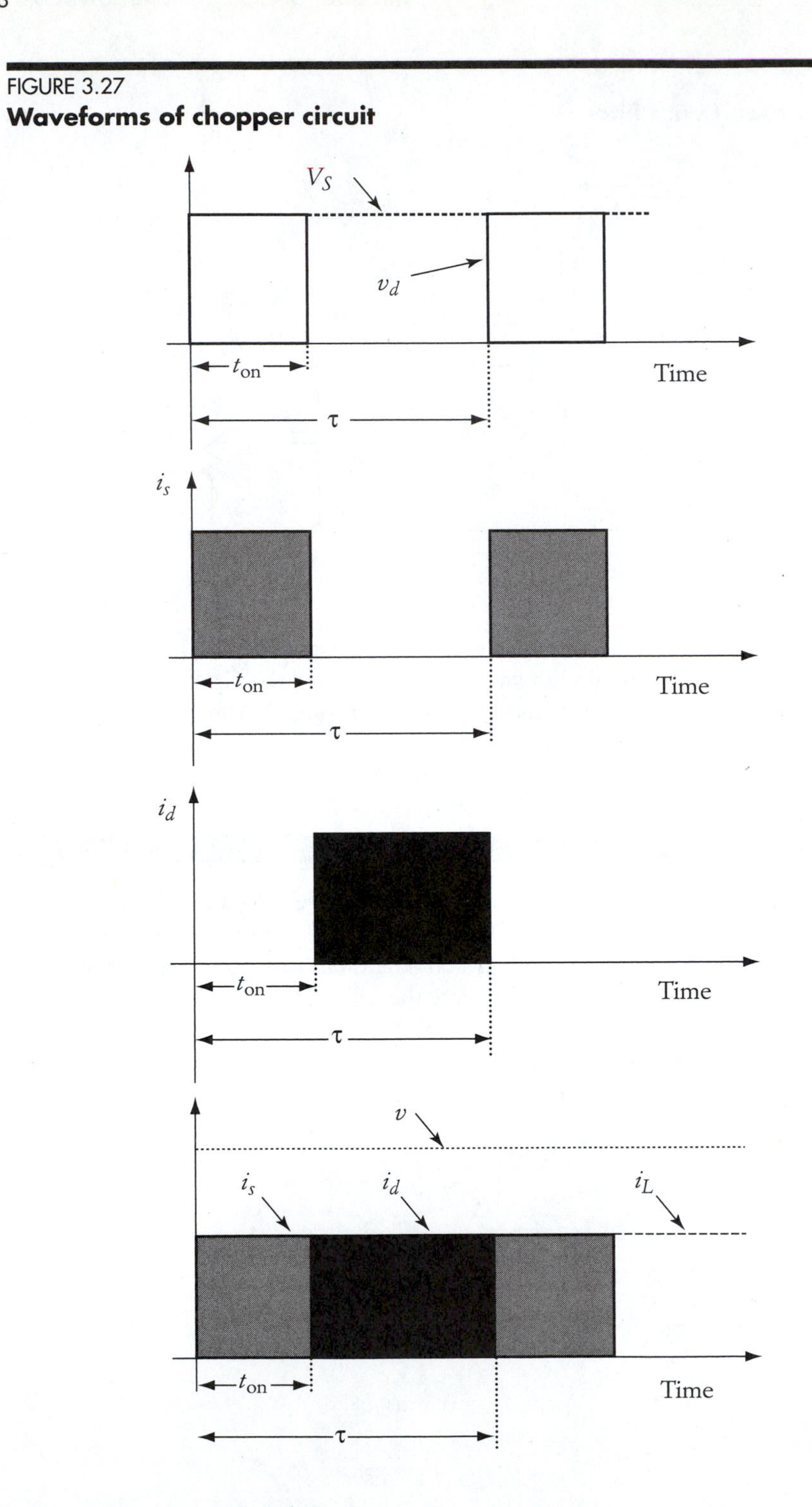

b. Use Equation (3.56) to compute the average voltage across the load.

$$V_{\text{ave}} = KV_s = 0.3\ (80) = 24\ \text{V}$$

c. The average voltage across the diode $V_{d\ \text{ave}}$ is the complement of the average voltage across the load.

$$V_{d\ \text{ave}} = \frac{1}{\tau}\int_{t_{\text{on}}}^{\tau} V_s\, dt = (1 - K)V_s = 0.7\ (80) = 56\ \text{V}$$

$V_{d\ \text{ave}}$ can also be computed by subtracting V_{ave} from the source voltage V_s.

d. The average load current is

$$I_{\text{ave}} = \frac{V_{\text{ave}}}{R} = \frac{24}{4} = 6\ \Omega$$

e. Since we assume that the current is ripple-free (no harmonics), and the load voltage is also ripple-free, the load power can be computed as

$$P = V_{\text{ave}}\, I_{\text{ave}} = 144\ \text{W}$$

3.9 dc/ac CONVERSION

This type of converter is also known as an inverter. Here, the input is a dc waveform and the output is an ac waveform. The inverter is widely used in uninterruptible power supplies, variable speed ac motors, and dc transmission lines.

Keep in mind that the term ac does not mean a perfect sinusoidal waveform; rather, it refers to a waveform that has positive and negative portions in each cycle. Furthermore, the ac waveform may have a small dc component; that is, the average value is not necessarily zero.

3.9.1 SINGLE-PHASE, dc/ac CONVERTER

Figure 3.28 shows a simple dc/ac inverter known as *H-bridge.* It consists of a dc source, four transistors, and a load. For simplicity, we shall assume that the load is resistive. At any period, only two transistors in opposite legs are turned on. To prevent short-circuiting the supply, any two transistors on the same leg cannot be turned on at the same time (either Q_1 and Q_2, or Q_3 and Q_4, are turned on simultaneously). Figure 3.28 also shows the voltage across the load. When Q_1 and Q_2 are closed, the current I_1 flows to the load. When Q_3 and Q_4 are closed, the current I_2 flows in the reverse direction to I_1. The load current in this case is alternating between positive and negative values. If the switching periods of all transistors are

FIGURE 3.28
dc/ac H-bridge and its ideal waveform

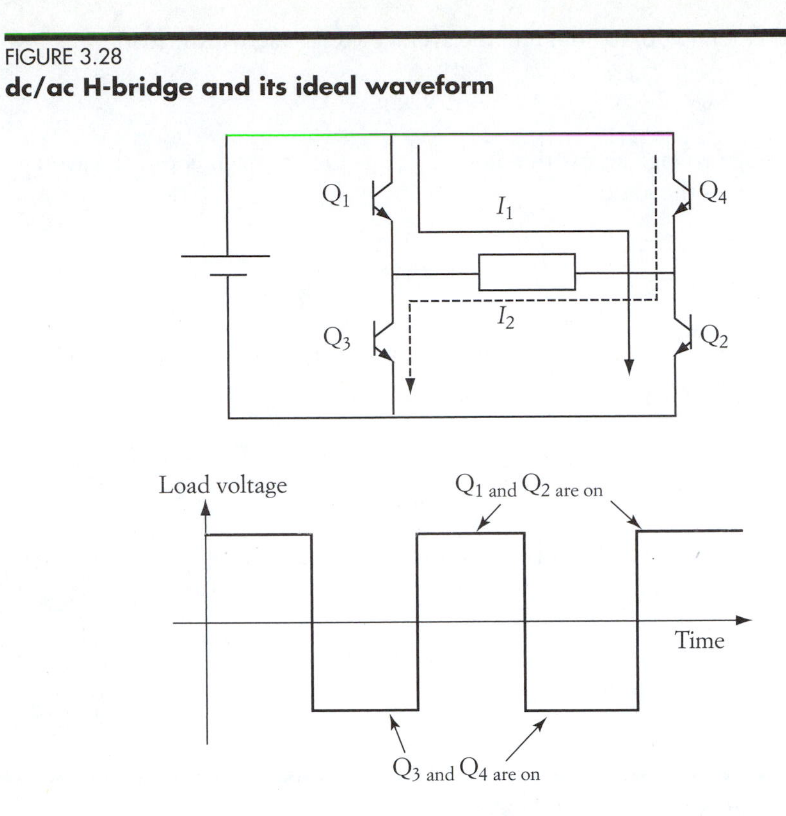

equal, the average component of the current is zero, and the waveform of the current or voltage is symmetrical around the time axis.

Notice that the waveform of Figure 3.28 is not sinusoidal. Nevertheless, it is an ac waveform. Reducing or prolonging the closing time of all transistors adjusts the frequency of the load voltage. The smaller the on time is, the higher is the frequency.

3.9.2 THREE-PHASE dc/ac CONVERTER

Three-phase waveforms can be obtained by using a relatively more elaborate dc/ac inverter similar to the one shown in Figure 3.29. The inverter is composed of six switches and six diodes. The diodes are used to allow the energy to flow back to the source. This is a requirement in most electric drive systems where the motor delivers energy back to the source under certain drive conditions. As discussed at the end of Chapter 2, when inductive load is interrupted, a large voltage builds up across the switch terminals to allow the energy stored in the inductive elements to return back to the source. The diodes in the figure provide a safe path for the flow of this energy back to the source, thus protecting the transistors from being damaged. The midpoint of each leg of the inverter is connected to the correspondingly-labeled load terminal. The load is shown separately on the right side of the figure.

FIGURE 3.29
Three-phase dc/ac inverter

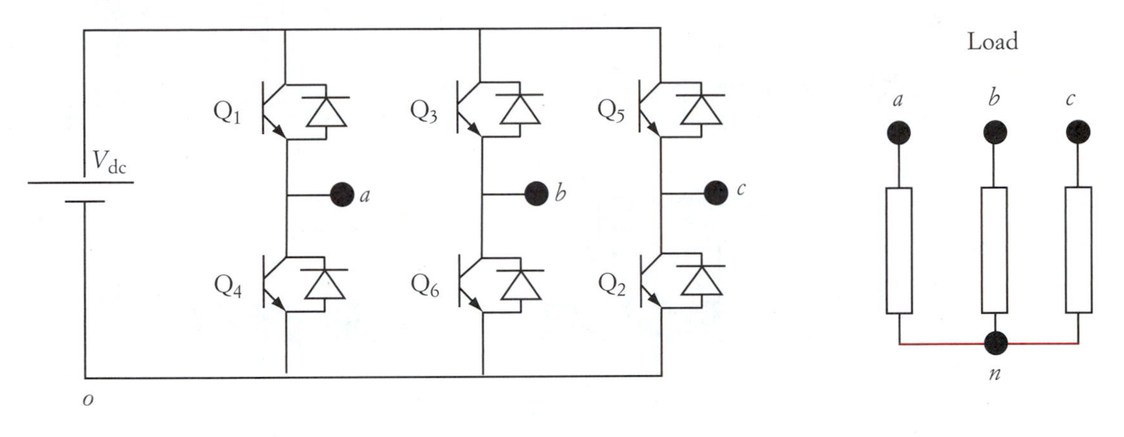

Before we continue any further, let us define the following terminology:

A *period* is the time of one complete cycle.

A *conduction period* is the time during which a transistor is closed.

A *switching interval* is one time segment. A conduction period can have more than one time segment.

The transistors of the three-phase dc/ac converter are switched in a specific sequence to generate three-phase waveforms and to prevent any two transistors on the same leg from being triggered—thus preventing the short-circuiting of the supply voltage. The switching sequence is shown in the top part of Figure 3.30. The figure also shows the line-to-line waveforms across the load terminals. Each cycle is divided into six time segments. Each segment is 60° long (electrical degrees), and each transistor is turned on for three time segments (180°). The switching of the transistors is based on their ascending order. For example, if transistor Q_1 is turned on, then after one time segment (60°) transistor Q_2 is turned on, and so on. Note that at any time segment, three transistors are closed, but only one transistor per leg is turned on.

Consider the first switching interval, where transistors Q_1, Q_5, and Q_6 are turned on. Because of this switching, the potentials of terminals a and c are positive, and the potential of terminal b is negative. Hence, the line-to-line voltages across the load during the first switching interval can be computed as follows:

$$\begin{aligned} v_{ab} &= v_a - v_b = V_{dc} \\ v_{bc} &= v_b - v_c = -V_{dc} \\ v_{ca} &= v_c - v_a = 0 \end{aligned} \tag{3.57}$$

FIGURE 3.30
Waveforms of the circuit in Figure 3.29

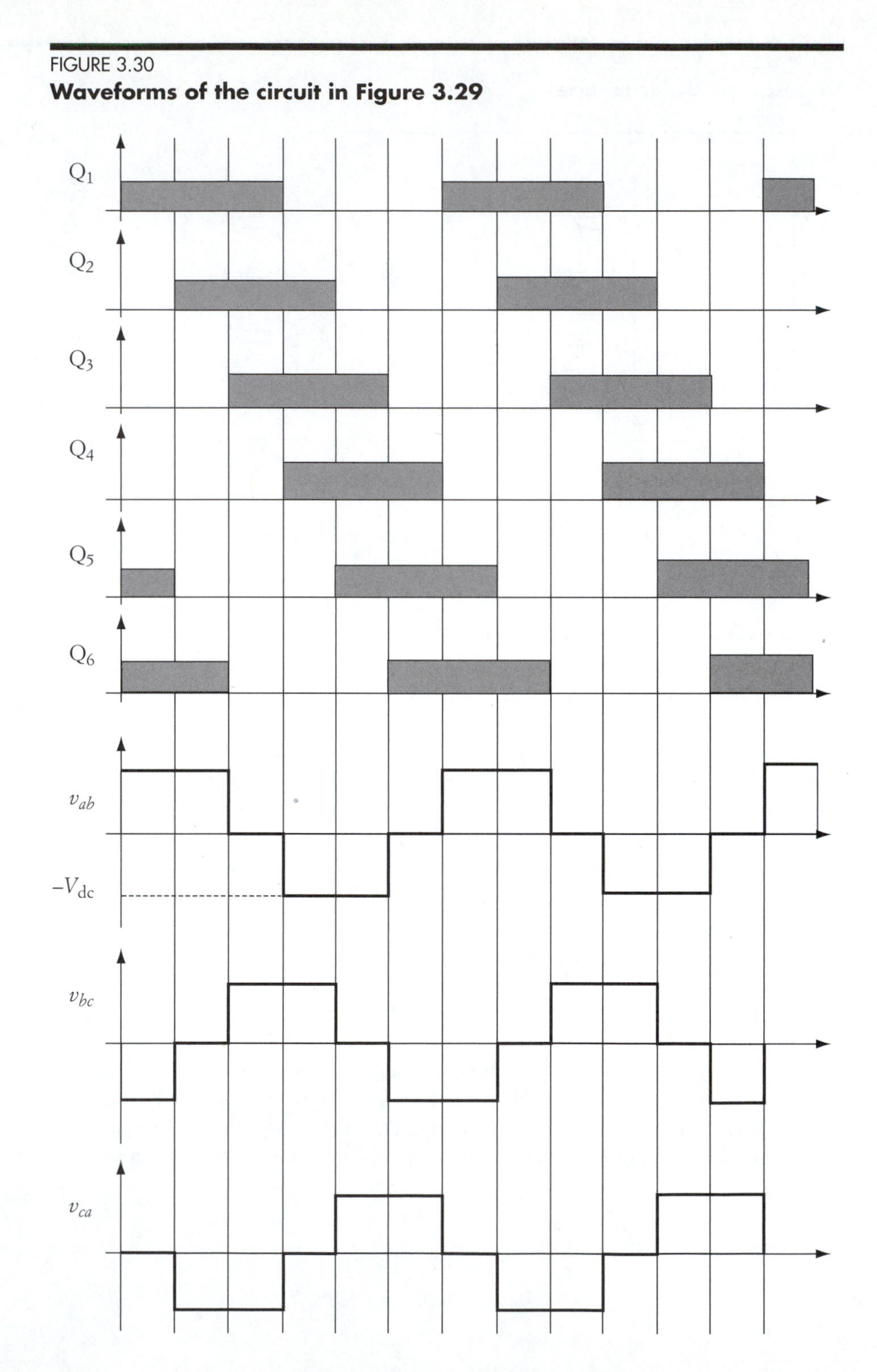

where V_{dc} is the voltage of the dc source. The line-to-line voltages for all other switching intervals can be computed using the same procedure. The waveforms are shown in the bottom part of Figure 3.30. Note that v_{ab} is leading v_{bc} by two time segments (120°), and v_{bc} is also leading v_{ca} by two time segments. The maximum voltage of any line-to-line voltage is equal to V_{dc}. Hence, the output of the inverter is a balanced, three-phase voltage.

The frequency of the load voltage can be adjusted by changing the period of the switching interval. The smaller the switching interval τ_{seg}, the higher the frequency. If the time of one switching interval is 2 μsec, the frequency of the load voltage is

$$f = \frac{1}{6\,\tau_{seg}} = \frac{1}{12} = 83.3 \text{ kHz}$$

The waveforms of the phase voltages (phase-to-neutral) are shown in Figure 3.31. These waveforms can be obtained by examining the switching status of the transistors in each switching interval. Assume that the load is connected in wye. The connection of the load to the source is changing every switching interval. Figure 3.32 shows the configuration of the load windings during the first three switching intervals. The load connection on the left side of the figure is for the first switching interval, when Q_1, Q_5, and Q_6 are closed and the rest of the transistors are in the open state. The middle part of the figure is for the load connection during the second interval (Q_1, Q_2, and Q_6 are closed). The third interval connection, when Q_1, Q_2, and Q_3 are closed, is shown on the right side of the figure.

As shown in Figure 3.32, during the first interval, when Q_1, Q_5, and Q_6 are closed, the potentials of terminals *a* and *c* are positive and that of terminal *b* is negative. If you assume that the load is balanced and the impedance Z is equal for each phase, then the potentials of phase *a, b,* and *c* are

$$v_{an} = v_{cn} = V_{dc}\frac{0.5\,Z}{1.5\,Z} = \frac{V_{dc}}{3} \tag{3.58}$$

$$v_{bn} = -V_{dc}\frac{Z}{1.5\,Z} = -\frac{2\,V_{dc}}{3} \tag{3.59}$$

The minus sign in Equation (3.59) is due to the direction of the current inside phase *b* as compared to that for phases *a* and *c*. In the second interval, phase *a* is positive potential, and *b* and *c* are negative potentials. The load voltages in this case are

$$v_{bn} = v_{cn} = -V_{dc}\frac{0.5\,Z}{1.5\,Z} = -\frac{V_{dc}}{3} \tag{3.60}$$

$$v_{an} = V_{dc}\frac{Z}{1.5\,Z} = \frac{2\,V_{dc}}{3} \tag{3.61}$$

FIGURE 3.31
Line-to-neutral voltage waveforms

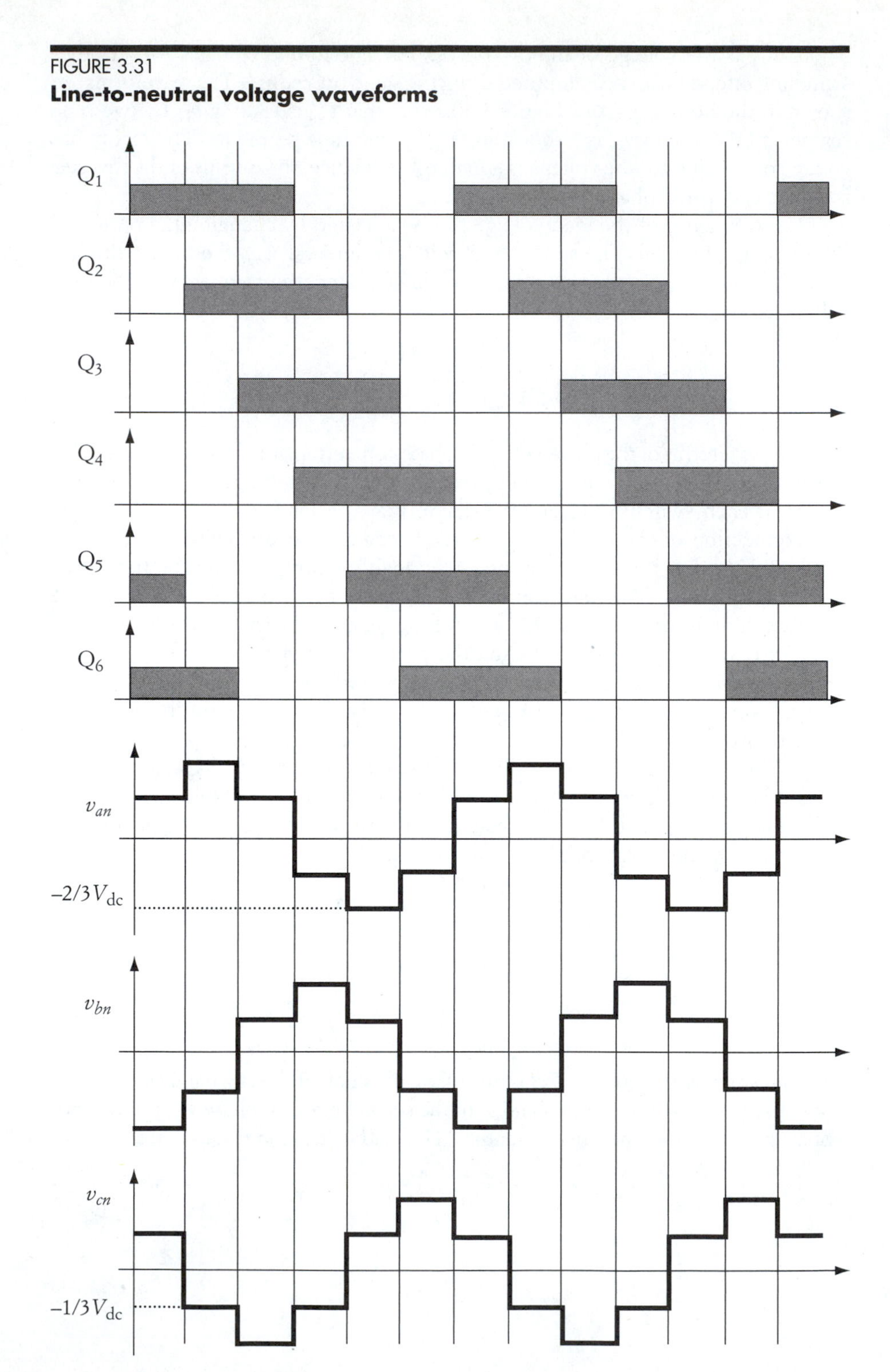

FIGURE 3.32
Connections of stator windings of an induction motor during the first three time intervals

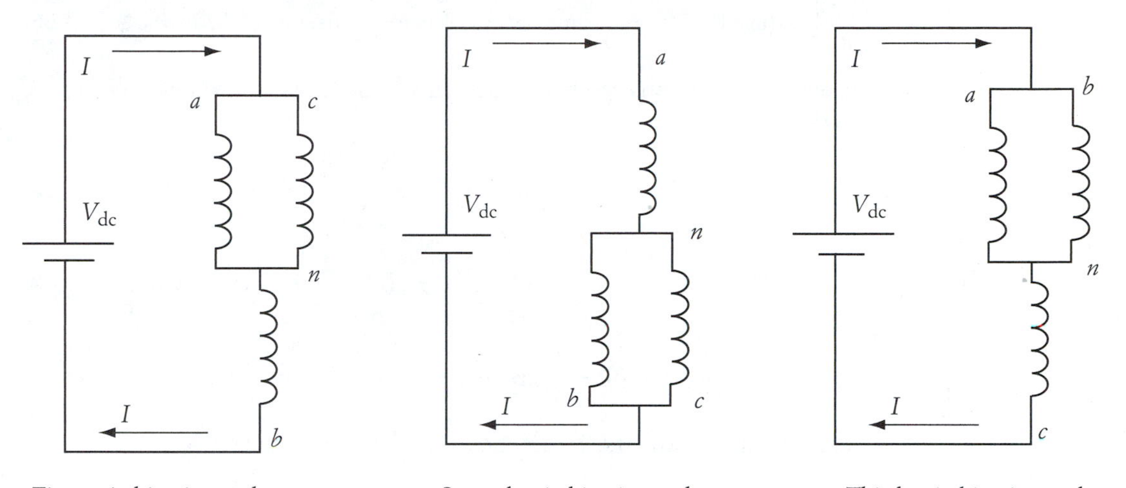

Similarly, during the third interval, the potentials of phases a and b are positive and that of c is negative. The winding potentials are

$$v_{an} = v_{bn} = V_{dc}\frac{0.5\,Z}{1.5\,Z} = \frac{V_{dc}}{3} \tag{3.62}$$

$$v_{cn} = -V_{dc}\frac{Z}{1.5\,Z} = -\frac{2\,V_{dc}}{3} \tag{3.63}$$

If you continue this process for the rest of the intervals, you can generate the waveforms in Figure 3.31. Note that the peak value of the phase voltage is equal to 2/3 of the supply voltage V_{dc}. Now examine the phase shift between the phase voltages. The shift is equal to two intervals. Since a complete cycle is six intervals, the phase shift is $2/6 \times 360° = 120°$. Also note that the voltage of phase a leads that of phase b by 120° and lags that of phase c by 120°. These are the main features of a balanced three-phase system.

The waveforms in Figure 3.31 contain several harmonic components. The fundamental component depends on the length of the switching interval. If you assume that the load is connected in wye, the harmonic components can be written as a Fourier expansion:

$$v_{an} = \frac{2V_{dc}}{\pi}\left(\sin \omega t + \frac{1}{3}\sin 3\omega t + \frac{1}{5}\sin 5\omega t + \ldots\right) \tag{3.64}$$

$$v_{bn} = \frac{2V_{dc}}{\pi}\left[\sin(\omega t - 120) + \frac{1}{3}\sin 3(\omega t - 120) + \frac{1}{5}\sin 5(\omega t - 120) + \ldots\right] \quad (3.65)$$

$$v_{cn} = \frac{2V_{dc}}{\pi}\left[\sin(\omega t + 120) + \frac{1}{3}\sin 3(\omega t + 120) + \frac{1}{5}\sin 5(\omega t + 120) + \ldots\right] \quad (3.66)$$

The phase voltage has no even harmonics. In addition, the line-to-line voltages are given by

$$v_{ab} = \sqrt{3}\,\frac{2V_{dc}}{\pi}\left[\sin(\omega t + 30) - \frac{1}{5}\sin 5(\omega t + 30) - \frac{1}{7}\sin 7(\omega t + 30)\ldots\right] \quad (3.67)$$

$$v_{bc} = \sqrt{3}\,\frac{2V_{dc}}{\pi}\left[\sin(\omega t - 90) - \frac{1}{5}\sin 5(\omega t - 90) - \frac{1}{7}\sin 7(\omega t - 90)\ldots\right] \quad (3.68)$$

$$v_{ca} = \sqrt{3}\,\frac{2V_{dc}}{\pi}\left[\sin(\omega t + 150) - \frac{1}{5}\sin 5(\omega t + 150) - \frac{1}{7}\sin 7(\omega t + 150)\ldots\right] \quad (3.69)$$

Note that the third harmonic does not exist in the line-to-line voltage.

EXAMPLE 3.11

Compute the rms voltage of the load for the circuit in Figure 3.29. The frequency of the fundamental component of the load voltage is 100 Hz.

SOLUTION

First, let us compute the period of the cycle τ and the period of one switching interval t_s.

$$\tau = \frac{1}{f} = 10 \text{ msec}$$

Since the cycle has six switching intervals,

$$t_s = \frac{\tau}{6} = 1.67 \text{ msec}$$

The general expression of the rms voltage is

$$V_{rms} = \sqrt{\frac{1}{\tau}\int_0^{\tau} v^2\,dt}$$

Divide the period into six switching intervals. The first interval ends at t_1, the second interval ends at t_2, and so on. Note that $t_2 = 2\,t_s$, $t_3 = 3\,t_s$, and so on.

$$V_{rms} = \sqrt{\frac{2}{\tau}\left[\int_0^{t_1} v^2\,dt + \int_{t_1}^{t_2} v^2\,dt + \int_{t_2}^{t_3} v^2\,dt\right]} = \sqrt{\frac{2}{\tau}\left[\int_0^{t_1}\left(\frac{V_{dc}}{3}\right)^2 dt + \int_{t_1}^{t_2}\left(\frac{2V_{dc}}{3}\right)^2 dt + \int_{t_2}^{t_3}\left(\frac{V_{dc}}{3}\right)^2 dt\right]}$$

$$V_{rms} = \sqrt{\frac{2V_{dc}^2}{9\tau}(3\,t_2 + t_3 - 3\,t_1)} = \sqrt{\frac{2V_{dc}^2}{9\tau}(6\,t_s)} = \sqrt{\frac{2V_{dc}^2}{90}(10)} = 0.47\,V_{dc}$$

Note that the rms value of the output voltage depends on the magnitude of the source voltage, and the fundamental component of the output frequency depends on the period of the switching interval.

3.9.3 VOLTAGE, FREQUENCY, AND SEQUENCE CONTROL

The switching of the transistors in the circuit of Figure 3.29 can be controlled to adjust the frequency, magnitude, or sequence of the load voltage. The control of these variables is essential for ac motor drives and many other applications. Many commercial triggering modules for the dc/ac converters have a built-in control circuit that allows adjustment of these variables.

3.9.3.1 FREQUENCY ADJUSTMENT

When a three-phase load is connected to a dc source via an inverter, adjusting the time of the switching interval can change the frequency of the load voltage.

$$f = \frac{1}{\tau} \tag{3.70}$$

where f is the frequency of the load voltage and τ is the period for one ac cycle. If τ is reduced, the frequency of the voltage waveform increases, and the reverse of this rule is also true.

EXAMPLE 3.12

A six-step inverter is used to supply a three-phase load using a dc voltage source. If the frequency at the load side is desired to be 500 Hz, calculate the conduction period of each transistor.

SOLUTION

The time for one cycle is

$$\tau = \frac{1}{f} = \frac{1}{500} = 2 \text{ msec}$$

The time of the switching segment t_{seg} is

$$t_{seg} = \frac{\tau}{6} = 0.33 \text{ msec}$$

The conduction period of each transistor t_{con} is three segments,

$$t_{con} = 3t_{seg} = 1 \text{ msec}$$

3.9.3.2 VOLTAGE ADJUSTMENT

The magnitude of the voltage across the terminals of the load can be adjusted by several techniques; one of them is fixed width modulation (FWM). A simple form of FWM is shown in Figure 3.33, where the conduction period of one transistor is shown. The rest of the transistors have similar conduction periods, but shifted as discussed earlier. Without the FWM, the transistor is continuously closed for the duration of its conduction period. By the FWM technique, the transistor is switched several times during its conduction period as shown in the lower part of the figure. The switching periods of the FWM techniques are called subintervals. Let us assume that the load voltage is at full value when the transistor is closed without the FWM technique. Hence, if the sum of the subintervals is less than the conduction period, the voltage across the load is less than the full voltage.

If the transistors in Figure 3.29 are switched without the FWM technique, the line-to-line voltage across the load will have the waveform shown in Figure 3.34. Examine the waveform in conjunction with the one shown in Figure 3.30. Notice that the line-to-line voltage across the load has a positive or negative duration of two switching intervals. The gap between the positive and negative durations is one switching interval. One cycle is equal to six switching intervals. The rms voltage across the motor terminals can be calculated by

$$V_{ab} = \sqrt{\frac{1}{6}\int_0^6 v_{ab}^2\,dx} \tag{3.71}$$

$$V_{ab} = \sqrt{\frac{2}{3}}\,V_{\mathrm{dc}} \tag{3.72}$$

where V_{ab} is the rms line-to-line voltage, v_{ab} is the instantaneous line-to-line voltage, and x is the interval. The integration is based on the duration rather than the

FIGURE 3.33
FWM of a single conduction period

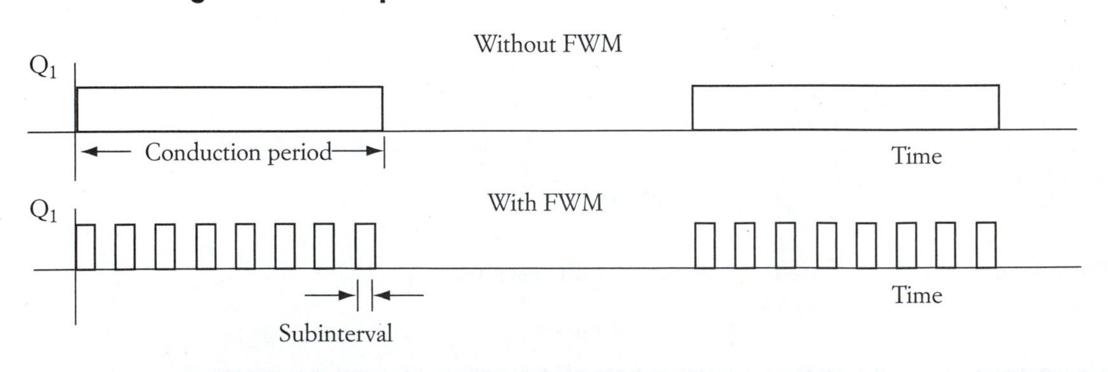

FIGURE 3.34
Line-to-line voltage without FWM

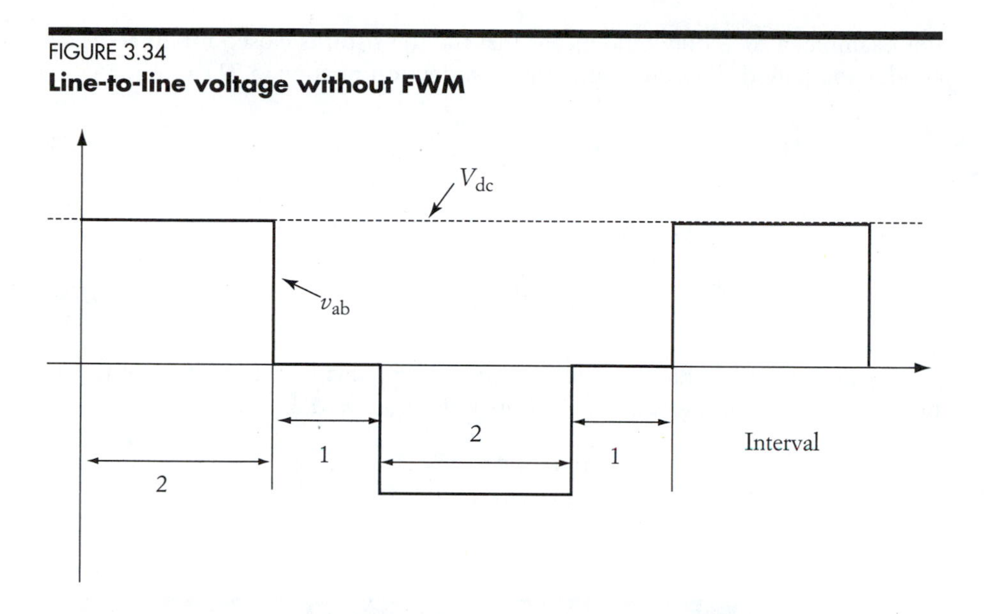

FIGURE 3.35
Line-to-line voltage with FWM

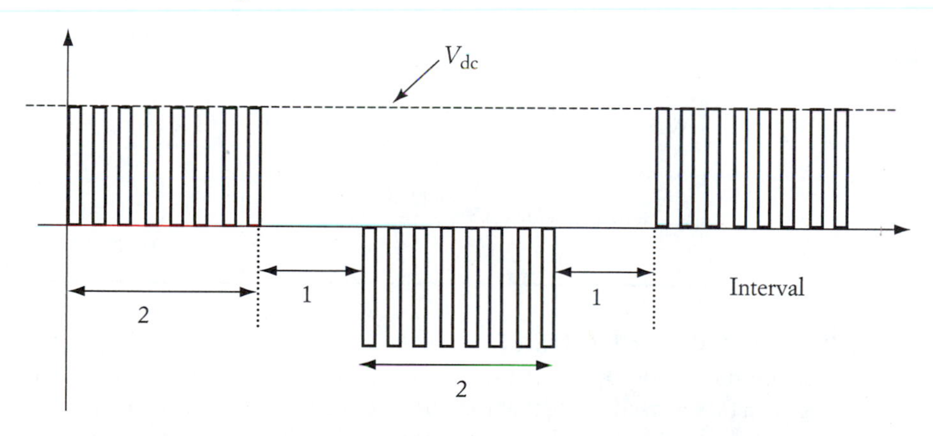

electrical angle. Since the time of all switching intervals is equal, for Equation (3.71) we use the number of intervals.

Now let us assume that due to the FWM, the load voltage will have the waveform shown in Figure 3.35. Identify the duty ratio d as

$$d = \frac{\Sigma \text{ subintervals of one conduction period}}{\text{conduction period}}$$

For example, a 20% duty ratio means that the transistor is closed for 20% of the conduction period. This duty ratio can be added to Equation (3.71) as follows:

$$V_{ab} = \sqrt{\frac{d}{6}\int_0^6 v_{ab}^2\,dx} \tag{3.73}$$

or

$$V_{ab} = \sqrt{\frac{2d}{3}}\,V_{dc} \tag{3.74}$$

The voltage reduction ratio VR can be computed by dividing the voltage of Equation (3.74) by the voltage with 100% duty ratio (no FWM).

$$VR = \frac{\text{voltage with FWM}}{\text{voltage without FWM}} = \sqrt{d} \tag{3.75}$$

EXAMPLE 3.13

An FWM with a duty ratio of 25% is used to reduce the voltage of the system described in Example 3.12. If the source voltage is 150 V, calculate the rms voltage applied to the motor windings with and without FWM.

SOLUTION

$$V_{ab}(\text{with FWM}) = \sqrt{\frac{2d}{3}}\,V_{\text{dc}} = \sqrt{\frac{2 \times 0.25}{3}} \times 150 = 61.24\text{ V}$$

$$V_{ab}(\text{without FWM}) = \frac{61.24}{\sqrt{d}} = 122.48\text{ V}$$

3.9.3.3 SEQUENCE ADJUSTMENT

By altering the succession of the transistor switching, the phase sequence of the load voltage can be reversed. Note that the phase sequence of the switching pattern in Figures 3.30 and 3.31 is *abc*. This sequence can be changed to *acb* simply by swapping the switching pattern of transistors Q_1 and Q_5, and also Q_2 and Q_4.

3.9.4 PULSE-WIDTH MODULATION (PWM)

Pulse-width modulation (PWM) is used to control the frequency and the magnitude of the ac voltage across the load and to reduce the harmonic contents in the output voltage or current. There are a number of PWM techniques, but the most common type is the sinusoidal PWM.

FIGURE 3.36
Control signals for PWM

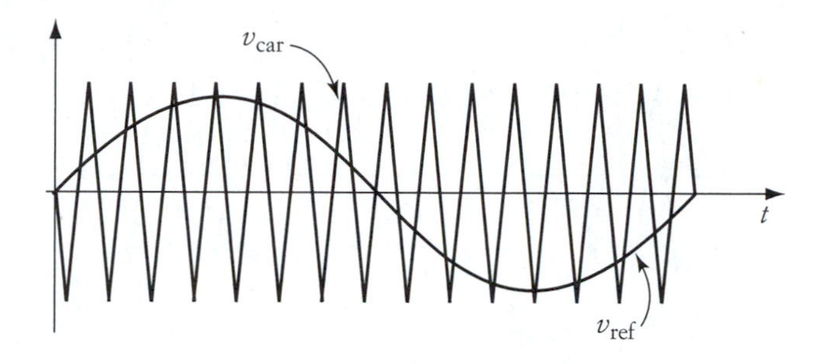

Figure 3.36 shows the basic idea of PWM for a voltage source inverter. Two control signals are used: a reference sinusoidal wave v_{ref} and a triangular carrier v_{car}. A control circuit at low voltage levels generates these two signals. They are used solely to create the triggering signals for two transistors on the same leg in the circuit shown in Figure 3.29. That is to say, they are for the terminal of one phase only. The control signals of the other two legs have the same triangular carrier, but their sinusoidal reference waves have the proper 120° shift associated with the balanced three-phase system. Thus, v_{ref} of phase b lags v_{ref} of phase a by 120°, and v_{ref} of phase c leads v_{ref} of phase a by 120°.

With the PWM technique, several parameters can be adjusted to generate the desired voltage and frequency at the load side. The basic parameters are the frequency and magnitude of the reference signal v_{ref}. The magnitude of the triangular carrier is usually kept constant, but its frequency can also vary. The upper limit of the frequency of the carrier is determined by the maximum switching frequency of the transistors. This frequency can be as high as 20 kHz.

Now let us see how the PWM works by examining Figure 3.36. We will assume that the figure is for the control signals of phase a only. Looking back at the circuit in Figure 3.29, you find that Q_1 and Q_4 are the two transistors switching in the leg of phase a. If Q_1 is closed and Q_4 is open, v_{ao} is positive. (v_{ao} is the potential of phase a with respect to point o.) Point o is just a reference point selected here to be the negative terminal of the input source V_{dc}. If Q_1 is open and Q_4 is closed, v_{ao} is zero. With PWM, the switching of Q_1 and Q_4 is based on the difference between the reference and carrier waveforms Δv:

$$\Delta v = v_{ref} - v_{car}$$

The switching conditions for any two transistors in one leg (say, for phase a) are as follows:

$\Delta v_a > 0$, Q_1 is closed and Q_4 is open

$\Delta v_a < 0$, Q_4 is closed and Q_1 is open

FIGURE 3.37
Potentials of phases *a* and *b* due to PWM

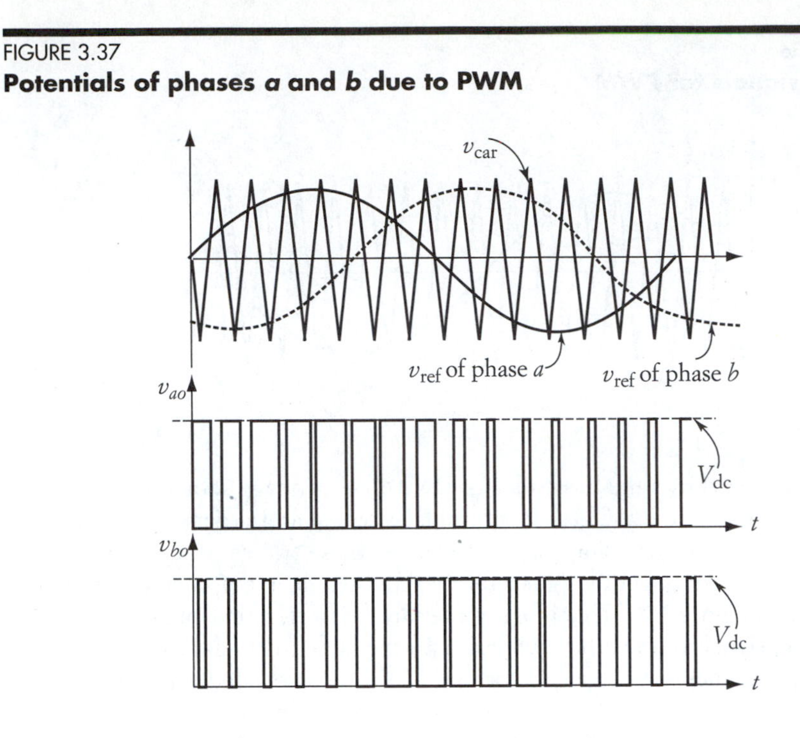

where

$$\Delta v_a = v_{ref\,a} - v_{car}$$

$v_{ref\,a}$ is the reference signal of phase a.

The reference signals for two phases are shown at the top of Figure 3.37. Using the rule stated in the previous paragraph, we can generate v_{ao} and v_{bo} shown in the figure. Note that these voltages have unequal switching intervals.

Figure 3.38 shows the line-to-line voltage v_{ab}, which is obtained by subtracting the potential of phase b from that of phase a:

$$v_{ab} = v_{ao} - v_{bo} \tag{3.76}$$

The line-to-line voltage consists of rectangular segments with different widths. It also has symmetrical positive and negative parts. Thus, it has a dominant component at the fundamental frequency. Using a harmonic analysis technique, the general expression of such a waveform can be written as

$$v_{ab}(t) = m_a \frac{V_{dc}}{2} \sin(2\pi f_s t) + \text{Bessel harmonic terms} \tag{3.77}$$

FIGURE 3.38
Line-to-line voltage and its fundamental component due to PWM

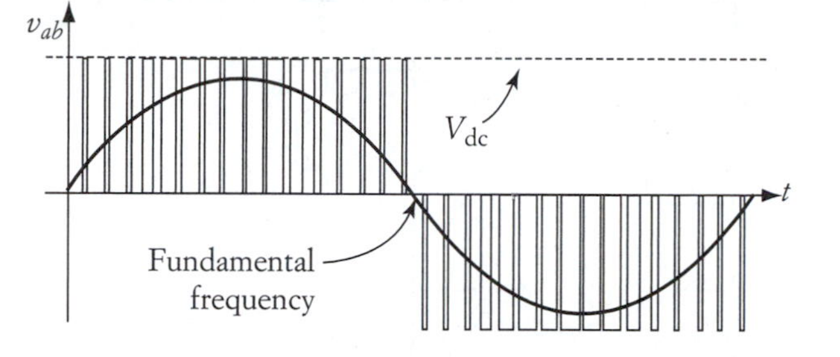

where f_s is the frequency of the reference signal and m_a is called the amplitude modulation, which is the ratio of the peak values of the reference signal to the carrier,

$$m_a = \frac{V_{\text{ref}}}{V_{\text{car}}} \tag{3.78}$$

By examining Equation (3.77), one can conclude that by adjusting the magnitude and frequency of the reference signal, the magnitude and frequency of the load voltage can be controlled. Assume that the carrier frequency and its magnitude are unchanged. When the magnitude of the reference signal increases, m_a increases, and so does the magnitude of the fundamental component of the load voltage v_{ab}. Also, since the frequency of the fundamental voltage across the load is the same as the frequency of the reference signal f_s, the frequency of the load voltage can be changed by changing the reference frequency. These are the major advantages of the PWM technique.

3.10 ENERGY RECOVERY SYSTEMS

In a number of applications, a two-way energy exchange between a source and a load is needed. This is particularly important for electric drive systems. The machine used in electric drives consumes electric energy when running as a motor, but it returns some energy back to the source when running as a generator. This process enhances the operation of the machine and improves the overall efficiency of the system. In later chapters, we will discuss this aspect in more detail.

Figure 3.39 shows an energy recovery circuit, where two sources are connected via two IGBT circuits in bridge configurations. Switches S_1 through S_4 are used to charge the battery. S_5 through S_8 are used for discharging. The charging and discharging operations are discussed in the following subsections.

FIGURE 3.39
Energy recovery circuit

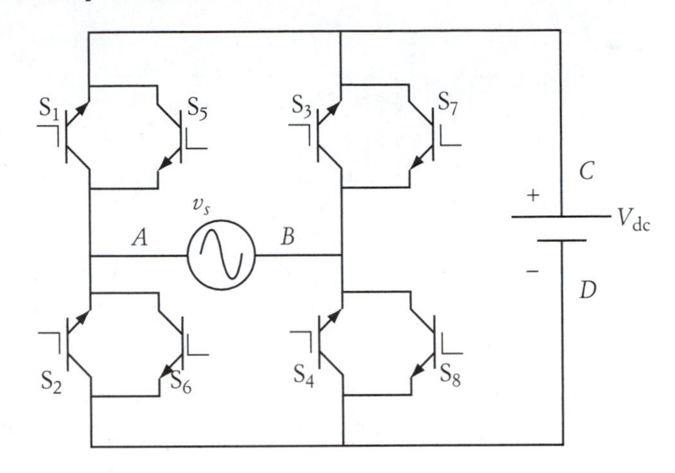

FIGURE 3.40
Waveforms of charging operation

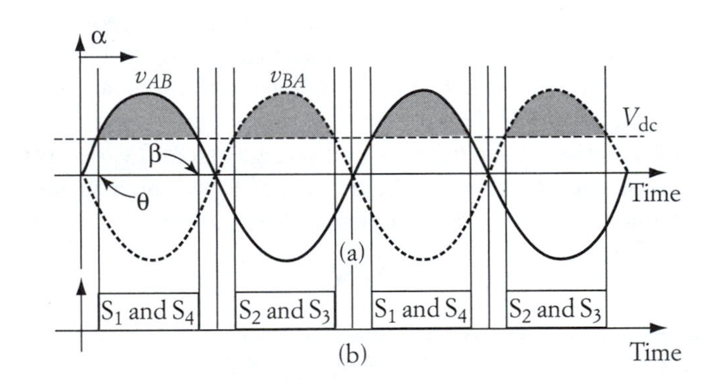

3.10.1 CHARGING OPERATION

When $v_{AB} > V_{dc}$, S_1 and S_4 can be triggered because their collector-to-emitter forward voltage is positive. The current loop in this case is *A*, S_1, *C*, *D*, S_4, and back to *B*. When point *B* is of higher potential than *A*, and $v_{BA} > V_{dc}$, S_2 and S_3 can be triggered. The current loop for this interval is *B*, S_3, *C*, *D*, S_2, and back to *A*.

Figure 3.40(a) shows the key waveforms of the system; Figure 3.40(b) shows the switching sequence and duration of the transistors. Only when $v_s > V_{dc}$ can the battery be charged. The shaded areas represent the maximum possible period for charging the battery.

FIGURE 3.41
Simplified equivalent circuit for charging and discharging

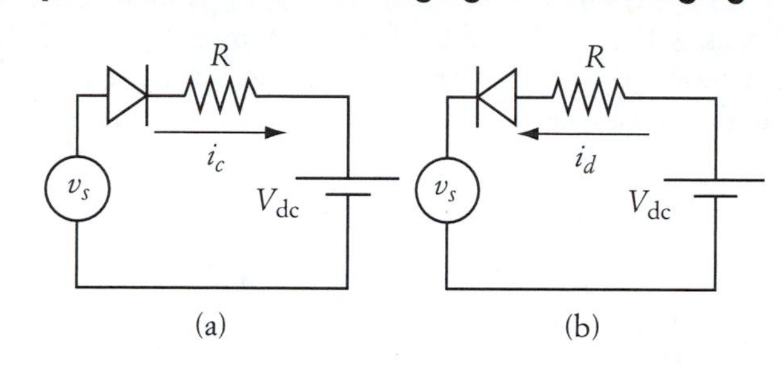

The fundamental equations for the charging operation can be obtained by considering the simplified equivalent circuit shown in Figure 3.41(a). In the figure, the resistance R represents the augmented value of the internal resistance of the battery, the forward resistance of the IGBTs while conducting, and the resistance of the cables. The diode is used to indicate the direction of thc charging current i_c.

The instantaneous voltage drop across the resistance can be written as

$$v_R = v_s - V_{dc} \quad (3.79)$$

At β,

$$v_s = V_{max} \sin \beta = V_{dc} \quad (3.80)$$

where $\beta > \pi$. The average voltage across the resistance can be expressed by

$$V_R = \frac{1}{\pi} \int_{\alpha}^{\beta} [V_{max} \sin(\omega t) - V_{dc}]\, d\omega t = \frac{V_{max}}{\pi} [\cos \alpha - \cos \beta] - V_{dc} \frac{\beta - \alpha}{\pi} \quad (3.81)$$

The average value of the charging current I_c is

$$I_c = \frac{V_R}{R}$$

Note that the minimum value of the triggering angle of the IGBTs (S_1 and S_4), as shown in Figure 3.40(a), is

$$\alpha_{min} = \theta$$

If the triggering angle is less than α_{min}, the collector–emitter voltage across these IGBTs is negative, and the transistors cannot close until $\omega t \geq \theta$.

EXAMPLE 3.14

For the circuit in Figure 3.39, assume that the voltage source is 110 V (rms) and the dc battery pack is 150 V. The value of the resistance between the battery and the ac source, including the internal resistance of the battery, is 1 Ω. Calculate the rms current and the power delivered to the battery during charging. Assume that α is 90°.

SOLUTION

Before we attempt to solve this problem, note that during charging, the triggering angle must be larger than the cross angle (when $v_s = V_{dc}$). The cross angle can be computed as

$$\theta = \sin^{-1}\frac{V_{dc}}{V_{max}} = \frac{150}{\sqrt{2}\,110} \approx 75^\circ$$

Any triggering angle less than 75° will result in no conduction. β is the first angle after α at which $v_s = V_{dc}$:

$$\beta = 180 - \theta = 105^\circ$$

The conduction period γ is then

$$\gamma = \beta - \alpha = 105 - 90 = 15^\circ$$

To compute the power delivered to the battery, we need to compute the rms current, the rms of the fundamental component of the current, and the phase shift of its fundamental component as given in Equations (3.15) through (3.25). The simplest way to compute the rms current is to compute the rms voltage across the resistance and divide it by the resistance itself.

$$V_{R\,rms} = \sqrt{\frac{1}{\pi}\int_\alpha^\beta v_R^2\, d\omega t} = \sqrt{\frac{1}{\pi}\int_\alpha^\beta (v_s - V_{dc})^2\, d\omega t}$$

$$V_{R\,rms} = \sqrt{\frac{1}{\pi}\left(V_{dc}^2\,\gamma + \int_\alpha^\beta v_s^2\, d\omega t - 2\,V_{dc}\int_\alpha^\beta v_s\, d\omega t\right)}$$

$$V_{R\,rms} = \sqrt{\frac{V_{dc}^2}{\pi}\gamma + \frac{V_{max}^2}{2\pi}\left[\gamma - \frac{\sin(2\beta) - \sin(2\alpha)}{2}\right] - \frac{2V_{dc}\,V_{max}}{\pi}(\cos\alpha - \cos\beta)} \qquad (3.82)$$

A direct substitution of the parameters into Equation (3.82) yields

$$V_{R\,rms} = 1.2 \text{ V}$$

The total rms current during charging $I_{c\ rms}$, which includes all the harmonics, is

$$I_{c\ rms} = \frac{V_{R\ rms}}{R} = 1.2\ \text{A}$$

As explained in Section 3.1.2.3, the power can only be computed by using currents and voltages of the same frequency. We cannot multiply by $I_{c\ rms}$ by V_{dc}. However, we can multiply the dc component of the current I_c by V_{dc}. Another method is to use the formula in Equation (3.17) to compute the power at the ac source side, then subtract the losses of the resistance R. The power at the ac source side is

$$P_s = VI_{1c\ rms} \cos \phi_1$$

where $I_{1c\ rms}$ is the rms value of the fundamental component of the charging current.

To calculate the power delivered by the ac source, we need to compute the phase shift of the fundamental component of the current with respect to the voltage of the ac source as given in Equation (3.20):

$$\phi_1 = \tan^{-1}\left(\frac{a_1}{b_1}\right)$$

$$a_1 = \frac{1}{\pi}\int_0^{2\pi} i(\omega t)\cos(\omega t)\, d\omega t = \frac{2}{R\pi}\int_\alpha^\beta v_R \cos(\omega t)\, d\omega t = \frac{2}{R\pi}\int_\alpha^\beta (v_s - V_{dc})\cos(\omega t)\, d\omega t$$

$$a_1 = \frac{2}{R\pi}\left[\int_\alpha^\beta V_{max}\sin(\omega t)\cos(\omega t)\, d\omega t - \int_\alpha^\beta V_{dc}\cos(\omega t)\, d\omega t\right]$$

$$a_1 = \frac{V_{max}}{R\pi}(\sin^2\beta - \sin^2\alpha) - \frac{2\,V_{dc}}{R\pi}(\sin\beta - \sin\alpha) = -0.063$$

$$b_1 = \frac{1}{\pi}\int_0^{2\pi} i(\omega t)\sin(\omega t)\, d\omega t = \frac{2}{R\pi}\int_\alpha^\beta v_R \sin(\omega t)\, d\omega t = \frac{2}{R\pi}\int_\alpha^\beta (v_s - V_{dc})\sin(\omega t)\, d\omega t$$

$$b_1 = \frac{2}{R\pi}\left[\int_\alpha^\beta V_{max}\sin^2(\omega t)\, d\omega t - V_{dc}\int_\alpha^\beta \sin(\omega t)\, d\omega t\right]$$

$$b_1 = \frac{V_{max}}{R\pi}\left[\gamma + \frac{\sin(2\alpha) - \sin(2\beta)}{2}\right] - \frac{2\,V_{dc}}{R\pi}(\cos\alpha - \cos\beta) = 0.628$$

$$\phi_1 = \tan^{-1}\left(\frac{a_1}{b_1}\right) = -5.73°$$

The rms value of the fundamental component of the charging current $I_{1c\ \text{rms}}$ is given by Equation (3.18)

$$I_{1c\ \text{rms}} = \frac{c_1}{\sqrt{2}} = \frac{\sqrt{a_1^2 + b_1^2}}{\sqrt{2}} = \frac{\sqrt{(-0.063)^2 + (0.628)^2}}{\sqrt{2}} = 0.446 \text{ A}$$

As given in Equation (3.17), the power of the ac source P_s is

$$P_s = VI_{1c\ \text{rms}} \cos \phi_1$$

$$= 110 \times 0.446 \times \cos(5.73) = 48.85 \text{ W}$$

The power losses in the resistance P_{loss} is

$$P_{\text{loss}} = I_{c\ \text{rms}}^2 R = 1.2^2 \times 1 = 1.44 \text{ W}$$

Note that the losses in the resistance are due to all harmonic components of the current. The power delivered to the battery pack P_{charge} is then

$$P_{\text{charge}} = P_s - P_{\text{loss}} = 47.41 \text{ W}$$

3.10.2 DISCHARGING OPERATION

The battery can return energy back to the ac source if $V_{\text{dc}} > v_{AB}$ and the proper IGBTs are triggered. The current in this case will flow in the opposite direction of the charging current as shown in Figure 3.41(b). The diode is used to indicate the direction of the discharging current i_d. The waveforms of the discharging operation are shown in Figure 3.42(a), and the triggering sequence is shown in Figure 3.42(b).

FIGURE 3.42
Waveforms of discharging operation

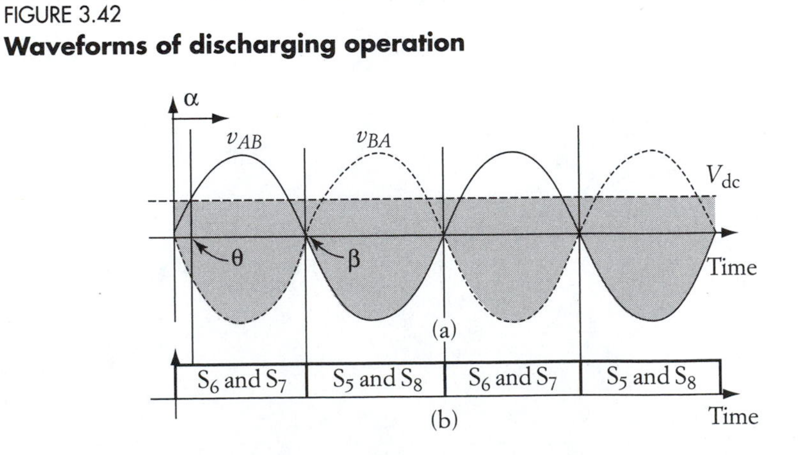

When $V_{dc} > v_{AB}$, S_5 and S_8 can be triggered, and the current loop is C, S_5, A, B, S_8, and back to D. When $V_{dc} > v_{BA}$, S_6 and S_7 can be triggered, and the current loop is C, S_7, B, A, S_6, and back to D. The equations of the system during discharging are similar to Equations (3.79) to (3.81). The main difference is that the voltage drop across the resistance is reversed.

$$v_R = V_{dc} - v_s$$

Let us analyze the first half of the cycle in Figure 3.42. In this period, v_{BA} is negative (point A has positive potential and B has negative) and is always less than V_{dc}. Thus, S_6 and S_7 are closed. The current will flow in the IGBTs until π. After π, point A turns positive with respect to B. Then, S_5 and S_8 are closed. The average component of the discharging current can be computed by using the average voltage across the resistance at any triggering angle α:

$$V_R = \frac{1}{\pi}\int_{\alpha}^{\pi} [V_{dc} - V_{max}\sin(\omega t)]\,d\omega t = V_{dc}\frac{\pi - \alpha}{\pi} - \frac{V_{max}}{\pi}(\cos\alpha + 1) \quad (3.83)$$

The minimum triggering angle α_{min} is achieved when $\omega t = 0$. The conduction period in this case is π, and the commutation angle is $\beta = \pi$. The average discharging current I_d is

$$I_d = \frac{V_R}{R}$$

The rms quantities of the discharging circuit are computed in the following example.

EXAMPLE 3.15

For the circuit in Example 3.14, calculate the rms current and the power delivered to the ac source during discharging. Assume that α is a minimum.

SOLUTION

The general expression for the rms voltage across the resistance while discharging is

$$V_{R\,rms} = \sqrt{\frac{1}{\pi}\int_{\alpha}^{\beta} v_R^2\,d\omega t} = \sqrt{\frac{1}{\pi}\int_{\alpha}^{\beta} (V_{dc} - v_s)^2\,d\omega t}$$

$$= \sqrt{\frac{1}{\pi}\left(V_{dc}{}^2(\gamma) + \int_{\alpha}^{\beta} v_s^2\,d\omega t - 2\,V_{dc}\int_{\alpha}^{\beta} v_s\,d\omega t\right)}$$

$$V_{R\,\text{rms}} = \sqrt{\frac{V_{dc}^2}{\pi}(\gamma) + \frac{V_{max}^2}{2\pi}\left[\gamma - \frac{\sin(2\beta) - \sin(2\alpha)}{2}\right] - \frac{2V_{dc}\,V_{max}}{\pi}(\cos\alpha - \cos\beta)}$$

where

$$\alpha_{min} = 0°$$

$$\beta = 180°$$

$$\gamma = \beta - \alpha_{min} = 180°$$

$$V_{R\,\text{rms}} = \sqrt{V_{dc}^2 + \frac{V_{max}^2}{2} - \frac{4V_{dc}V_{max}}{\pi}}$$

$$V_{R\,\text{rms}} = \sqrt{150^2 + 110^2 - \frac{4(150)(\sqrt{2}\,110)}{\pi}} = 70 \text{ V}$$

The total rms current (including all harmonics) during discharging $I_{d\,\text{rms}}$ is

$$I_{d\,\text{rms}} = \frac{V_{R\,\text{rms}}}{R} = 70 \text{ A}$$

To calculate the power delivered to the ac source, we need to compute the rms value of the fundamental component of the current, and the phase shift as given in Equations (3.16) through (3.25).

$$a_1 = \frac{1}{\pi}\int_0^{2\pi} i(\omega t)\cos(\omega t)\,d\omega t = \frac{2}{R\pi}\int_\alpha^\beta v_R\cos(\omega t)\,d\omega t = \frac{2}{R\pi}\int_\alpha^\beta (V_{dc} - v_s)\cos(\omega t)\,d\omega t$$

$$a_1 = \frac{2\,V_{dc}}{R\pi}(\sin\beta - \sin\alpha) - \frac{V_{max}}{R\pi}(\sin^2\beta - \sin^2\alpha)$$

$$a_1 = 0$$

Since a_1 is zero, there is no phase shift between the ac source voltage and the fundamental component of the current.

$$b_1 = \frac{1}{\pi}\int_0^{2\pi} i(\omega t)\sin(\omega t)\,d\omega t = \frac{2}{R\pi}\int_\alpha^\beta v_R\sin(\omega t)\,d\omega t = \frac{2}{R\pi}\int_\alpha^\beta (V_{dc} - v_s)\sin(\omega t)\,d\omega t$$

$$b_1 = \frac{2\,V_{dc}}{R\pi}(\cos\alpha - \cos\beta) - \frac{V_{max}}{R\pi}\left[\gamma + \frac{\sin(2\alpha) - \sin(2\beta)}{2}\right]$$

$$b_1 = \frac{4\,V_{dc}}{R\pi} - \frac{V_{max}}{R} = 35.42$$

$$\phi_1 = \tan^{-1}\left(\frac{a_1}{b_1}\right) = 0°$$

The rms value of the fundamental component of the current is given by

$$I_{1d\,\mathrm{rms}} = \frac{c_1}{\sqrt{2}} = \frac{\sqrt{a_1^2 + b_1^2}}{\sqrt{2}} = \frac{35.42}{\sqrt{2}} = 25\ \mathrm{A}$$

As given in Equation (3.17), the power delivered to the ac source P_{ac} can be computed by

$$P_{\mathrm{ac}} = VI_{1d\,\mathrm{rms}} \cos \phi_1 = 110 \times 25 \times \cos(0) = 2.75\ \mathrm{kW}$$

3.11 THREE-PHASE ENERGY RECOVERY SYSTEMS

In three-phase systems, the energy recovery circuit is similar to the circuit of the three-phase ac/dc converter in Figure 3.21. However, the switches are oriented to allow the current to flow from the dc source to the ac source. A typical circuit is shown in Figure 3.43. It consists of six IGBTs, a dc source, and an ac source. The current I can flow from the dc side to the ac side under the following conditions:

When $v_{ab} < V_{\mathrm{dc}}$, and S_1 and S_6 are triggered
When $v_{bc} < V_{\mathrm{dc}}$, and S_2 and S_3 are triggered
When $v_{ca} < V_{\mathrm{dc}}$, and S_4 and S_5 are triggered

Figure 3.44 shows the waveforms of the circuit and a switching pattern. The bottom part of the figure shows the voltage difference between the ac and dc sources. Note that this switching sequence provides a balanced energy recovery for all three phases. The triggering sequence of the IGBTs does not result in a waveform at the ac side similar to those in Figures 3.30 or 3.31, because the ac side is already a voltage source and its waveforms cannot be altered.

FIGURE 3.43
Energy recovery circuit

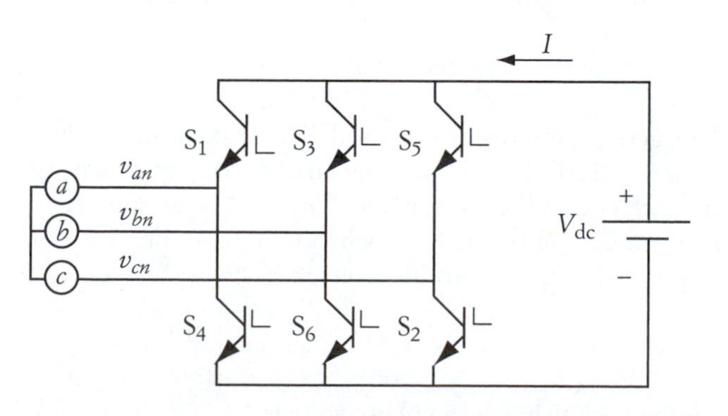

FIGURE 3.44
Waveforms of switching patterns for energy recovery

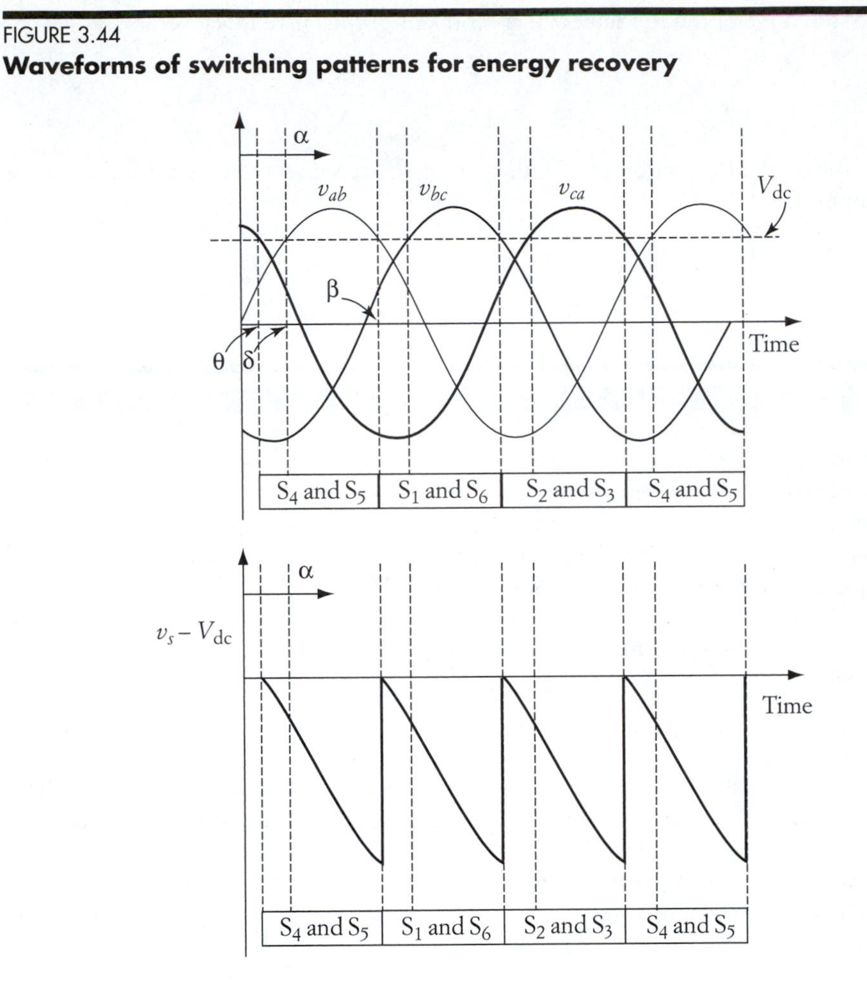

The single-phase equivalent circuit is similar to that given in Figure 3.41. The equations governing the discharging of the battery can be expressed by

$$v_R = V_{dc} - v_{ab}$$

Several triggering patterns can be used. The maximum conduction period of any transistor is $\gamma = 120°$. In the following analysis, $\alpha_{min} = \theta$, whereby the voltage of the battery pack equals the voltage of the line-to-line ac source.

β can be computed at the moment when the line-to-line waveform equals the dc voltage. Assume that the line-to-line voltage is

$$v_{ab} = \sqrt{3}V_{max} \sin(\omega t)$$

where V_{max} is the peak value of the phase voltage.

Hence,

$$\sqrt{3}V_{\max}\sin(\delta) = V_{\mathrm{dc}}$$

$$\beta = 180° - \sin^{-1}\left(\frac{V_{\mathrm{dc}}}{\sqrt{3}V_{\max}}\right) \quad (3.84)$$

$$\theta = \beta - 120°$$

$$\theta = 60° - \sin^{-1}\frac{V_{\mathrm{dc}}}{\sqrt{3}\,V_{\max}}$$

The average voltage across the resistance during discharging is

$$V_R = \frac{3}{2\pi}\int_{\alpha}^{\beta}[V_{\mathrm{dc}} - \sqrt{3}\,V_{\max}\sin(\omega t)] \quad (3.85)$$

$$V_R = V_{\mathrm{dc}}\frac{3\gamma}{2\pi} + \frac{3\sqrt{3}\,V_{\max}}{2\pi}(\cos\beta - \cos\alpha)$$

EXAMPLE 3.16

The three-phase circuit shown in Figure 3.43 is used to discharge a battery bank of 250 V. The line-to-line ac voltage is 208 V. The system resistance between the battery bank and the source during conduction is 3 Ω. Calculate the following:

a. Minimum triggering angle and the associated conduction period
b. Average charging current for the minimum triggering angle

SOLUTION

a. The minimum triggering angle

$$\alpha_{\min} = \theta = 60° - \sin^{-1}\frac{V_{\mathrm{dc}}}{\sqrt{3}V_{\max}} = 2°$$

$$\gamma = \beta - \alpha_{\min} = 120°$$

b. To compute the average charging current, we need to compute the voltage across the system resistance.

$$V_R = V_{\mathrm{dc}}\frac{3\gamma}{2\pi} + \frac{3\sqrt{3}\,V_{\max}}{2\pi}(\cos\beta - \cos\alpha)$$

$$V_R = V_{\mathrm{dc}} + \frac{9V_{\max}}{2\pi}\cos(\alpha + 150)$$

$$V_R = 250 + 9\frac{169}{2\pi}\cos(152) = 36.26 \text{ V}$$

$$I_{R\text{ ave}} = \frac{V_R}{R} = 12.09 \text{ A}$$

3.12 CURRENT SOURCE INVERTER

The current source inverter (CSI) has several advantages over the voltage source inverters; among these advantages are the following:

1. The load current is constant, even when the load impedance changes.
2. When misfiring occurs and two switches on the same leg conduct, the supply of the voltage source inverter is shorted and the switches are damaged. However, for the CSI, the current through these switches is controlled to stay below the damage level.
3. When a commutation circuit is needed, the current source inverter demands a much simpler commutation than the voltage source inverter.

During the conduction period, the current source inverter (CSI) is designed to maintain the current of the load constant, while the voltage is allowed to fluctuate. To explain the operation of the CSI, consider the dc/ac inverter shown in Figure 3.45. The figure shows a variable dc source, which could be the output of an ac/dc converter. The magnitude of the source voltage v_s is continually adjusted to maintain the current in the inductor L constant. A large enough inductance is chosen

FIGURE 3.45
Current source inverter

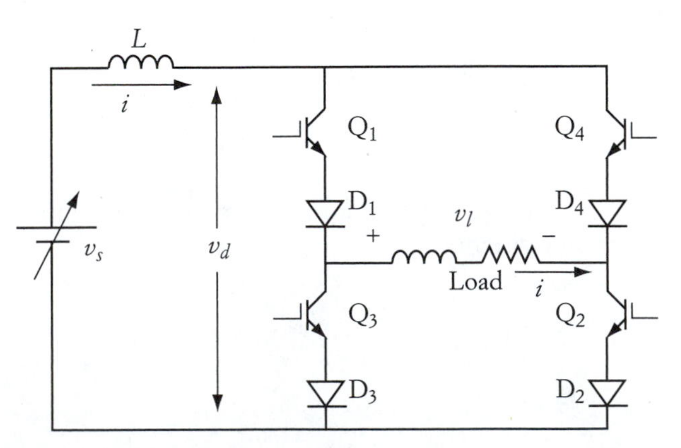

so as to reduce current fluctuations. The dc/ac bridge circuit is similar to that shown in Figure 3.28, except for the four additional diodes D_1 to D_4. The function of these diodes will become apparent after we discuss the operation of this circuit. The four switches (Q_1 to Q_4) could be GTO, bipolar transistors, IGBT, or any other switching device.

Since the current is maintained constant, two transistors must be in conduction at any time. When Q_1 and Q_2 are conducting, the current flows in the direction shown in Figure 3.45. The load current reverses its direction when Q_3 and Q_4 are conducting.

The voltage equation of the circuit can be written as

$$v_d = v_s - L\frac{di}{dt} \tag{3.86}$$

Also,

$$v_d = v_{sw} + iR_l + L_l\frac{di_l}{dt} \tag{3.87}$$

where R_l and L_l are the resistance and inductance of the load, respectively. v_{sw} is the voltage drop across the switches in conduction, including the conducting diodes. Note that in Equation (3.86), we used di/dt, and in Equation (3.87), we used di_l/dt. Although the currents in the load and inductor L are the same during the steady state, they can have different values for di/dt. When two of the switches are in the process of closing, the other two must be in the process of opening. This opening and closing normally happens very rapidly, so the load current reverses its direction, while the inductor current remains relatively unchanged.

The rapid reversal of the load current results in a high di_l/dt. For a highly inductive load with small resistance, Equations (3.86) and (3.87) can be rewritten as

$$v_{sw} = v_d - L_l\frac{di_l}{dt} = v_s - L\frac{di}{dt} - L_l\frac{di_l}{dt} \tag{3.88}$$

If the load inductance is large enough and the switches are high-speed devices, $L_l\,(di_l/dt)$ during the current reversal could become much larger than v_d. Depending on the sign of the di_l/dt, v_{sw} could become negative. As we know from the property of transistors, an excessive reverse voltage can damage the transistors. Therefore, by inserting a diode, with good reverse-voltage property, in series with the switches, most of the negative voltage of v_{sw} will be on the diodes, and the switches are saved.

Figure 3.46 shows a simple dc/ac current source inverter with a commutation circuit. It has four diodes (D_1 to D_4), four SCRs (Q_1 to Q_4), an inductor (L) and two capacitors (C_1 and C_2). The four SCRs are switched in pairs; Q_1 and Q_2 are closed during one half-cycle, and Q_3 and Q_4 are closed during the other half. This circuit is similar to that in Figure 3.28, but has more components and its switches are SCRs instead of transistors.

FIGURE 3.46
Commutation of current source inverter

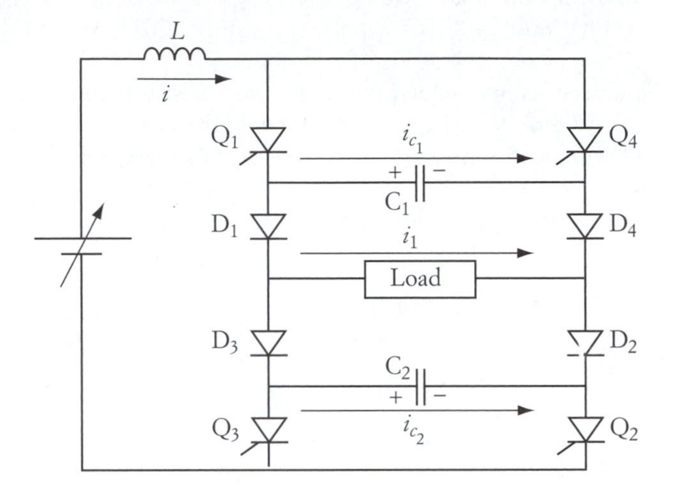

FIGURE 3.47
Waveform of current source circuit

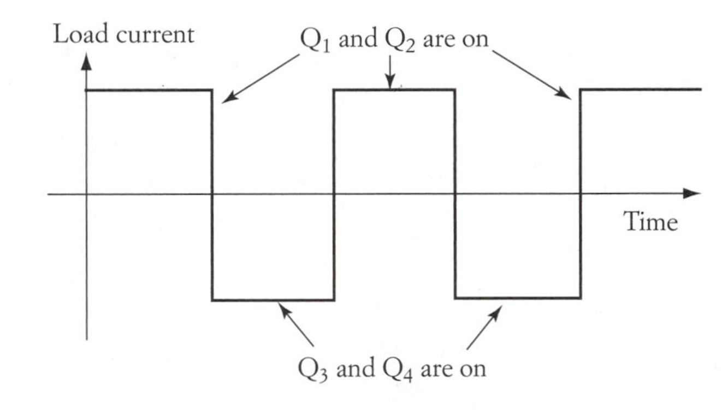

For now, ignore the capacitors and diodes and assume that the bridge has only four SCRs. When Q_1 and Q_2 are switched on, the current flows in the load in one direction. When these switches are turned off, and Q_3 and Q_4 are turned on, the current flows in the opposite direction. If we ignore the transition period between the turn-off and turn-on of the switches, one would expect the load current to have the waveform shown in Figure 3.47. If the load changes during a conduction period (180°), the current of the load is maintained constant due to the presence of the inductor L.

As you know, the SCR cannot be commutated unless its current falls below its holding value, which can happen when the terminal voltage of the SCR is reversed.

This circuit can be commutated by using the capacitors and diodes. When Q_1 and Q_2 are switched on, the source current flows in three paths:

$$i = i_1 + i_{c_1} + i_{c_2}$$

The loop of the capacitor current i_{c1} is from the source through Q_1, C_1, D_4, D_2, Q_2, and back to the source. Similarly, the i_{c2} loop is through Q_1, D_1, D_3, C_2, and Q_2. The loop of the load current is Q_1, D_1, D_2, and Q_2. The currents i_{c1} and i_{c2} flow until the capacitors are fully charged. The capacitance of C_1 and C_2 are selected sufficiently large so as to fully charge them in less than 180°. (The maximum conduction period of any pair of the SCRs is 180°.) After the capacitors are fully charged, only i_1 continues to flow. Now let us assume that Q_1 and Q_2 are to be turned off, and Q_3 and Q_4 turned on. All we have to do is to trigger Q_3 and Q_4. By doing this, C_1 becomes in parallel with Q_1, and the voltage across Q_1 is negative. Hence, Q_1 is turned off. Similarly, C_2 becomes in parallel with Q_2, and it turns Q_2 off. This process is repeated for the second half of the cycle.

One other function for the diodes is to isolate the capacitors from the load. In this case, when the load voltage varies, the capacitors remain at constant voltage after they are charged. Thus, the capacitor voltage will be available and ready for commutation when needed. The waveform of this circuit is shown in Figure 3.47 and is similar to that in Figure 3.28. The only difference is that the waveform here is for current instead of voltage.

CHAPTER 3 PROBLEMS

3.1 A half-wave, single-phase ac-to-dc converter is loaded by an impedance of 10 mH inductance in series with 10 Ω resistance. The ac voltage is 110 V (rms). For α equal to 30° and 90°, calculate the following:
   a. Conduction period
   b. Average current of the load
   c. Average voltage of the load
   d. dc power

3.2 Repeat Problem 3.1 for $\alpha = 10°$, assuming that a freewheeling diode is used.

3.3 Assume that an additional inductance can be inserted in series with the load. Also assume that the converter has no freewheeling diodes. Calculate the added inductance that leads to a conduction period of 180° when $\alpha = 30°$.

3.4 Calculate the average current, average voltage, and the power of the load for the case described in Problem 3.3.

3.5 A single-phase, half-wave SCR circuit is used to control the power consumption of an inductive load. The resistive component of the load is 5 Ω. The source voltage is 120 V (rms). When the triggering angle is adjusted to 60°, the average current of the load is 6 A. Calculate the following:
   a. Average voltage across the load
   b. Conduction period in degrees

3.6 A dc/dc converter consists of a 100 V dc source in series with a 10 Ω load resistance and a bipolar transistor. Assume that the transistor is an ideal switch. In each cycle, the transistor is turned on for 100 μs and turned off for 300 μs. Calculate the following:
   a. Switching frequency of the converter
   b. Average voltage across the load
   c. Average load current
   d. rms voltage across the load
   e. rms current
   f. rms power consumed by the load

3.7 A 120 V (rms), 60 Hz source is connected to a full-wave bridge as shown in Figure 3.48. The load is an arc welding machine that can be represented by a resistance of 1 Ω in series with an inductive reactance of 3 Ω. At a triggering angle of 60°, the current of the load is continuous. Calculate the following:
   a. Average voltage across the load
   b. Average voltage across the resistive element of the load
   c. Average current of the load

FIGURE 3.48

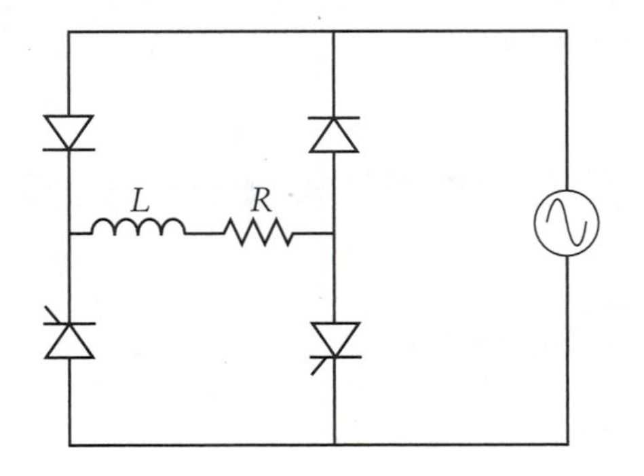

3.8 An inductive load consists of a resistance and an inductive reactance connected in series. The circuit is excited by a full-wave, ac/dc SCR converter. The ac voltage (input to the converter) is 120 V (rms), and the circuit resistance is 5 Ω. At a triggering angle of 30°, the load current is continuous. Calculate the following:
   a. Average voltage across the load
   b. Average load current
   c. rms voltage across the load

3.9 A purely inductive load of 10 Ω is connected to an ac source of 120 V (rms) through a half-wave SCR circuit.
   a. If the SCR is triggered at 90°, calculate the angle at the maximum instantaneous current.

b. If the triggering angle is changed to 120°, calculate the angle at the maximum instantaneous current.
c. Calculate the conduction period for the case in (b).

3.10 A resistive load of 5 Ω is connected to an ac source of 120 V (rms) through an SCR circuit.
a. If the SCR circuit consists of a single SCR, and if the triggering angle is adjusted to 30°, calculate the power consumption of the load.
b. If the SCR circuit consists of two back-to-back SCRs, calculate the power consumption of the load assuming that the triggering angle is kept at 30°.

3.11 An inductive load that has a resistive component of 4 Ω is connected to an ac source of 120 V (rms) through a half-wave SCR circuit. When the triggering angle of the SCR is 50°, the conduction period is 160°. Calculate the following:
a. Average voltage across the load
b. Average voltage across the resistive element of the load
c. rms voltage across the load
d. Average current of the load
e. If a freewheeling diode is connected across the load, calculate the load rms voltage. Assume that the current of the diode flows for a complete half-cycle.

3.12 The full-wave, ac/dc converter shown in Figure 3.49 is operating under continuous current (conduction period = 180°). The source voltage is 120 V (rms), and the load resistance is 2 Ω. For an average load current of 40 A, calculate the triggering angle of the SCRs.

FIGURE 3.49

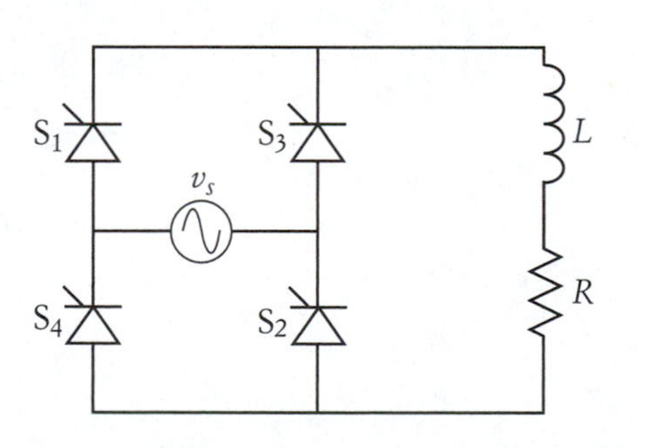

3.13 Draw the waveforms of the load voltage for the circuit in Figure 3.21, assuming that the triggering angle is −30°.

3.14 A three-phase, ac/dc converter is excited by a three-phase source of 480 V (rms and line-to-line). Compute the following:
a. The rms voltage across the load when the triggering angle is 30°

b. The average voltage across the load when the triggering angle is 140°. Keep in mind that the conduction is incomplete when the triggering angle is greater than 140°.

3.15 The three-phase circuit shown in Figure 3.43 is used to discharge a battery bank of 250 V. The line-to-line ac voltage is 208 V. The system resistance between the battery bank and the source during conduction is 3 Ω. Compute the triggering angle of the IGBTs that limits the average current to 5 A.

4

Joint Speed–Torque Characteristics of Electric Motors and Mechanical Loads

Electric motors exhibit a variety of speed–torque characteristics that are suitable for a wide range of load demands. A single motor can exhibit different speed–torque characteristics based on its winding configuration or the characteristics of the electric supply. As seen in Chapter 1, loads also have a wide range of speed–torque characteristics depending on their mechanical properties.

When an electric motor is connected to a mechanical load, the system operates at a speed–torque status that matches the characteristic of the motor as well as the mechanical load. Let us explain this by examining Figure 4.1. The figure shows three speed–torque characteristics of an electric motor (CC_1, CC_2, and CC_3). Assume that these characteristics can be obtained by adjusting the voltage across the terminals of the motor where CC_1 requires higher voltage as compared to CC_2 or CC_3. Assume also that the motor is driving an elevator (hoist). As we explained in Chapter 1, the load torque of a hoist is independent of speed. Let us assume that the motor voltage is adjusted so that its speed–torque characteristic is CC_1. The system operating point in this case is H_1—the coordinates of point H_1 determine the speed and torque of the system. Now assume that the motor voltage is reduced to the level of characteristic CC_2. The new system operating condition in this case is H_2, and so on. Note that the torque of the system is unchanged because of the hoist's characteristic.

Now we assume that the same motor is loaded by a blower (fan), and the fan characteristic is the one shown in Figure 4.1. The operating points of the system with the fan are F_1, F_2, and F_3, depending on the motor voltage. Note that the speed and torque of the system are changing for the fan load.

From the preceding assumptions, we conclude that the speed of the system is not determined by the motor only, but is also heavily dependent on the load characteristics. Hence, the characteristics of the load cannot be ignored when designing an effective electric drive system.

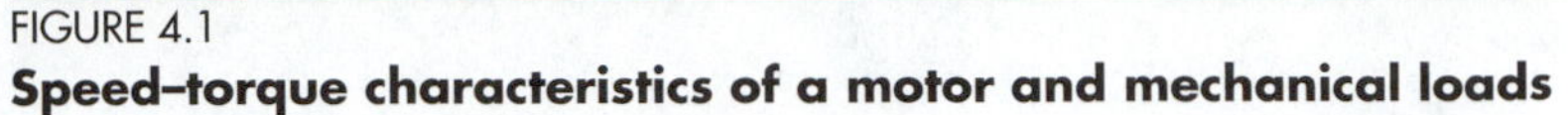

FIGURE 4.1
Speed–torque characteristics of a motor and mechanical loads

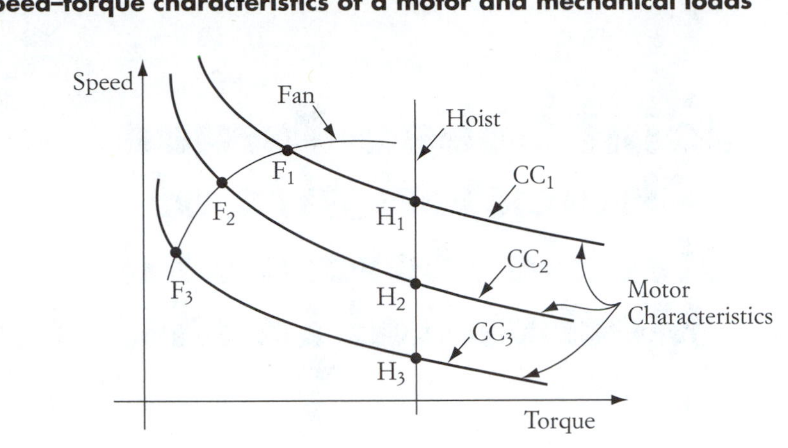

4.1 BIDIRECTIONAL ELECTRIC DRIVE SYSTEMS

One of the basic laws of physics is the theory of equilibrium developed by Isaac Newton in 1686. The essence of Newton's third law of motion is that *whenever one body exerts a force on another, the second exerts a force on the first that is equal in magnitude, opposite in direction, and has the same line of action.* This is also known as the *action–reaction* theory.

Now consider the case of an electric motor driving a mechanical load in a steady-state operation. A force exerted by either part of the drive system (load or motor) is opposed by a force equal in magnitude and opposite in direction from the other. If a frictional force is present, it is a part of the load force; this is true for any drive system even in standstill.

In drive applications, classifying the action and reaction forces is not always self-evident. Either part of the drive system can produce an action force depending on the nature of the operation. It is imperative to know the part of the drive system (motor or load) that produces the action force before any worthwhile analysis can start.

Consider the two examples given in Figures 4.2 and 4.3. The first, shown in Figure 4.2, represents an electric bus driven uphill, then downhill. To simplify the system, assume that the electric motor is directly mounted on the front wheels of the bus. Let us first study the system motion in the uphill direction. The force of the load is divided into two components: one is perpendicular to the road, F, producing the frictional force, and the other, F_l, is parallel to the road and represents the load torque exerted on the motor. The direction of F_l depends on the orientation of the road with respect to the gravitational force. F_l always pulls the bus toward the base of the hill. If frictional forces are ignored, the load torque seen by the motor is F_l multiplied by the radius of the wheel. This load torque must be matched by a motor torque F_m in the opposite direction to F_l.

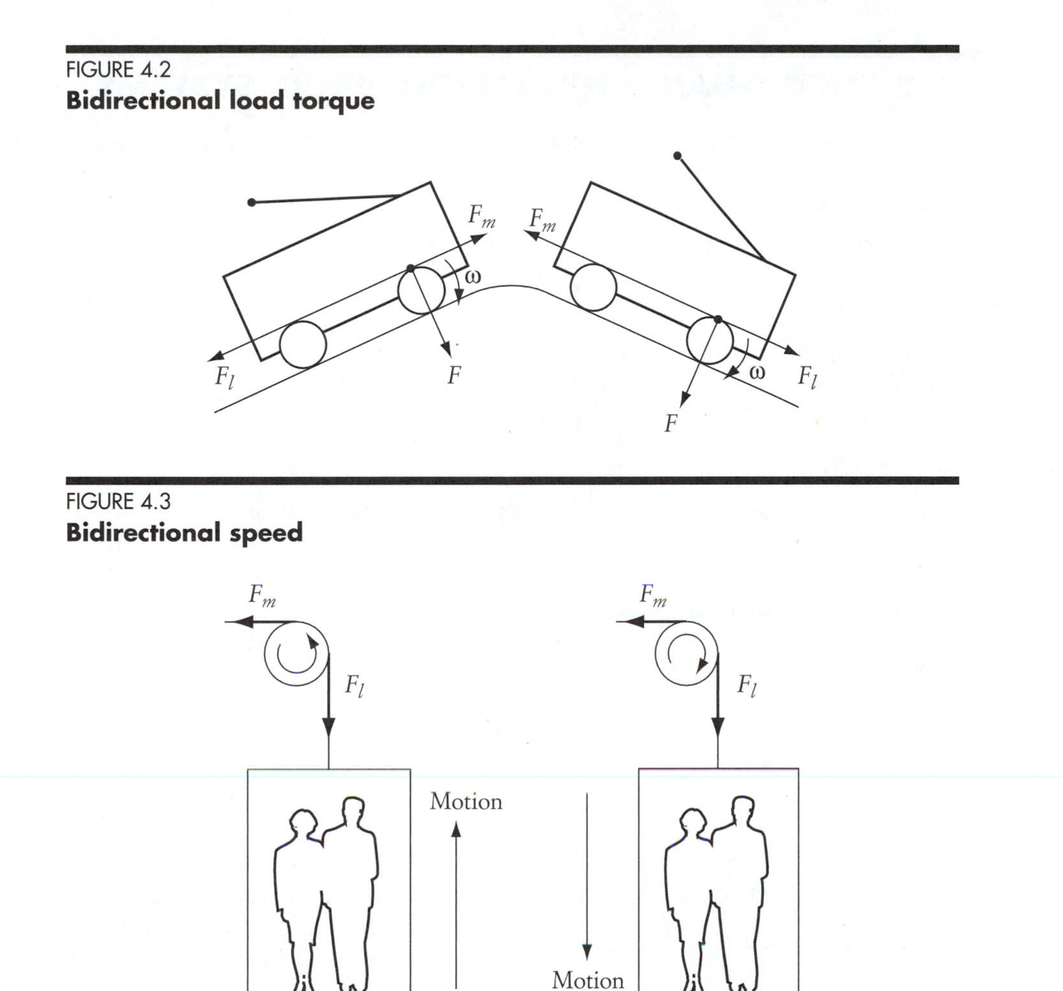

FIGURE 4.2
Bidirectional load torque

FIGURE 4.3
Bidirectional speed

Now let us assume that the bus is in the downhill direction. Because of the gravitational force, F_l still pulls the bus toward the bottom of the hill. However, as seen by the motor, the load force is reversed. The motor torque *must* then change its direction to counterbalance the torques of the load as described by Newton's laws.

Note that the motor speed is unidirectional in the uphill and downhill motions. Only the torques of the system are reversed.

The second example is shown in Figure 4.3. An elevator is moving passengers in both directions (up and down). For simplicity, let us assume that the elevator does not have a counterweight. In the upward and downward directions, the motor sees the load force F_l, which is a function of the weight of the passengers plus elevator cabin, cables, and so on. Since the weight and F_l are unidirectional, the motor force F_m is also unidirectional. The speed of the motor in this operation is bidirectional.

4.2 FOUR-QUADRANT ELECTRIC DRIVE SYSTEMS

The following conventions govern the power flow analysis of electric drive systems.

1. When the torque of an electric machine is in the same direction as the system speed, the machine consumes electric power from the electric source and delivers mechanical power to the load. The electric machine is then operating as a motor.
2. If the speed and torque of the machine are in opposite directions, the machine is consuming mechanical power from the load and delivering electric power to the source. In this case, the electric machine is acting as a generator.

Figure 4.4 shows the four quadrants of the speed–torque characteristic that cover all possible combinations of any electric drive system. Let us define the first quadrant as the reference. In this quadrant, the torque of the electric machine is in the same direction as the speed. The load torque, of course, is opposite to the machine torque. The electric machine in this case is operating as a motor. The flow of power is from the machine to the mechanical load.

FIGURE 4.4
Four-quadrant drives

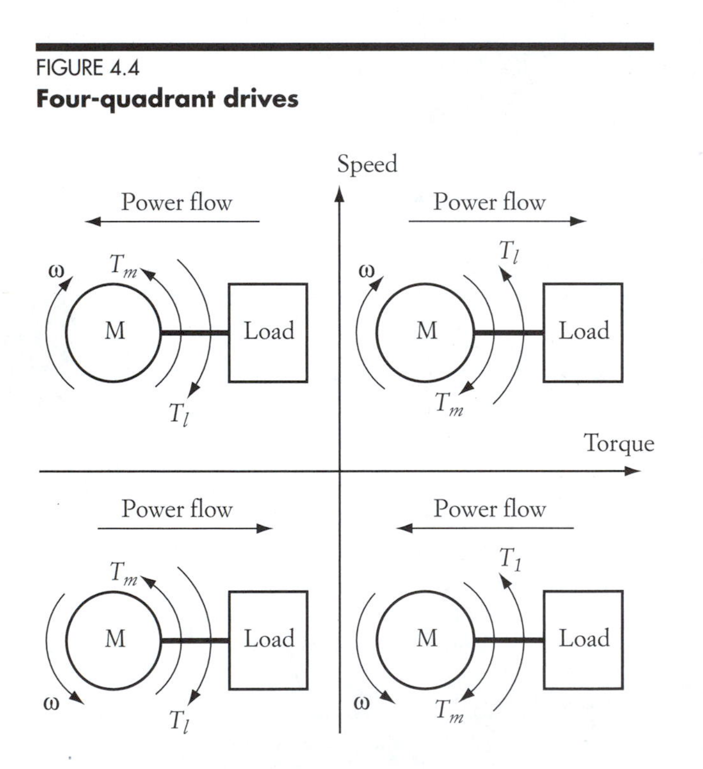

In the second quadrant, the speed direction of the system is unchanged, while the torque of the load and the motor torque are reversed. Since the load torque is in the same direction as the speed, the mechanical load is delivering power to the machine. The machine then receives this mechanical energy, converting it to electric energy and returning it back to the electric source. The electric machine in this case is acting as a generator.

Note that the example of Figure 4.2 represents the operation of the drive system in the first and second quadrants. The first quadrant represents the bus going uphill, whereas the second quadrant represents the bus going downhill.

Compared to the first quadrant, the system speed and torque are reversed in the third quadrant. Since the machine torque and speed are in the same direction, the flow of power is from the machine to the load. The machine is therefore acting as a motor rotating in the reverse direction to the speed of the first quadrant.

A bidirectional grinding machine is a good example of the first and third quadrant operation. The direction of the load torque of a grinding load is reversed when the speed is reversed (third quadrant). A horizontal conveyor belt is another example of this type of operation.

In the fourth quadrant, the torques remain unchanged as compared to the first quadrant. The speed, however, changes direction. From the load perspective, the load torque and the speed are in the same direction. Hence, the electric power flow is from the load to the machine. The machine in this case is operating as a generator delivering the electric power to the source. An example of the first and fourth quadrant operation is shown in Figure 4.3. The first quadrant may represent an elevator in the upward direction. When the elevator is going in the downward direction, the speed of the motor is reversed, but the torques are still unidirectional (fourth quadrant).

Any electric drive system operates in more than one quadrant. In fact, most versatile systems operate in all four quadrants. The converters of these systems must be designed to allow the electric power to flow in both directions.

Keep in mind that when an electric machine operates as a generator, it delivers electric power to the source. A large amount of power for a short period can be delivered to the source, but such energy must be fully absorbed by the source for a drive system to be efficient. If the source is composed of batteries, the rate by which they can absorb the energy is often slow due to the chemical process. The energy of the generator in this case is often wasted in resistive elements, which is why electric cars cannot, so far, utilize this energy to its fullest.

EXAMPLE 4.1

The mass of the electric bus in Figure 4.2 is 5000 kg, including the passengers. A single motor mounted on the front wheels drives the bus. The wheel diameter is 1 m. The bus is going uphill at a speed of 50 km/hr. The slope of the hill is 30°. The friction coeffiecient of the road surface at a given weather condition is 0.4. Ignore the motor losses, and compute the power consumed by the motor.

SOLUTION

To compute the electric power consumed by the motor, you must first calculate the total force exerted by the system on the motor. Consider the system forces in Figure 4.5. When the bus is moving uphill, the weight of the bus is divided into two components: one perpendicular to the road surface, which is responsible for the friction force, F_r, and the other, F_l, which pulls the bus toward the bottom of the hill. The direction of the friction force is always opposite to the direction of motion of the bus. The motor force must equal all forces in the opposite direction.

$$F_m = F_l + F_r$$

FIGURE 4.5
Forces acting on an electric bus moving uphill

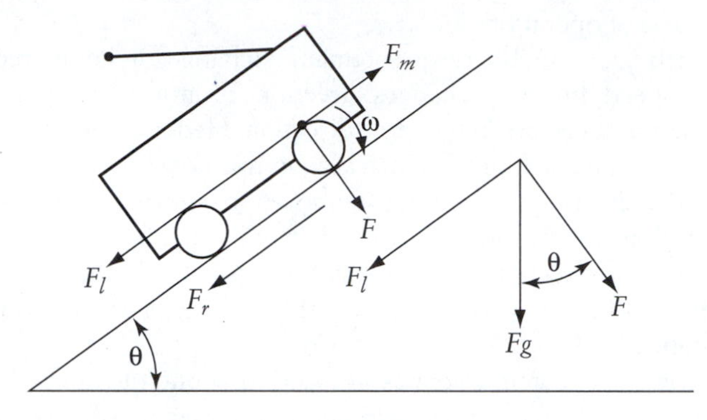

All these forces are dependent on the gravitational force F_g. Consider the force diagram on the right side of the figure. The normal force F and the load-pulling force F_l can be computed by

$$F = F_g \cos \theta$$

$$F_l = F_g \sin \theta$$

where θ is the slope of the hill.

The gravitational force F_g is

$$F_g = mg$$

where m is the total mass of the bus and passengers and g is the gravitational acceleration.

$$F_g = 5000 \times 9.8 = 49{,}000 \text{ N}$$

Hence,

$$F = F_g \cos \theta = 49{,}000 \times \cos 30 = 42{,}435.25 \text{ N}$$

$$F_l = F_g \sin \theta = 24{,}500 \text{ N}$$

The friction force is

$$F_r = \mu F$$

where μ is the coefficient of friction.

$$F_r = \mu F = 0.4 \times 42{,}435.25 = 16{,}974 \text{ N}$$

The total force seen by the motor is

$$F_m = F_l + F_r = 24{,}500 + 16{,}974 = 41{,}474 \text{ N}$$

The torque seen by the motor is equal to the total force multiplied by the radius of the wheel:

$$T_m = F_m r = 41{,}474 \times 0.5 = 20{,}737 \text{ Nm}$$

The power of the motor is the torque multiplied by the angular speed:

$$P_m = T_m \omega = T_m \frac{v}{r} = 20{,}737 \frac{50}{0.5} = 2073.7 \text{ kW}$$

CHAPTER 4 PROBLEMS

4.1 The mass of an electric car is 900 kg including the passengers. A single motor mounted on the front wheels drives the car, and the radius of the wheel is 0.3 m. The car is going downhill at a speed of 50 km/hr, and the slope of the hill is 30°. The friction coefficient of the road surface at a given weather condition is 0.8. Ignore the motor losses and compute the power generated by the electric machine.

4.2 Assume that the electric car in Problem 4.1 is moving at a constant speed. It takes the car 1 min to reach the bottom of the hill. Calculate the energy generated by the electric machine.

4.3 If the energy generated by the electric machine in Problem 4.2 is totally consumed by the batteries of the car, compute the charging rate of the batteries. Is it possible to charge a battery pack at this rate?

4.4 The electric drive system shown in Figure 4.6 consists of a motor, a pulley, a rigid belt, and a stage. The motor moves the stage in either direction. Explain the motion of system and the power flow using the four-quadrant drive concept.

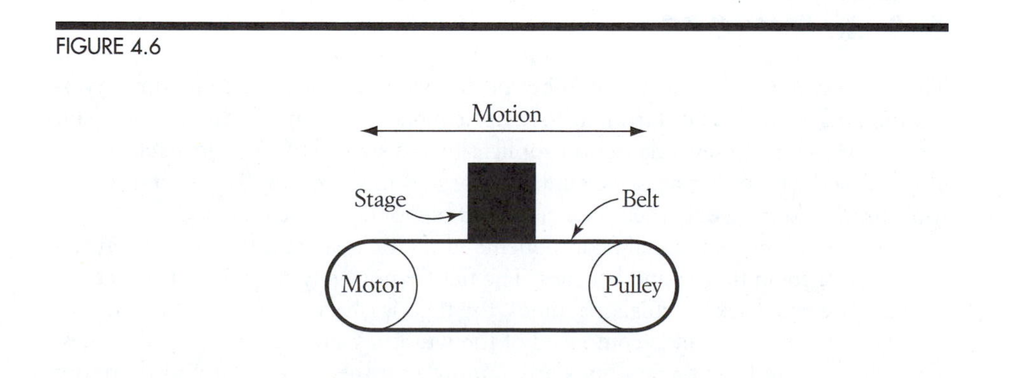

FIGURE 4.6

5

Speed–Torque Characteristics of Electric Motors

Electric motors have a variety of speed–torque characteristics during steady-state and transient operations. For a given drive application, engineers often selected motors with characteristics matching the needed operation, which could be driven by existing power sources. Due to advances in power electronic devices and circuits, such stringent restrictions no longer exist. The characteristics of most motors can now be altered to match the desired performance when external power converters are used and advanced control strategies are employed.

In this chapter, the speed–torque characteristics of major types of electric motors are presented. Models and formulas of speed equations as related to the torque are explained from the electric drive perspective. These characteristics form the basis for the speed control and braking of electric motors that are discussed in following chapters. Three types of electric motors are discussed here: dc, induction, and synchronous. Although there are several other types of motors, such as the brushless, reluctance, linear, and stepper motors, they all share common features with the three presented here. For example, the brushless machine can be considered a special form of a synchronous machine switched to imitate a dc motor. The linear induction motor is also considered a special form of the induction motor.

5.1 dc MOTORS

The dc machine is popular in a number of drive applications due to its simple operation and control. The starting torque of dc machines is large, which is the main reason for using it in several traction applications. A special form of dc machine can also be used with either ac or dc supply. A large number of appliances and power tools used at home, such as circular saws and blenders, are dc machines.

Figure 5.1 shows the main components of the dc machine: field circuit, armature circuit, commutator, and brushes. The field is normally an electric magnet fed by a dc power source. In small machines, the field is often a permanent magnet.

The armature circuit is composed of the windings, commutator, and brushes. The windings and the commutator are mounted on the rotor shaft and therefore

FIGURE 5.1
Main components of a dc machine

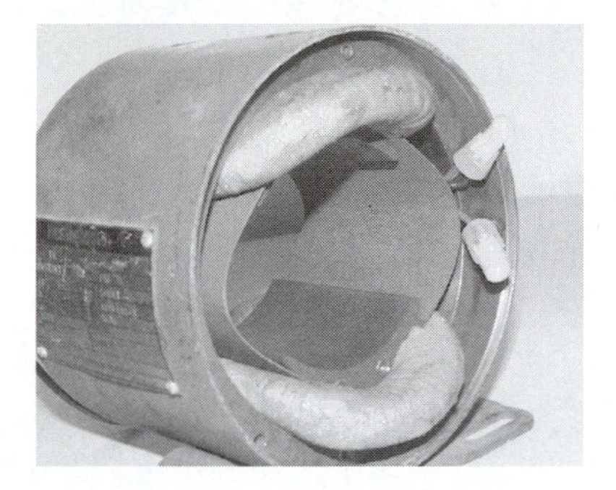

Field windings mounted on stator

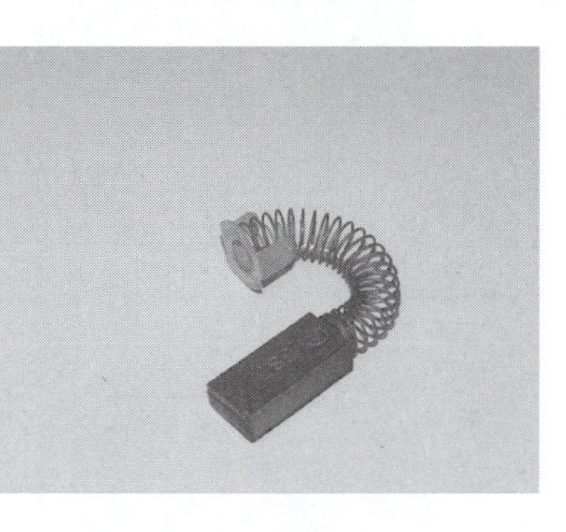

Brush

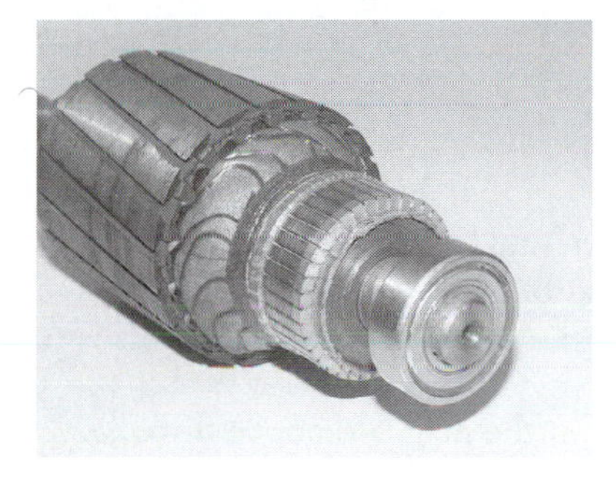

Rotating armature with commutator

Rotating armature with commutator and brushes

rotate. The brushes are mounted on the stator and are stationary, but in contact with the rotating commutator segments. The rotor windings are composed of several coils; each has two terminals connected to the commutator segments on opposite sides. The commutator segments are electrically isolated from one another. The segments are exposed, and the brushes touch two opposing segments. The brushes allow the commutator segments to be connected to an external dc source.

The diagram in Figure 5.2 illustrates the operation of a typical dc machine. The stator field produces flux ϕ from the N pole to the S pole. The brushes touch the terminals of the rotor coil under the pole. When the brushes are connected to an external dc source of potential V, a current I enters the terminal of the rotor coil under the N pole and exits from the terminal under the S pole. The presence of the stator flux and rotor current produces a force F on the coil known as the Lorentz force. The direction of F is shown in the figure. This force produces torque that rotates the armature counterclockwise. The coil that carries the current moves away

FIGURE 5.2
Operation of a typical dc machine

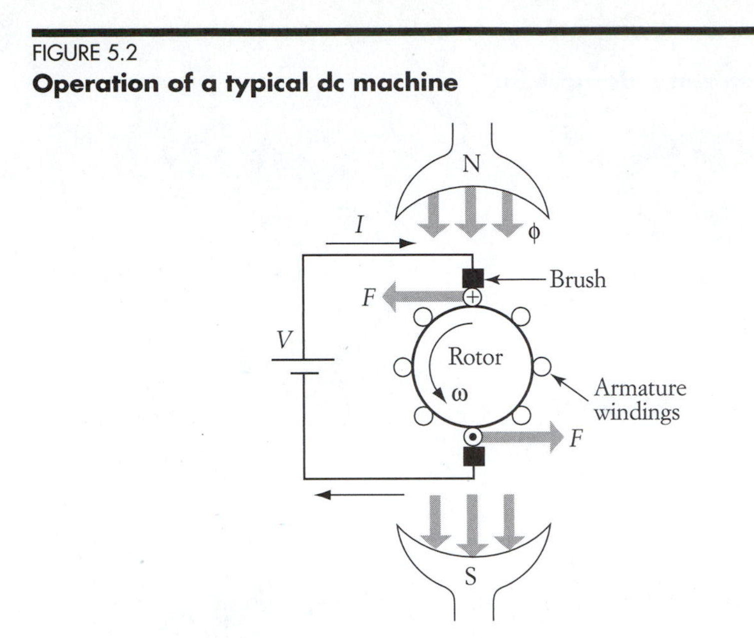

from the brush and is disconnected from the external source. The next coil moves under the brush and carries the current *I*. This produces a continuous force *F* and continuous rotation. Note that the function of the commutator and brushes is to switch the coils mechanically.

The rotation of the machine is dependent on the magnetomotive force *MMF* of the field circuit, which is described by

$$MMF \approx NI$$

where *N* is the number of turns and *I* is the field current. The desired *MMF* can be achieved by the design of the field windings. There are basically two types of field windings: the first has a large number of turns and low current, and the second type has a small *N* and high current. Both types achieve the desired range of *MMF*. Actually, any two different windings can produce identical amounts of *MMF* if their current ratio is inversely proportional to their turns ratio. The first type of winding can handle higher voltage than the second type. Moreover, the cross section of the wire is smaller for the first type since it carries a smaller current.

Direct current motors can be classified into four groups based on the arrangement of their field windings. Motors in each group exhibit distinct speed–torque characteristics and are controlled by different means. These four groups are:

1. *Separately excited machines.* The field winding is composed of a large number of turns with small cross-section wire. This type of field winding is designed to withstand the rated voltage of the motor. The field and armature circuits are excited by separate sources.

2. *Shunt machines.* The field circuit is the same as that for separately excited machines, but the field winding is connected in parallel with the armature circuit. A common source is used for the field and armature windings.
3. *Series machines.* The field winding is composed of a small number of turns with a large cross-section wire. This type is designed to carry large currents and is connected in series with the armature winding.
4. *Compound machines.* This type uses the shunt and series windings.

5.1.1 SEPARATELY EXCITED MOTORS

The equivalent circuit of a separately excited motor is shown in Figure 5.3. The motor consists of two circuits: field and armature. The field circuit is mounted on the stator of the motor and is energized by a separate dc source of voltage V_f. The field has a resistance R_f and a high inductance L_f. The field inductance has no impact in the steady-state analysis, since the source is a dc type. The field current I_f can then be represented by

$$I_f = \frac{V_f}{R_f} \tag{5.1}$$

For small motors (up to a few hundred watts), the field circuit is a permanent magnet. In such a case, the flux of the field is constant and cannot be adjusted.

The armature circuit, mounted on the rotor, is composed of a rotor winding and commutator segments. An external source of voltage V_t is connected across the armature to provide the electric energy needed to drive the load. The source is connected to the armature circuit via the commutator segments and brushes. The direction of the current in the armature winding is dependent on the location of the winding with respect to the field poles.

FIGURE 5.3
Equivalent circuit of a dc motor in steady-state operation

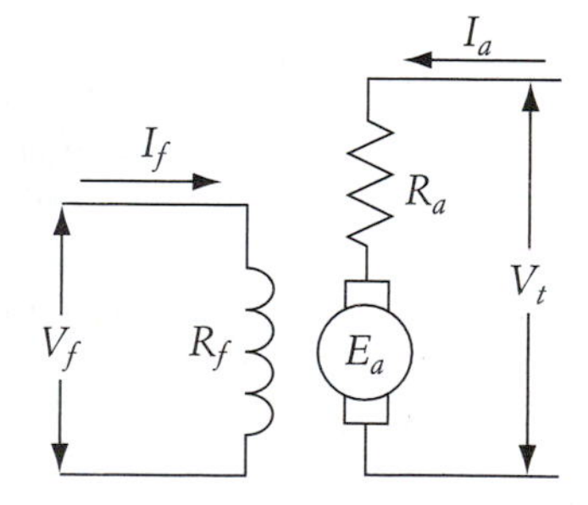

Relative to the field circuit, the armature carries a much higher current. Therefore, the wire cross section of the armature winding is much larger than that for the field circuit. The armature resistance R_a is, therefore, much smaller than the field resistance R_f. R_a is in the range of a few ohms and is smaller for larger horsepower motors. The field resistance is a hundred times larger than the armature resistance. The field current is usually in the neighborhood of 1% to 10% of the rated armature current. The field voltage is usually in the same order of magnitude as the armature voltage.

The back electromagnetic force E_a shown in Figure 5.3 is equal to the voltage of the source minus the voltage drop due to the armature resistance. The armature current I_a can then be expressed by

$$I_a = \frac{V_t - E_a}{R_a} \tag{5.2}$$

The multiplication of I_a by E_a represents the developed power P_d. In mechanical representation, the developed power is also equal to the developed torque multiplied by the angular speed.

$$P_d = E_a I_a = T_d \omega \tag{5.3}$$

The developed power P_d is equal to the output power consumed by the mechanical load plus rotational losses (frictional and windage). Similarly, the developed torque T_d is equal to the load torque plus the rotational torque. The angular speed ω in Equation (5.3) is in radians/second.

Using Faraday's law and the Lorentz force expressions, the relationships that govern the electromechanical motion are

$$e = Blv$$

$$F = Bli$$

where B is the flux density, l is the length of a conductor carrying the armature current, v is the speed of the conductor relative to the speed of the field, and i is the conductor current. F and e are the force and the induced voltage on the conductor, respectively. If we generalize these equations by including all conductors, using the torque expression instead of the force F, and using the angular speed instead of v, we can rewrite E_a and T_d as

$$E_a \sim e$$

$$E_a = K\phi\omega \tag{5.4}$$

$$T_d \sim F$$

$$T_d = K\phi I_a \tag{5.5}$$

where ϕ is the flux, which is almost proportional to I_f for separately excited motors. The constant K is dependent on design parameters such as the number of poles, number of conductors, and number of parallel paths.

The speed–torque equation can be obtained by first substituting I_a of Equation (5.2) into Equation (5.5).

$$T_d = K\phi \frac{V_t - E_a}{R_a} \tag{5.6}$$

Then, by substituting E_a of Equation (5.4) into Equation (5.6), we get

$$T_d = K\phi \frac{V_t - K\phi\omega}{R_a} \tag{5.7}$$

or

$$\omega = \frac{V_t}{K\phi} - \frac{R_a}{(K\phi)^2} T_d \tag{5.8}$$

The speed–current equation can be obtained if $\frac{T_d}{K\phi}$ of Equation (5.8) is replaced by I_a.

$$\omega = \frac{V_t}{K\phi} - \frac{R_a I_a}{K\phi} \tag{5.9}$$

If we ignore the rotational losses, the developed torque T_d is equal to the shaft torque, and the no-load armature current is equal to zero. Hence, the no-load speed can be calculated from Equation (5.8) or (5.9) by setting the armature current and load torque equal to zero.

$$\omega_0 = \frac{V_t}{K\phi} \tag{5.10}$$

In reality, the mass of the drive system and the rotational losses are the base load of the motor. The no-load speed ω_0 is therefore slightly smaller than the value computed in Equation (5.10). Nevertheless, Equation (5.10) is an acceptable approximation.

In the steady state, the developed torque T_d is equal to the load torque T_m. At a given value of load torque T_m, the speed of the motor drops by an amount of $\Delta\omega$ that is equal to the second term on the right side of Equation (5.8).

$$\Delta\omega = \frac{R_a}{(K\phi)^2} T_m \tag{5.11}$$

The speed of the motor can then be expressed by using the no-load and speed drop.

$$\omega = \omega_0 - \Delta\omega \tag{5.12}$$

Figures 5.4 and 5.5 show the speed–torque and speed–current characteristics when the field and armature voltages are kept constant.

For large motors (greater than 10 hp), the armature resistance R_a is very small, because the armature carries higher currents, and the cross section of the wire must then be larger. For these motors, the speed drop $\Delta\omega$ is small, and the motors can be considered constant-speed machines.

The developed torque at starting T_{st} and the starting armature current I_{st} can be calculated from Equations (5.8) and (5.9) by setting the motor speed to zero.

$$T_{st} = K\phi \frac{V_t}{R_a} \tag{5.13}$$

$$I_{st} = \frac{V_t}{R_a} \tag{5.14}$$

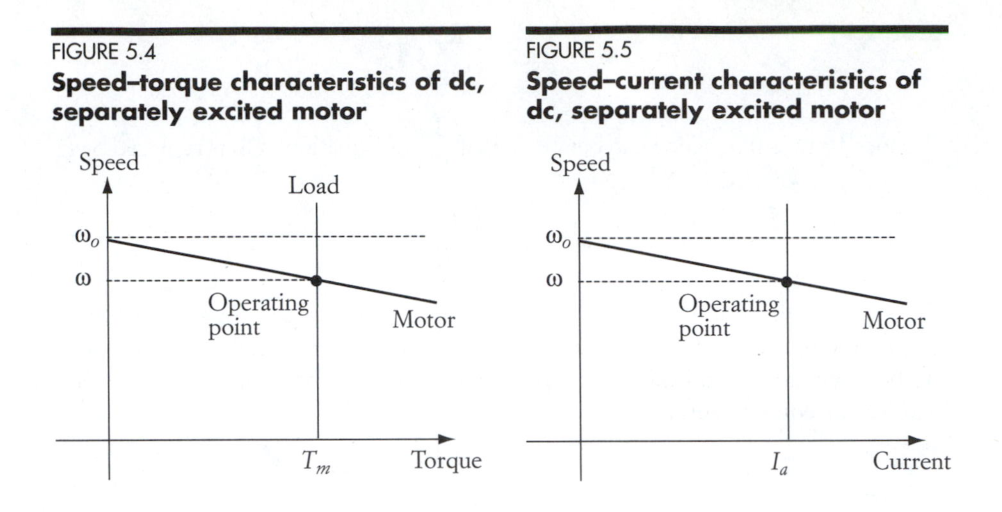

FIGURE 5.4
Speed–torque characteristics of dc, separately excited motor

FIGURE 5.5
Speed–current characteristics of dc, separately excited motor

Equations (5.13) and (5.14) provide important information about the starting behavior of the dc, separately excited motor. As we stated earlier, R_a is usually small. Hence, the starting torque of the motor is very large when the source voltage is equal to the rated value. This is an advantageous feature, and is highly desirable when motors start under heavy loading conditions. A problem, however, will arise from the fact that the starting current is also very large, as seen in Equation (5.14). Large currents at starting might have a damaging effect on the motor windings. Excessive currents flowing inside a winding will result in large losses due to the winding resistance. These losses, when accumulated over a period of time, may result in excessive heat that could melt the insulations of the winding, causing an eventual short circuit. This is illustrated by the next example.

EXAMPLE 5.1

A dc, separately excited motor has the following data:

$K\phi = 3.0$ V sec (volt second)

$V_t = 600.0$ V

$R_a = 2.0\ \Omega$

$I_a = 5.0$ A (armature current at full load)

Calculate the rated torque, starting torque, and starting current at full voltage.

SOLUTION

$$\text{Rated torque} = T_d = K\phi I_a = 3 \times 5 = 15 \text{ Nm}$$

$$\text{Starting torque} = T_{st} = V_t \frac{K\phi}{R_a} = 600.0 \frac{3.0}{2.0} = 900.0 \text{ Nm}$$

$$\text{Starting current} = I_{st} = \frac{V_t}{R_a} = 300.0 \text{ A}$$

As seen from these results, the starting torque is 60 times the rated torque, and the starting current is also 60 times the rated current. Such a high current over a period of time is damaging to the motor winding.

One important parameter missing in this example is the inductance of the armature winding. This inductance reduces the value of the current during transient conditions such as starting or braking. Nevertheless, the starting current under full voltage conditions is excessively large, and methods must be implemented to bring this current to a lower and safer value.

By examining Equation (5.14), the starting current can be reduced by lowering the terminal voltage or inserting a resistance in the armature circuit.

Let us assume that the starting current must be limited to six times the rated value. This can be achieved by reducing the terminal voltage at starting to

$$V_{st} = I_{st} R_a = 6 \times 5.0 \times 2.0 = 60.0 \text{ V}$$

Figure 5.6 illustrates the effect of reducing the terminal voltage during starting. When the voltage is reduced from V_{t_1} to V_{t_2}, the slope of the speed–current characteristic remains unchanged, whereas the no-load speed is reduced. Note that the starting current I_{st_2} is less than I_{st_1}.

FIGURE 5.6
Effect of reducing source voltage at starting

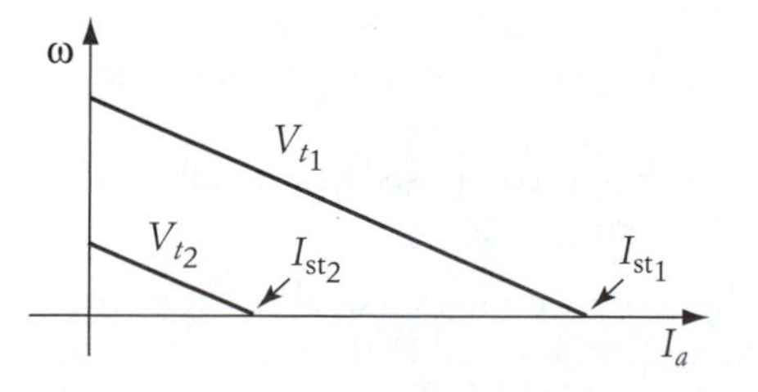

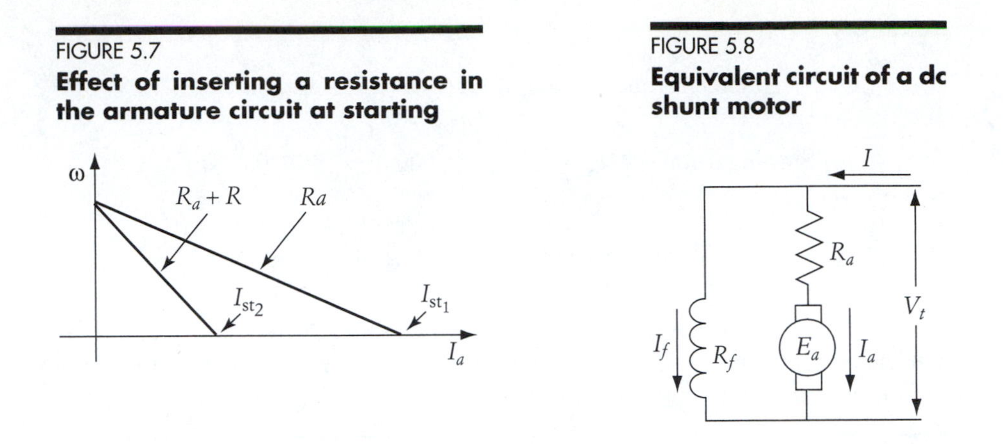

FIGURE 5.7
Effect of inserting a resistance in the armature circuit at starting

FIGURE 5.8
Equivalent circuit of a dc shunt motor

Another method to reduce the starting current is to add a resistance R to the armature circuit.

$$R + R_a = \frac{V_t}{I_{st}}$$

$$R = \frac{V_t}{I_{st}} - R_a = \frac{600.0}{30.0} - 2.0 = 18\ \Omega$$

Figure 5.7 illustrates the effect of reducing the starting current by adding a resistance to the armature circuit. The resistance increases the slope of the speed–current characteristic but keeps the no-load speed unchanged.

5.1.2 SHUNT MOTORS

A shunt motor has its field winding connected across the same voltage source used for the armature circuit, as shown in Figure 5.8. The current of the source I is equal to the sum of the armature current I_a and the field current I_f. The shunt motor exhibits characteristics identical to those of the separately excited motor.

5.1.3 SERIES MOTORS

The field winding of a series motor is connected in series with the armature circuit, as shown in Figure 5.9. There are several distinct differences between the field winding of a series machine and that of a shunt machine; among them are

1. The series field winding is composed of a small number of turns as compared to the shunt field winding.
2. The current of the series winding is equal to the armature current, whereas the current of the shunt field is equal to the supply voltage divided by the field resistance. Hence, the series field winding carries a much larger current than the shunt field winding.

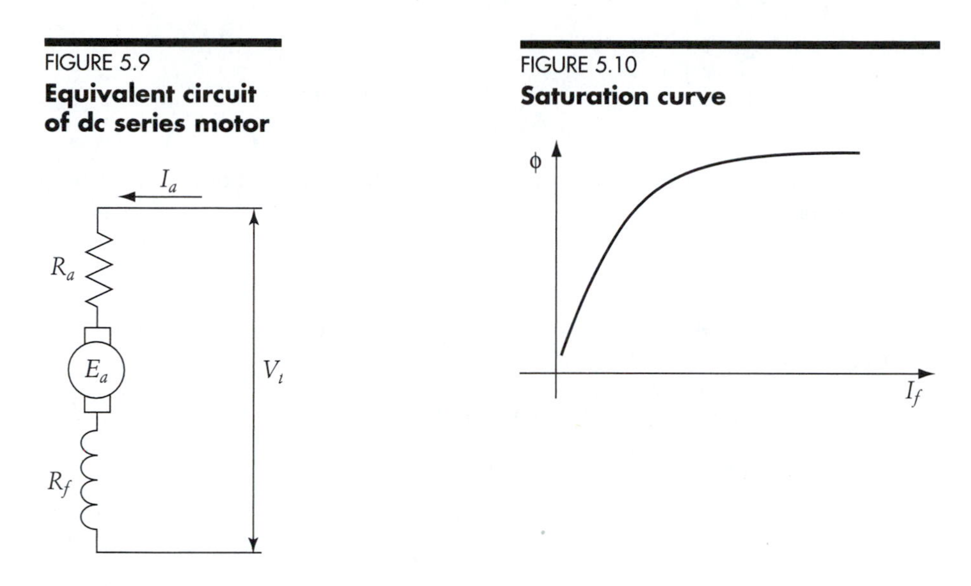

FIGURE 5.9
Equivalent circuit of dc series motor

FIGURE 5.10
Saturation curve

3. The field current of the shunt machine is constant regardless of loading conditions (armature current). The series machine, on the other hand, has a field current varying with the loading of the motor—the heavier the load, the stronger the field. At light or no-load conditions, the field of the series motor is very small.

When analyzing series machines, one should keep in mind the effect of flux saturation due to high field currents. A flux saturation curve is shown in Figure 5.10. The field coil is wound around the metal core of the stator. The current of the field winding produces the flux inside the core. When the current increases, the flux increases in a linear proportion unless the core is saturated. At saturation, the flux tends to increase at a progressively diminishing rate when the field current increases.

The series motor has the same basic equations used for shunt motors: Equations (5.4) and (5.5). The armature current is calculated by using the loop equation of the armature circuit.

$$I_a = \frac{V_t - E_a}{R_a + R_f} \tag{5.15}$$

Note that R_f is present in Equation (5.15). A similar process to the one used in Equation (5.7) can compute the torque of the machine.

$$T_d = K\phi \frac{V_t - E_a}{R_a + R_f} \tag{5.16}$$

$$T_d = K\phi \frac{V_t - K\phi\omega}{R_a + R_f} \tag{5.17}$$

or

$$\omega = \frac{V_t}{K\phi} - \frac{R_a + R_f}{(K\phi)^2} T_d \tag{5.18}$$

Let us assume that the motor operates in the linear region of the saturation curve; that is,

$$\phi = CI_a \tag{5.19}$$

where C is a proportionality constant. The developed torque in this case can be represented by

$$T_d = K\phi I_a = KCI_a^2 \tag{5.20}$$

Substituting Equations (5.19) and (5.20) into Equation (5.18) yields

$$\omega = \frac{V_t}{KCI_a} - \frac{R_a + R_f}{KC} \tag{5.21}$$

Equation (5.21) can also be obtained as a function of the developed torque.

$$\omega = \frac{V_t}{\sqrt{KCT_d}} - \frac{R_a + R_f}{KC} \tag{5.22}$$

Equations (5.21) and (5.22) show that the speed at no load or light loads is excessively high. Such a high speed may be damaging due to excessive centrifugal forces exerted on the rotor. For this reason, series motors must always be connected to a mechanical load.

The speed–torque characteristic of a series motor is shown in Figure 5.11. Note that the speed of the motor is rapidly decreasing when the load torque increases. This can be explained by Equation (5.22), where the motor speed is inversely proportional to the square root of the load torque.

The starting current of a series motor is calculated by setting E_a equal to zero in Equation (5.15), since ω is equal to zero.

FIGURE 5.11
Speed–torque characteristic of dc series motor

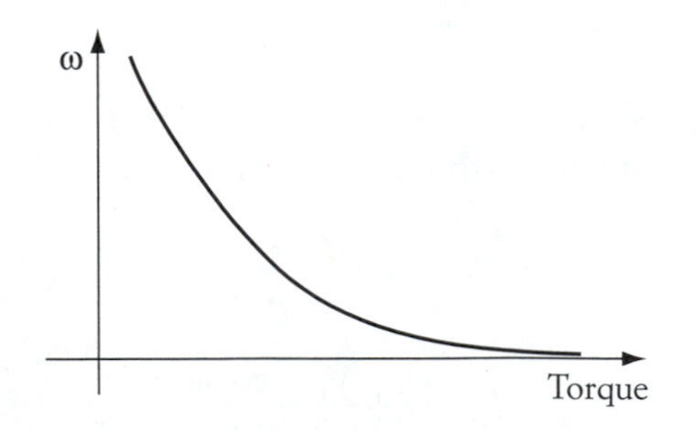

$$I_{st} = \frac{V_t}{R_a + R_f} \tag{5.23}$$

Compare Equation (5.23) to Equation (5.14). Note that, for the same terminal voltage, the starting current of a series motor is smaller than the starting current of the shunt motor due to the presence of R_f in Equation (5.23).

If we ignore the core saturation, the starting torque of the series motor is

$$T_{st} = K\phi I_{st} = KCI_{st}^2 = KC\left(\frac{V_t}{R_a + R_f}\right)^2 \tag{5.24}$$

To compare the starting torque of a series motor to that of a shunt motor, let us rewrite Equation (5.13), assuming that the flux is proportional to the field current.

$$T_{st\ shunt} = K\phi \frac{V_t}{R_a} = KC\frac{V_t}{R_{f\ shunt}}\frac{V_t}{R_a} = KC\frac{V_t^2}{R_a R_{f\ shunt}} \tag{5.25}$$

where $R_{f\ shunt}$ is the resistance of the shunt field winding and is usually a few hundred times larger than the resistance of the series field R_f. If we assume that KC in Equations (5.24) and (5.25) are of comparable value, one can conclude that the starting torque of a series motor is much larger than that for a shunt motor. Also, keep in mind that the starting current of a series motor is lower than that for a shunt motor. These features make the series motor a popular machine in such applications as traction and transportation. A trolley bus, for example, requires a high starting torque, especially when loaded with passengers.

Another great feature of series motors is their ability to be directly driven by ac supplies. To explain this, let us examine Figure 5.9, where E_a is equal to the source voltage minus the voltage drop across the armature and field resistances. When the source voltage reverses its polarity, E_a follows. Since the field and armature inductances of series motors are small, E_a reverses its polarity without any tangible delay. Hence, E_a is always in phase with the supply voltage, and the field current is also in phase with the supply voltage. Since

$$\omega = \frac{E_a}{K\phi}$$

the speed of the motor remains unchanged when both E_a and ϕ reverse their polarities. Because of this important feature, we can find dc series motors used in household appliances and tools such as blenders, food processors, washing machines, drills, and circular saws. Note that high starting torque—another good feature of series motors—is also needed in all these applications.

5.1.4 COMPOUND MOTORS

A compound motor is composed of shunt and series windings. Two types of compound configurations can be used. One is called a cumulative compound, where

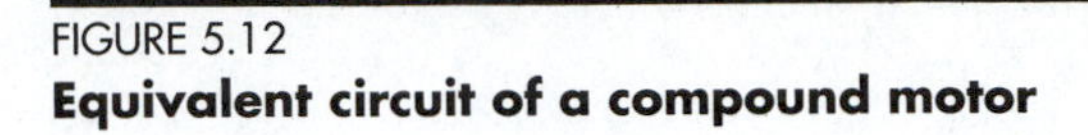

FIGURE 5.12
Equivalent circuit of a compound motor

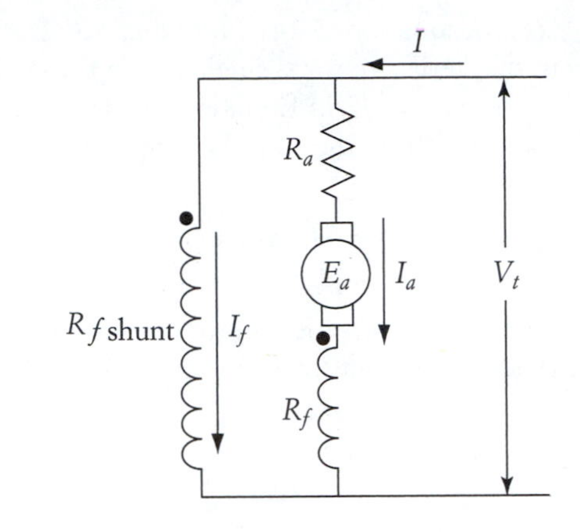

the airgap flux is the sum of the flux of the two field windings. The second is a subtractive compound, where the airgap flux is the difference between the flux of the two field windings. The subtractive compound may result in a very low flux in the airgap, leading to excessive speeds. This type is therefore considered unstable in operation and is not widely used.

The cumulative compound motor (hereafter called the compound motor) has the schematic shown in Figure 5.12. The direction of the currents with respect to the windings' dots represents flux polarities that are cumulative.

Equations (5.4) and (5.5) are also valid for compound motors. The flux in these equations can represent the compound machine by setting

$$\phi = \phi_{\text{series}} + \phi_{\text{shunt}}$$

The speed equation of a compound machine is similar to that given in Equation (5.9), but the resistive term and the flux are modified to reflect the parameters of the compound machine.

$$\omega = \frac{V_t}{K(\phi_{\text{series}} + \phi_{\text{shunt}})} - \frac{(R_a + R_f)I_a}{K(\phi_{\text{series}} + \phi_{\text{shunt}})} \tag{5.26}$$

Assuming that the terminal voltage and ϕ_{shunt} are constant, and

$$\phi_{\text{series}} = CI_a$$

then the speed–current equation can be modified as follows:

$$\omega = \frac{V_t}{KCI_a + K\phi_{\text{shunt}}} - \frac{(R_a + R_f)I_a}{KCI_a + K\phi_{\text{shunt}}} \tag{5.27}$$

FIGURE 5.13
Characteristics of compound, series, and shunt motors

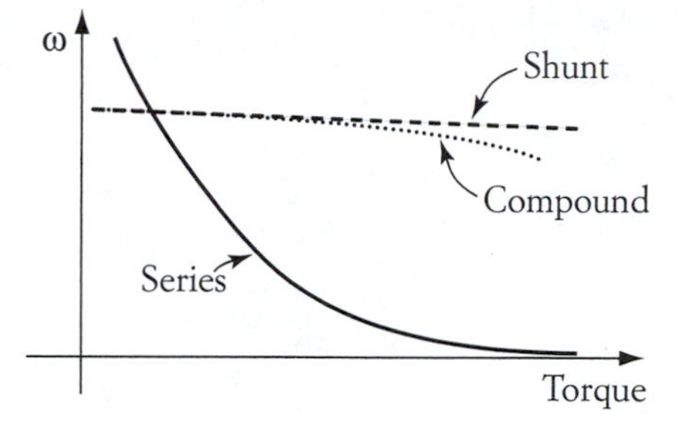

Also, since the motor torque is a function of the armature current and the total flux, it can be represented by

$$T_d = K(\phi_{\text{series}} + \phi_{\text{shunt}})I_a \tag{5.28}$$

$$\omega = \frac{V_t}{K(\phi_{\text{series}} + \phi_{\text{shunt}})} - \frac{(R_a + R_f)T_d}{[K(\phi_{\text{series}} + \phi_{\text{shunt}})]^2}$$

Note that at no load ($T_d = 0$), the armature current is zero, and ϕ_{series} is also zero. In this case, the no-load speed of the compound motor is

$$\omega_0 = \frac{V_t}{K\phi_{\text{shunt}}} \tag{5.29}$$

which is the same as the no-load speed of the shunt machine. By using the compound connection, the excessive no-load speed of the series motor is avoided.

The speed–torque characteristic of the compound motor is shown in Figure 5.13. For comparison purposes, the figure also shows the characteristics of the shunt and series motors. The starting current of the armature circuit of the compound machine can be calculated using the circuit in Figure 5.12.

$$I_{\text{st}} = \frac{V_t}{R_a + R_f} \tag{5.30}$$

The starting current of the compound motor is the same as that for the series motor. The starting torque of the compound motor is

$$\begin{aligned} T_{\text{st}} &= K(\phi_{\text{series}} + \phi_{\text{shunt}})I_{\text{st}} \\ &= KC\left(\frac{V_t}{R_a + R_f}\right)^2 + K\phi_{\text{shunt}}\left(\frac{V_t}{R_a + R_f}\right) \end{aligned} \tag{5.31}$$

which is higher than the starting torque of the series motor, given in Equation (5.24).

FIGURE 5.14
Components of an induction motor

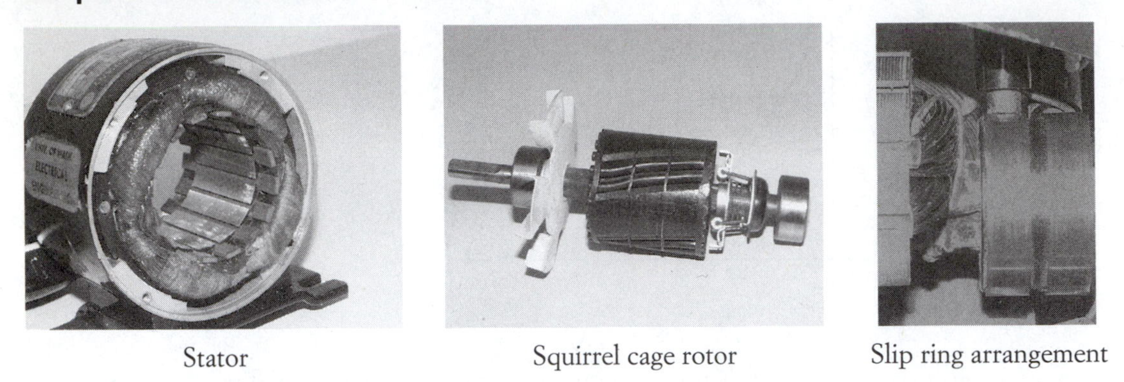

Stator   Squirrel cage rotor   Slip ring arrangement

5.2 INDUCTION MOTORS

About 65% of the electric energy in the United States is consumed by electric motors. In the industrial sector alone, about 75% of the total energy is consumed by motors, and over 90% of them are induction machines. The main reasons for the popularity of the induction machines are that they are rugged, reliable, easy to maintain, and relatively inexpensive. Their power densities (output power to weight) are higher than those for dc motors.

An induction machine is shown in Figure 5.14. The induction machine is composed of a stator circuit and a rotor circuit. The stator circuit has three sets of coils. In its simplest arrangement, the coils are separated by 120° and are excited by a three-phase supply. A conceptual representation is shown in Figure 5.15. The rotor circuit is also composed of three-phase windings that are shorted internally (within the rotor structure) or externally (through slip rings and brushes). The rotor with internal short is called a *squirrel cage rotor.* It consists of wire bars slanted and shorted on both ends of the rotor. The *slip rings* type of rotor is also shown in Figure 5.14. The terminals of the rotor windings in this type are connected to rings mounted on the rotor shaft. These slip rings are electrically isolated from one another. Most rotor windings are connected in wye, and the three terminals are connected to three slip rings. Carbon brushes mounted on the stator are continuously touching the slip rings to achieve the connectivity of the rotor windings with any external equipment. Unlike the commutator of the dc machine, the slip rings allow the brushes to be connected to the same coil regardless of the rotor position.

FIGURE 5.15
Conceptual representation of an induction motor

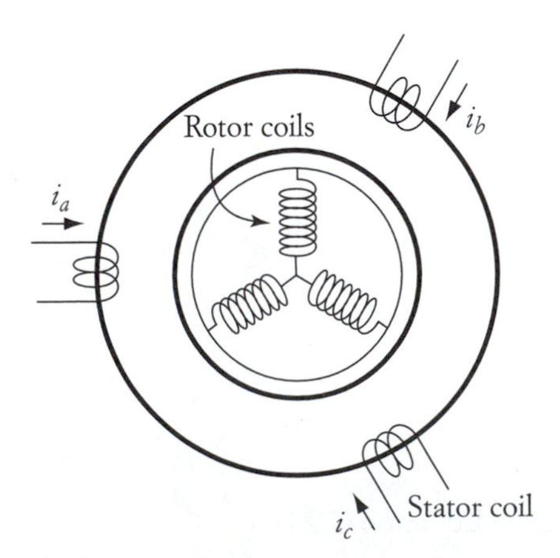

Before we explain how the induction machine rotates, we need to understand the concept of rotating fields. The three-phase stator windings are excited by a three-phase source with sinusoidal waveforms separated by 120°. The currents of the three phases produce a three-phase flux, as shown in Figure 5.16. Because of the arrangements of the stator windings, the flux of each phase travels along the windings' axes, as shown in Figure 5.17. The airgap flux is the resultant of all flux produced by the three windings.

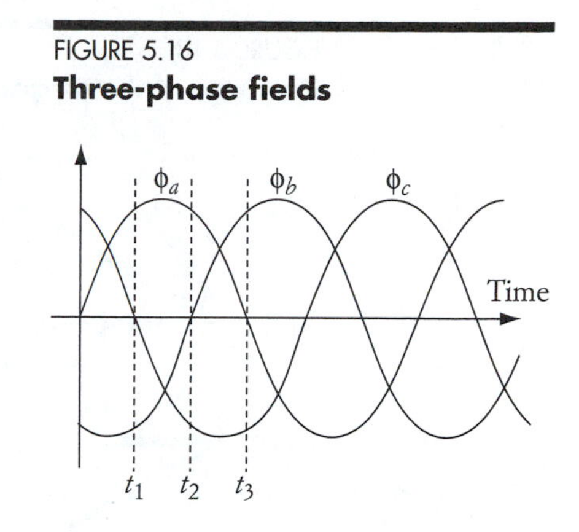

FIGURE 5.16
Three-phase fields

Now let us consider any three time instances such as those given in Figure 5.16 (t_1, t_2, and t_3). At t_1, the flux of phase a is $(\sqrt{3}/2)\,\phi_{max}$, the flux of phase b is $-(\sqrt{3}/2)\,\phi_{max}$, and the flux of phase c is zero. These flux phases are depicted in Figure 5.18. The resultant airgap flux is the phasor sum of all flux present in the airgap. Hence, at t_1,

$$\bar{\phi}(t_1) = \bar{\phi}_a(t_1) + \bar{\phi}_b(t_1) + \bar{\phi}_c(t_1) = \frac{\sqrt{3}}{2}\,\phi_{max}\angle 0 + \frac{\sqrt{3}}{2}\,\phi_{max}\angle 60^\circ + 0 = \frac{3}{2}\,\phi_{max}\angle 30^\circ$$

At t_2, the flux of phase a is $(\sqrt{3}/2)\,\phi_{max}$, the flux of phase b is zero, and the flux of phase c is $-(\sqrt{3}/2)\,\phi_{max}$. The total airgap flux at t_2 is

$$\bar{\phi}(t_2) = \bar{\phi}_a(t_2) + \bar{\phi}_b(t_2) + \bar{\phi}_c(t_2) = \frac{\sqrt{3}}{2}\,\phi_{max}\angle 0 + 0 + \frac{\sqrt{3}}{2}\,\phi_{max}\angle -60^\circ = \frac{3}{2}\,\phi_{max}\angle -30^\circ$$

Similarly, at t_3, the flux of phase a is zero, the flux of phase b is $(\sqrt{3}/2)\,\phi_{max}$, and the flux of phase c is $-(\sqrt{3}/2)\,\phi_{max}$. The total airgap flux at t_3 is

$$\bar{\phi}(t_3) = \bar{\phi}_a(t_3) + \bar{\phi}_b(t_3) + \bar{\phi}_c(t_3) = 0 + \frac{\sqrt{3}}{2}\,\phi_{max}\angle -120 + \frac{\sqrt{3}}{2}\,\phi_{max}\angle -60^\circ = \frac{3}{2}\,\phi_{max}\angle -90^\circ$$

FIGURE 5.17
Direction of the airgap field of each phase

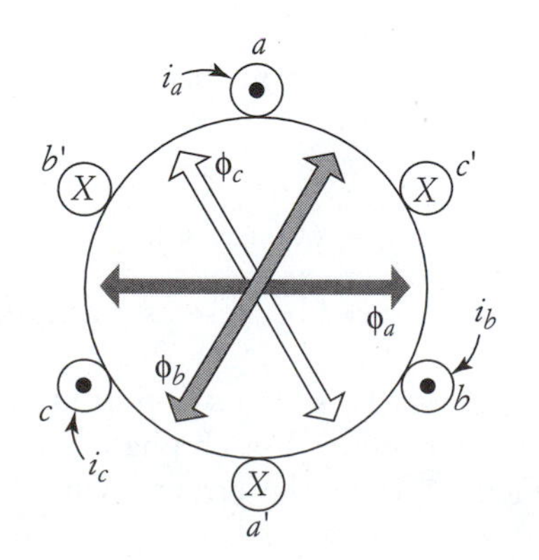

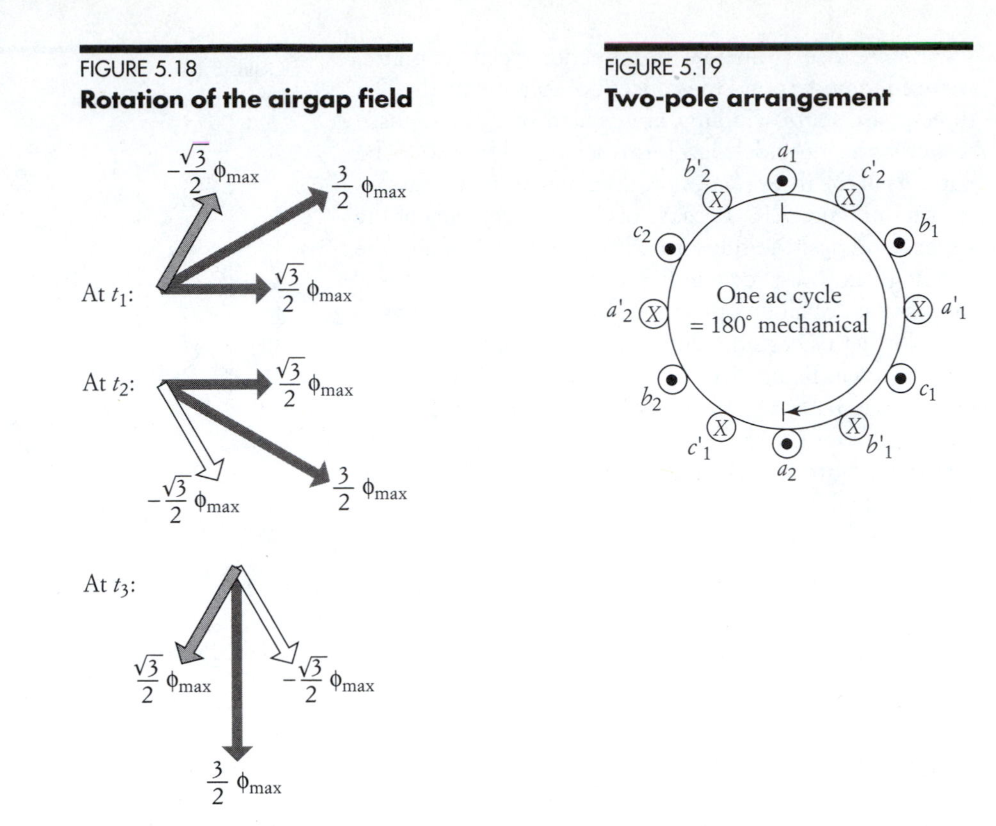

FIGURE 5.18
Rotation of the airgap field

FIGURE 5.19
Two-pole arrangement

These three equations show that the airgap flux has a constant magnitude of (3/2) ϕ_{max}, but its angle is changing. The airgap flux is then rotating in the clockwise direction. This rotating flux is one of the main advantages of the three-phase systems used in power distribution.

The speed of the airgap flux is one revolution per ac cycle. The time of one ac cycle $\tau = 1/f$, where f is the frequency of the supply voltage. Thus, the speed of the airgap n_s is

$$n_s = f \text{ rev/sec}$$

or

$$n_s = 60f \text{ rev/min}$$

n_s is known as the synchronous speed because its magnitude is synchronized with the supply frequency.

The arrangement in Figure 5.17 is for a two-pole machine. (Every coil has two poles, one north and the other south.) If each phase has two coils, the machine is four-pole, as shown in Figure 5.19. In this arrangement, the rotor moves 180° mechanical for every one complete ac cycle. Hence, the mechanical speed of the air-gap flux is

$$n_s = \frac{60f}{pp} = 120\frac{f}{p} \text{ rpm} \tag{5.32}$$

where pp is the number of pole-pairs, and p is the number of poles ($p = 2\,pp$).

The rotation of the induction motor can be explained using Faraday's law and the Lorentz force equations. Assuming that a conductor is carrying current in a uniform magnetic field, the relationships that govern the electromechanical motion are depicted in the following equations:

$$e = Blv$$

$$F = Bli$$

where B is the flux density, l is the length of the current-carrying conductor, v is the speed of the conductor relative to the speed of the field, and i is the conductor current. F and e are the force and the induced voltage on the conductor, respectively. If we generalize these equations for the rotating field, we can rewrite them in the following form:

$$e = f(\phi,\Delta n) \tag{5.33}$$

$$T = f(\phi,i) \tag{5.34}$$

where f (·) is a function notation, ϕ is the flux, T is the torque developed by the current-carrying conductors, and Δn is the relative speed between the conductor and the airgap flux. Now, let us assume that the rotor is at standstill. When a three-phase voltage is applied to the stator windings, a rotating flux is generated in the airgap. The speed of this flux is the synchronous speed n_s. The relative speed Δn is equal to the synchronous speed (the rotor is stationary). A voltage e is then induced in the rotor windings according to Equation (5.33). Since the rotor windings are shorted, a rotor current i flows. This current produces a Lorentz force F and torque T that spins the rotor.

The steady-state operation is achieved when the motor, on a continuous basis, provides the torque needed by the load. Assuming that the flux has a fixed magnitude, the rotor current is adjusted so that the Lorentz force and torque given in Equation (5.34) meet the load torque demand. The magnitude of the rotor current requires an induced voltage e in the rotor windings that is equal to the rotor current multiplied by the rotor impedance. This voltage in turn requires a certain speed deviation Δn as given in Equation (5.33). Hence, the steady-state speed of the rotor must always be slightly less than the synchronous speed to maintain the desired magnitude of the developed torque. If the rotor speed is equal to the synchronous speed ($\Delta n = 0$), the rotor current is dropped to zero, and so is the developed torque. Thus, the rotor cannot sustain the synchronous speed and the machine slows down to lower speed.

The difference between the rotor speed (n or ω) and the synchronous speed (n_s or ω_s) is known as the slip s,

$$s = \frac{\Delta n}{n_s} = \frac{\Delta\omega}{\omega_s} = \frac{n_s - n}{n_s} = \frac{\omega_s - \omega}{\omega_s} \tag{5.35}$$

where $\omega = 2\pi\,(n/60)$, n is in revolutions/minute (rpm), and ω is in radians/second. Note that the slip at starting, when the motor speed is zero, is equal to one. At no load, when the motor speed is very close to synchronous speed, the slip is about zero.

5.2.1 EQUIVALENT CIRCUIT

A single-phase equivalent circuit can be developed for the induction motor by first separating the stator and rotor circuits. The equivalent circuit of the stator is shown in Figure 5.20. The stator is a set of windings made of copper material mounted on the core. The windings have a resistance R_1 and inductive reactance X_1. The core, which is made of steel alloy, can be represented by a linear combination of a parallel resistance and a reactance (R_m and X_m). This core representation approximately models the hysteresis and eddy current effects. The sum of currents in R_m and X_m is called the magnetizing current I_m. R_m and X_m are each of a high ohmic value. The number of turns of the stator windings is N_1, and its effective voltage drop E_1 is equal to the source voltage V minus the drop across the winding impedance.

$$\overline{E}_1 = \overline{V} - \overline{I}_1(R_1 + jX_1) \tag{5.36}$$

The magnetizing current I_m is a small fraction of I_1 and can be ignored for heavily loaded motors.

The rotor circuit needs a special analysis. First, let us assume that the rotor is at standstill. In this case, the induction machine is behaving similarly to the transformer. The rotor can be represented by a winding impedance composed of a re-

FIGURE 5.20
Equivalent circuit of the stator

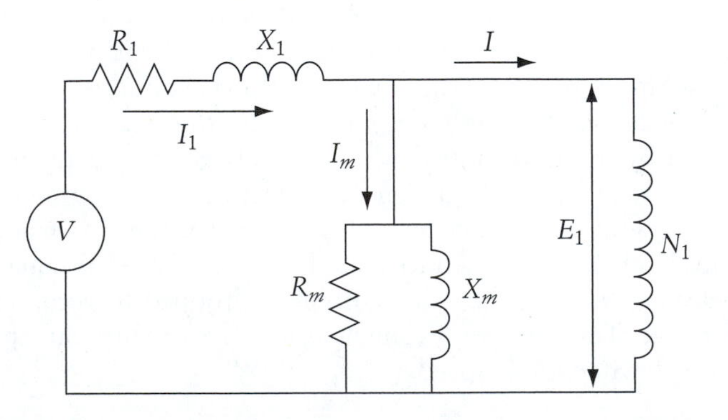

FIGURE 5.21
Equivalent circuit of the rotor at standstill

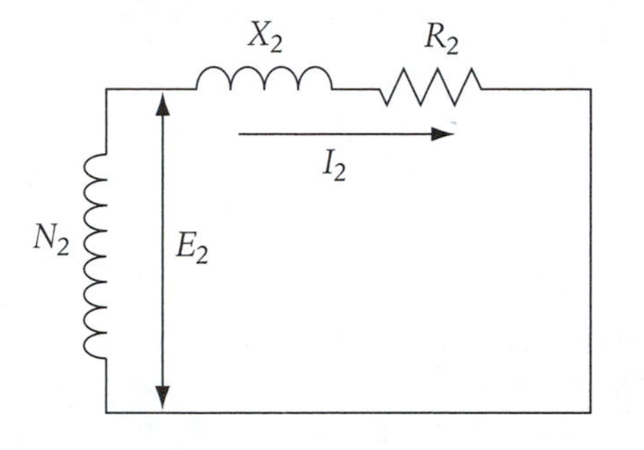

sistance R_2 and an inductive reactance X_2, as shown in Figure 5.21. The number of turns of the rotor windings is N_2, and its terminals are shorted. The induced voltage across the rotor windings at standstill E_2 is

$$\frac{E_2}{E_1} = \frac{N_2}{N_1} \tag{5.37}$$

Now let us assume that the rotor is spinning at speed n. In this case, the induced voltage across the rotor E_r is proportional to the relative speed Δn between the rotor and the field as given in Equation (5.33). Keep in mind that the induced voltage at standstill E_2 is proportional to the synchronous speed ($\Delta n = n_s$).

$$E_2 \sim n_s$$

$$E_r \sim n_s - n \tag{5.38}$$

Hence, the rotor voltage E_r , at any speed n, is

$$\frac{E_r}{E_2} = \frac{n_s - n}{n_s}$$

$$E_r = sE_2 \tag{5.39}$$

The frequency of the rotor current is also dependent on Δn. At standstill ($\Delta n = n_s$), the frequency of E_2 or I_2 is the same as the stator's supply frequency f. At any other speed, the frequency of the rotor current depends on the rate by which the rotor windings cut through the field. Hence, it depends on the relative speed Δn. At standstill, the rotor frequency f_{ss} is

$$f_{ss} \sim \Delta n, \quad \Delta n = n_s$$

$$f_{ss} = f \tag{5.40}$$

At any other speed, the rotor frequency f_r is

$$f_r \sim \Delta n, \quad \Delta n = n_s - n \tag{5.41}$$

Hence,

$$\frac{f_r}{f_{ss}} = \frac{f_r}{f} = \frac{n_s - n}{n_s}$$

$$f_r = sf \tag{5.42}$$

Equations (5.39) and (5.42) change the equivalent circuit of the rotor to that shown in Figure 5.22, which is a more general circuit for any rotor speed. The rotor inductive reactance in this circuit is

$$X_r = 2\pi f_r L_2 = 2\pi s f L_2 = s(2\pi f L_2) = sX_2 \tag{5.43}$$

where L_2 is the inductance of the rotor windings, and X_2 is the inductive reactance of the rotor at standstill. The rotor current of the induction motor I_r at any speed can be represented by

$$\bar{I}_r = \frac{s\bar{E}_2}{R_2 + jsX_2} \tag{5.44}$$

which can be modified to

$$\bar{I}_r = \frac{\bar{E}_2}{\frac{R_2}{s} + jX_2} \tag{5.45}$$

Equation (5.45) can lead to the modified rotor circuit shown in Figure 5.23. Now, let us put the stator and rotor equivalent circuits together, as shown in Figure 5.24(a). The equivalent circuit can be simplified by eliminating the turns ratio by means of referring all the parameters and variables to the stator, as shown in Figure 5.24(b).

FIGURE 5.22
Equivalent circuit of the rotor at any speed

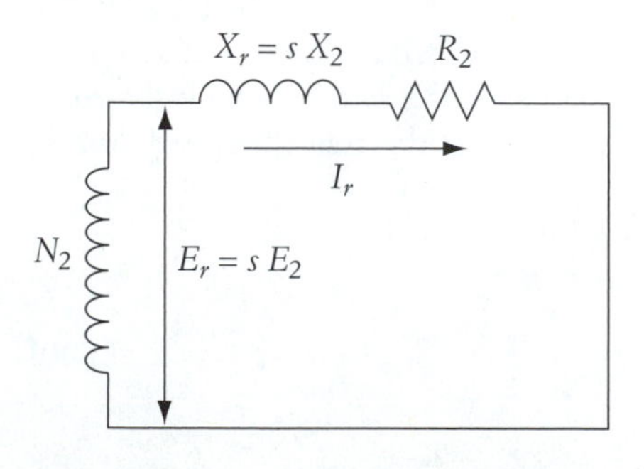

FIGURE 5.23
Modified equivalent circuit of the rotor at any speed

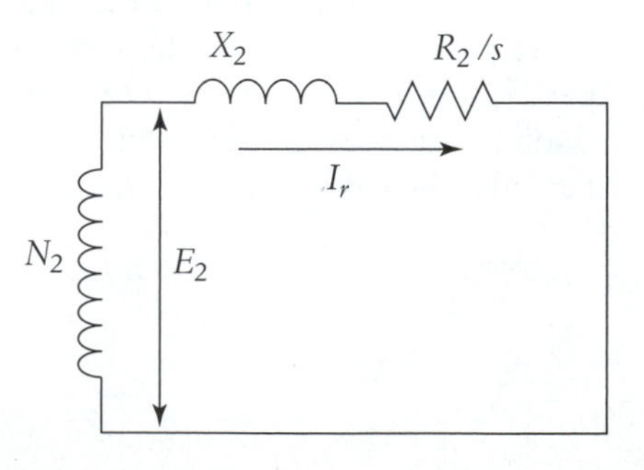

FIGURE 5.24
Development of approximate equivalent circuit for an induction motor

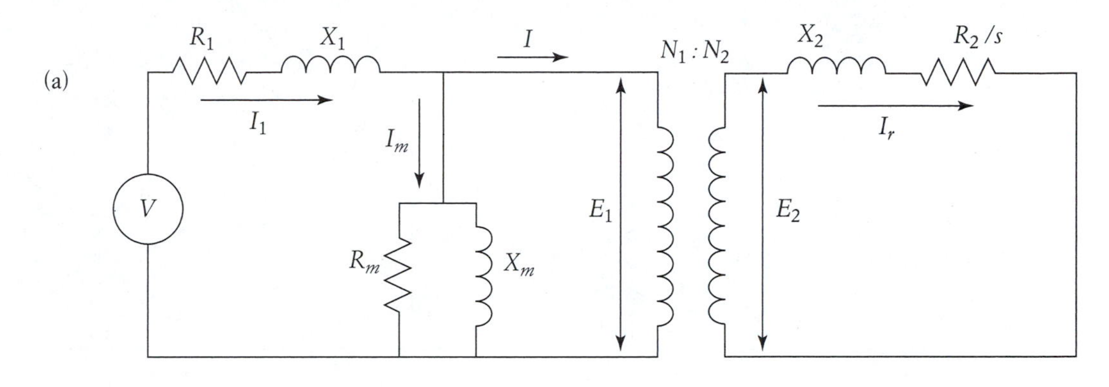

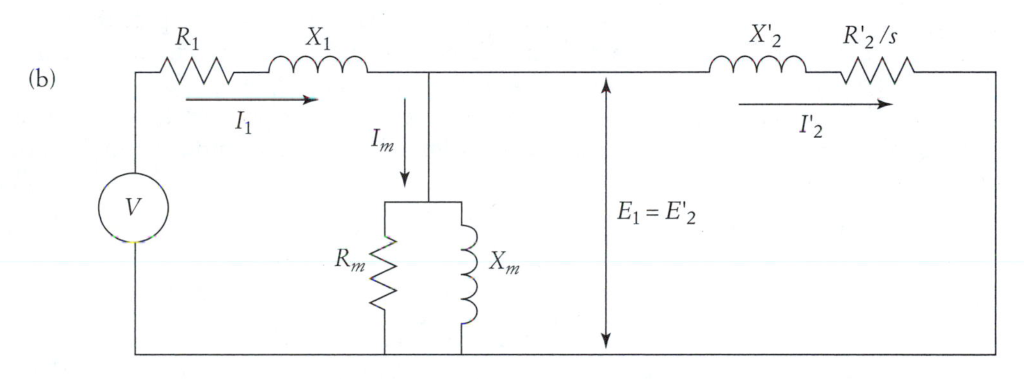

The resistance R'_2 and inductive reactance X'_2 of the rotor winding referred to the stator circuit are computed as follows:

$$R'_2 = R_2\left(\frac{N_1}{N_2}\right)^2$$

$$X'_2 = X_2\left(\frac{N_1}{N_2}\right)^2$$

where N_1 and N_2 are the number of turns of the stator and rotor windings, respectively. The rotor current referred to the stator circuit I'_2 can be computed as

$$I'_2 = I_r\left(\frac{N_2}{N_1}\right)$$

To conveniently analyze the rotor circuit, let us divide $\frac{R'_2}{s}$ into two components:

$$\frac{R'_2}{s} = R'_2 + \frac{R'_2}{s}(1 - s)$$

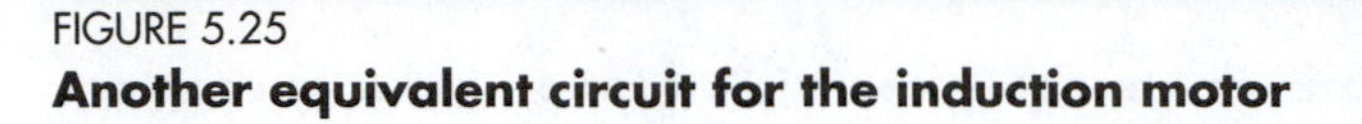
FIGURE 5.25
Another equivalent circuit for the induction motor

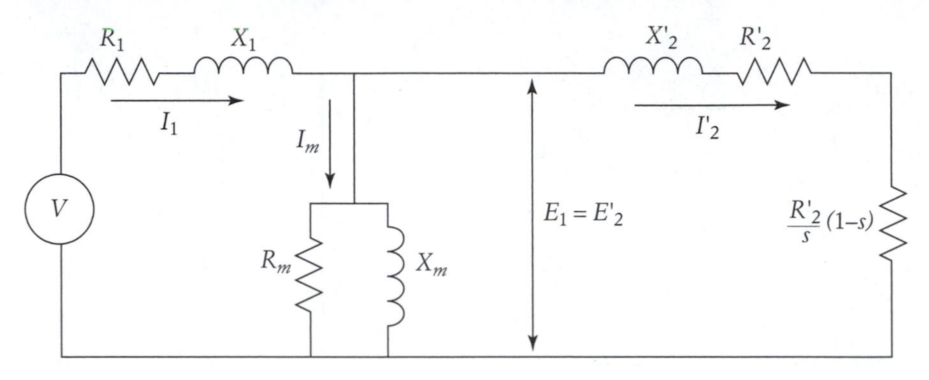

With the rotor resistance components divided this way, we can compute the losses of the rotor windings separately from the developed power, as will be explained later. The equivalent circuit can now be represented by Figure 5.25.

We can further modify the equivalent circuit by assuming that $I_m << I_1$. This makes $I_1 \cong I'_2$, and we can assume that the impedances of the stator and rotor windings are in series, as shown in Figure 5.26(a).

R_{eq} and X_{eq} of Figure 5.26(b) are defined as

$$R_{eq} = R_1 + R'_2$$

$$X_{eq} = X_1 + X'_2$$

The resistive element $(R'_2/s)\,(1-s)$ represents the load of the motor, which includes the mechanical and rotational loads. Rotational loads include the friction and windage. Note that the value of the load resistance is dependent on the motor speed. At no load, when the slip is close to zero, the load resistance is very large. At starting, when the slip is unity, the load resistance is zero.

5.2.2 POWER FLOW

The diagram in Figure 5.27 represents the power flow of the induction motor. Part of the input power to the motor P_{in} is consumed in the stator circuit in the form of winding losses $P_{cu\,1}$ and core losses P_{iron}. The rest of the power P_g passes through the airgap to the rotor circuit. This power is called the airgap power. P_g enters the rotor circuit, where part of it is consumed in the rotor resistance as copper losses $P_{cu\,2}$. The rest is called the developed power P_d. Part of the developed power is rotational losses $P_{\text{rotational}}$ due to friction, windage, and so on. The rest is the output power P_{out} consumed by the load.

The input power can be computed as

$$P_{\text{in}} = 3VI_1 \cos \theta_1 \tag{5.46}$$

FIGURE 5.26
More equivalent circuits for the induction motor

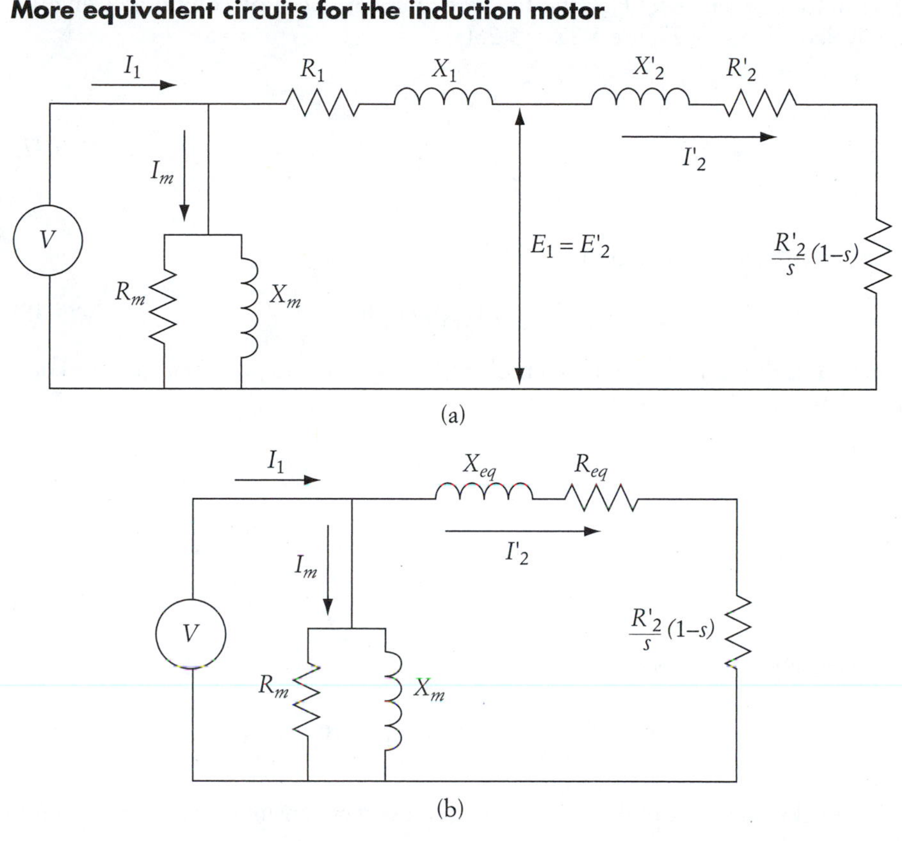

FIGURE 5.27
Power flow of the induction motor

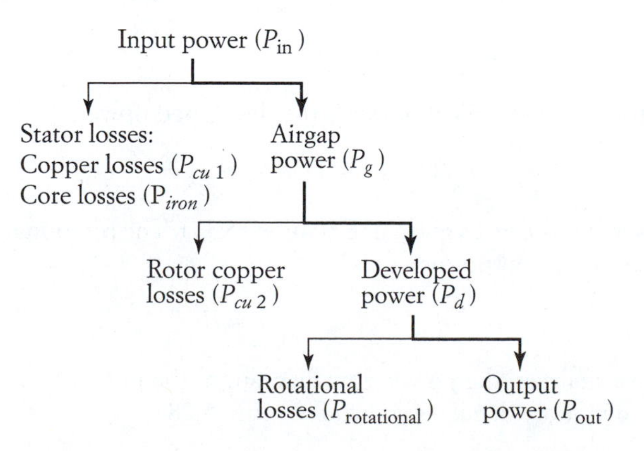

where V is the phase voltage of the source and θ_1 is the phase angle of the current. The stator copper losses P_{cu1} and the core losses P_{iron} can be computed using the equivalent circuit of Figure 5.25 or 5.26(a).

$$P_{cu\,1} = 3\, I_1^2 R_1$$

$$P_{iron} \cong 3\frac{V^2}{R_m} \tag{5.47}$$

The airgap power can be computed by

$$P_g = 3\, E'_2 I'_2 \cos\theta_2 \tag{5.48}$$

where θ_2 is the phase angle between E'_2 and I'_2. The airgap power can also be computed as

$$P_g = 3(I'_2)^2\frac{R'_2}{s} \tag{5.49}$$

The rotor losses are

$$P_{cu\,2} = 3(I'_2)^2 R'_2 = sPg \tag{5.50}$$

The developed power is

$$P_d = P_g - P_{cu\,2} = 3(I'_2)^2\frac{R'_2}{s}(1 - s) = P_g(1 - s) \tag{5.51}$$

The developed power of the motor is the shaft power consumed by the mechanical load plus the rotational losses.

The powers of the induction motor can be represented by mechanical terms such as torque and speed. The first form of mechanical power is airgap power, which is equal to the developed torque T_d exerted by the flux (Lorentz force) times the speed of the flux ω_s.

$$P_g = T_d\omega_s \tag{5.52}$$

The second form of mechanical power is the developed power,

$$P_d = P_g(1 - s) = T_d\omega_s(1 - s) = T_d\omega \tag{5.53}$$

where ω is the rotor speed, as given in Equation (5.35). The rotational losses reduce the torque; hence, the output power is

$$P_{\text{out}} = T\omega, \qquad T < T_d \tag{5.54}$$

Based on these analyses, the power flow diagram of the induction motor can now be represented in more detail, as shown in Figure 5.28.

FIGURE 5.28
Detailed power flow of the induction motor

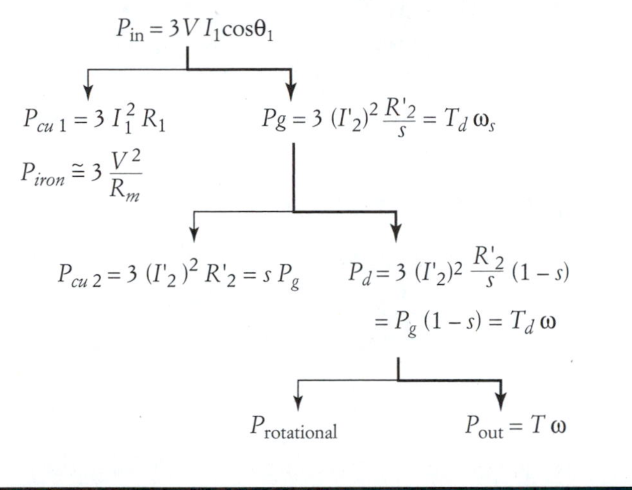

EXAMPLE 5.2

A 50 hp, 60 Hz, three-phase, Y-connected induction motor operates at full load at a speed of 1764 rpm. The rotational losses of the motor are 950 W, the stator copper losses are 1.6 kW, and the iron losses are 1.2 kW. Compute the motor efficiency.

SOLUTION

The output power at full load is 50 hp.

$$P_{out} = \frac{50}{1.34} = 37.3 \text{ kW}$$

$$P_d = P_{out} + P_{rotational} = 37.3 + 0.95 = 38.25 \text{ kW}$$

We need to calculate the slip before we can proceed. We have the actual motor speed, but the synchronous speed and the number of poles are not given. Nevertheless, we know from the principal of operation that the motor speed at full load is just slightly less than the synchronous speed. Since the number of poles is always even, this machine must be four-pole with a synchronous speed of 1800 rpm (see Equation (5.23)). Hence,

$$s = \frac{n_s - n}{n_s} = \frac{1800 - 1764}{1800} = 0.02$$

Then,

$$P_g = \frac{P_d}{1 - s} = \frac{38.25}{0.98} = 39 \text{ kW}$$

$$P_{in} = P_g + P_{cu\,1} + P_{core} = 39 + 1.6 + 1.2 = 41.8 \text{ kW}$$

The motor efficiency η is

$$\eta = \frac{P_{out}}{P_{in}} = \frac{37.3}{41.8} = 0.89 \text{ or } 89\%$$

5.2.3 TORQUE CHARACTERISTICS

To establish the speed–torque relationship, let us use the equivalent circuit in Figure 5.26(b) to compute the rotor current.

$$I'_2 = \frac{V}{\sqrt{\left(R_1 + \frac{R'_2}{s}\right)^2 + X_{eq}^2}} \tag{5.55}$$

The developed torque of the motor is computed by dividing the developed power by the rotor speed:

$$T_d = \frac{P_d}{\omega} = \frac{3}{\omega}(I'_2)^2 \frac{R'_2}{s}(1 - s) = \frac{3V^2 R'_2(1 - s)}{s\omega\left[\left(R_1 + \frac{R'_2}{s}\right)^2 + X_{eq}^2\right]} \tag{5.56}$$

From Equation (5.35), $\omega = \omega_s(1 - s)$. Hence,

$$T_d = \frac{P_d}{\omega} = \frac{3V^2 R'_2}{s\omega_s\left[\left(R_1 + \frac{R'_2}{s}\right)^2 + X_{eq}^2\right]} \tag{5.57}$$

V is the phase voltage and Equation (5.57) represents the motor torque due to the three phases.

The slip-torque (or speed–torque) characteristic of the induction motor using Equation (5.57) is shown in Figure 5.29. At starting, when the motor speed is zero (slip is unity), the rotor current produces a starting torque T_{st}. If the starting torque is greater than the entire load torque, including inertia torques, the motor shaft spins. When the speed of the motor increases, so does the motor torque. The maximum torque T_{max} occurs at slip s_{max}. Since in normal steady-state operation the rotor speed is close to synchronous speed (the slip is about 2% to 7%), the motor speed continues to increase until it reaches a steady-state value in the linear region of the characteristic.

The speed–torque characteristic can be divided into three major regions, as shown in Figure 5.30: large slip, small slip and maximum torque. In the large slip region, which is also known as the starting region, the torque equation of the motor can be approximated by assuming that

$$\left(R_1 + \frac{R'_2}{s}\right)^2 << X_{eq}^2$$

FIGURE 5.29
Speed–torque characteristics of the induction motor

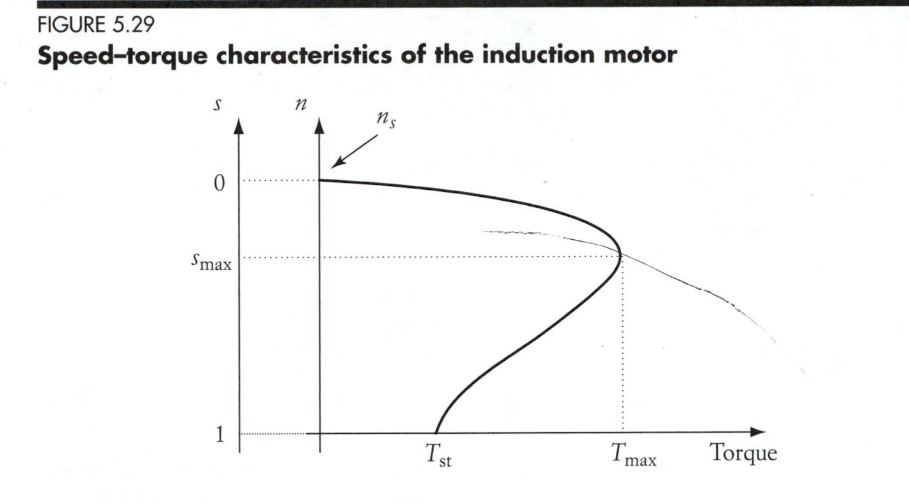

FIGURE 5.30
Main regions of the speed–torque characteristic

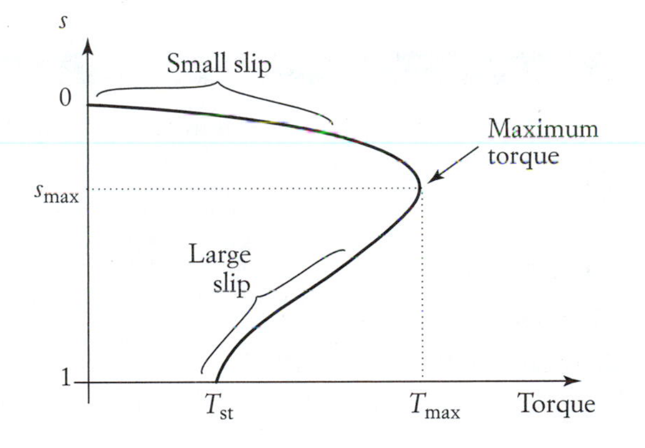

Hence,

$$T_d \approx \frac{3V^2R'_2}{s\omega_s X_{eq}^2} \tag{5.58}$$

By setting $s = 1$ in the large slip approximation, we can compute the starting torque:

$$T_{st} \approx \frac{3V^2R'_2}{\omega_s X_{eq}^2} \tag{5.59}$$

For the small slip region, when the rotor speed is close to synchronous, the motor torque can be approximated, assuming that

$$R_1 << \frac{R'_2}{s} >> X_{eq}$$

Hence,

$$T_d \approx \frac{3V^2 s}{\omega_s R'_2} \tag{5.60}$$

To compute the maximum torque and the slip at maximum torque s_{max}, the first derivative of Equation (5.57) with respect to slip must be set equal to zero. Doing that results in the following equations:

$$s_{max} = \frac{R'_2}{\sqrt{R_1^2 + X_{eq}^2}} \tag{5.61}$$

$$T_{max} = \frac{3V^2}{2\omega_s [R_1 + \sqrt{R_1^2 + X_{eq}^2}]} \tag{5.62}$$

Note that the slip at maximum torque s_{max} is linearly proportional to the rotor resistance, whereas the magnitude of the maximum torque is independent of the rotor resistance. For motors with large rotor resistance, the maximum torque occurs at low speeds.

EXAMPLE 5.3

A 50 hp, 440 V, 60 Hz, three-phase, four-pole induction motor develops a maximum torque of 250% at slip of 10%. Ignore the stator resistance and rotational losses. Calculate the following:

a. Speed of the motor at full load
b. Copper losses of the rotor
c. Starting torque of the motor

SOLUTION

a. *Motor speed.* Using the small slip approximation of Equation (5.60), we can write the motor torque at full load as

$$T \approx \frac{3V^2 s}{\omega_s R'_2}$$

The maximum-torque equation is given by (5.62). The equation can be rewritten to ignore the effect of R_1 by assuming $R_1 << X_{eq}$:

$$T_{max} = \frac{3V^2}{2\omega_s X_{eq}}$$

Then

$$\frac{T_{max}}{T} = \frac{R'_2}{2sX_{eq}}$$

Now let us modify Equation (5.61) by ignoring the effect of the stator resistance.

$$s_{max} = \frac{R'_2}{X_{eq}}$$

Then

$$\frac{T_{max}}{T} = \frac{s_{max}}{2s}$$

$$s = \frac{T}{T_{max}} \frac{s_{max}}{2} = \frac{1}{2.5} \frac{0.1}{2} = 0.02$$

The motor speed at full load is

$$n = n_s(1 - s) = 120 \frac{f}{p} (1 - s) = 120 \frac{60}{4} (1 - 0.02) = 1764 \text{ rpm}$$

b. *Copper losses of the rotor.* Since the rotational losses are ignored, the developed power is equal to the output power.

$$P_d = \frac{50}{1.34} = 37.3 \text{ kW}$$

Since

$$P_d = P_g(1 - s)$$

and

$$P_{cu\,2} = sP_g$$

then

$$\frac{P_{cu\,2}}{P_d} = \frac{s}{1 - s}$$

$$P_{cu\,2} = P_d\left(\frac{s}{1 - s}\right) = 37.3 \frac{0.02}{0.98} = 760 \text{ W}$$

c. *Starting torque.* The starting torque can be obtained by the large slip approximation when $s = 1$.

$$T_{st} \approx \frac{3V^2 R'_2}{\omega_s X^2_{eq}}$$

The full load torque represented by the small slip approximation is

$$T \approx \frac{3V^2 s}{\omega_s R'_2}$$

Hence,

$$\frac{T_{st}}{T} = \frac{(R'_2)^2}{sX_{eq}^2} = \frac{s_{max}^2}{s}$$

$$T_{st} = \frac{s_{max}^2}{s} T = \frac{s_{max}^2}{s} \frac{P_{out}}{\omega} = \frac{(0.1)^2}{0.02} \frac{37300}{2\pi \frac{1764}{60}} = 101 \text{ Nm}$$

5.2.4 STARTING PROCEDURE

In many cases, induction motors do not need a special starting procedure because the starting current is generally limited to tolerable values by the winding impedance. However, for large motors with small winding resistance, the starting current could be excessive and a starting mechanism must be used.

If we ignore the magnetizing current at starting, the starting current I'_{st_2} can be computed using Equation (5.55). In this equation, the slip is set equal to one.

$$I'_{st_2} = \frac{V}{\sqrt{(R_1+R'_2)^2 + X_{eq}^2}} \tag{5.63}$$

To reduce the starting current of an induction motor, several methods can be used. The common ones are based on reducing the terminal voltage or inserting a resistance in the rotor circuit.

Figure 5.31 shows the speed–torque characteristics of the induction motor under different voltage levels. Voltage reduction results in a linearly proportional reduction of the starting current. However, the starting torque and the maximum torque of the motor will also be reduced. Note that the torque is proportional to the square of the voltage. Hence, a 20% reduction in the voltage reduces the starting current by 20%, but also reduces the starting torque and the maximum torque by 36% each. If the motor is heavily loaded, the starting torque may not be adequate to spin the shaft.

FIGURE 5.31
Speed–torque characteristics at different voltage levels

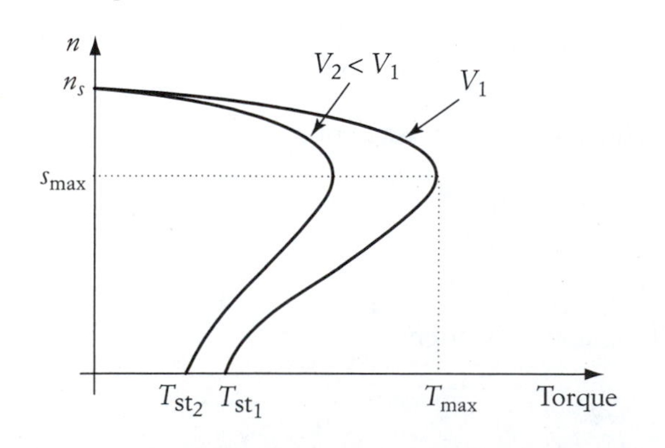

The other starting method is based on adding a resistance to the rotor circuit, as shown in Figure 5.32. Notice that according to Equations (5.59) and (5.61), when a resistance is added to the rotor circuit, the starting torque and the slip at maximum torque increase. In fact, if the added resistance makes Equation

(5.61) equal to one, the maximum torque occurs at starting. This is a very good starting method for heavily-loaded machines.

The insertion of rotor resistance is only possible if the rotor is accessible through brushes and a slip ring arrangement. For squirrel-cage motors, adding a resistance is not possible since the rotor is fully enclosed. However, some types of squirrel cage motors have rotor windings made of alloys that exhibit skin effects at 60 Hz. Since the rotor frequency at starting is 60 Hz, the starting rotor resistance is high due to the skin effect. Once the speed of the motor increases, the rotor frequency is reduced, and the skin effect is diminished. The rotor resistance is then reduced.

FIGURE 5.32
Speed–torque characteristics when a resistance is added to the rotor circuit

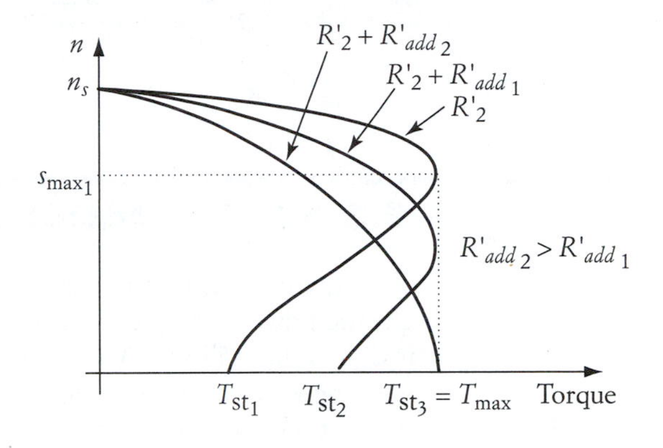

EXAMPLE 5.4

An induction motor has a stator resistance of 3 Ω, and the rotor resistance referred to the stator is 2 Ω. The equivalent inductive reactance $X_{eq} = 10\ \Omega$. Calculate the change in the starting torque if the voltage is reduced by 10%. Also, compute the resistance that should be added to the rotor circuit to achieve the maximum torque at starting.

SOLUTION

Using the large slip approximation of Equation (5.59), we can compute the starting torque by setting $s = 1$.

$$T_{st} \approx \frac{3V^2 R'_2}{\omega_s X_{eq}^2}$$

If T_A is the starting torque at full voltage and T_B is the starting torque at the reduced voltage, then

$$\frac{T_A}{T_B} = \left(\frac{V}{0.9\ V}\right)^2$$

$$T_B = 0.81\ T_A$$

Hence, the reduction of the starting torque is 19%.

To compute the value of the inserted resistance in the rotor circuit for maximum torque at starting, we can use Equation (5.61). s_{max} must then be set equal to one.

$$s_{max} = 1 = \frac{R'_2 + R'_{add}}{\sqrt{R_1^2 + X_{eq}^2}}$$

$$R'_{add} = \sqrt{R_1^2 + X_{eq}^2} - R'_2 = \sqrt{9 + 100} - 2 = 7.54\ \Omega$$

5.3 SYNCHRONOUS MOTORS

The synchronous machine is used mainly for power generation. Over 97% of all electric power generated worldwide is produced by synchronous generators. This is due to the ability of synchronous generators to produce ac power directly without a need for conversion, and the effective and simple control of its voltage and power flow. The frequency of the generated power is directly proportional to the speed of the machine. Hence, the speed of the generator must be maintained constant at synchronous speed at all times.

The synchronous machine is also used as a motor. Several applications that demand fixed speeds regardless of load changes employ synchronous machines. The motor can also be used as an effective tool for reactive power and voltage controls.

A synchronous machine, as the name implies, operates at the synchronous speed n_s. The machine, as shown in Figures 5.33 and 5.34, is composed of a stator and a rotor. The stator of a synchronous machine is similar to that of an induction motor. The stator has three phase windings connected to a three-phase source. The stator windings generate a rotating magnetic field ϕ_s in the airgap, as shown in Figure 5.18. The speed of ϕ_s is the synchronous speed, which is a function of the supply frequency as given in Equation (5.32). For small machines, the rotor could be a permanent magnet. For larger machines, the rotor is an electrical magnet excited externally by a dc source known as the exciter. The winding of the rotor circuit is connected to slip rings mounted on the rotor shaft. Brushes are used to connect the rotor circuit to the exciter. Because of the slip ring arrangement, the rotor winding does not reverse its polarities. Hence, the rotor magnetic field ϕ_f is stationary relative to the rotor shaft.

The airgap of the synchronous machine has two fields: one is ϕ_s rotating at a synchronous speed due to the stator excitation, and the other ϕ_f is due to the rotor excitation and is stationary with respect to the rotor. These two fields must be aligned at all times (provided that the fields are strong enough). Therefore, the rotor field ϕ_f must also rotate at the synchronous speed of ϕ_s. Since the rotor field is stationary with respect to the rotor, the rotor will also rotate at the synchronous speed n_s.

Using the schematic of Figure 5.35, the equivalent circuit of the synchronous machine can be developed. The figure shows the rotor circuit excited by a dc source V_f. The excitation current I_f produces a field ϕ_f that is stationary with respect to the rotor. Now let us look at the windings of one phase in the stator circuit. Assume

FIGURE 5.33
Synchronous machine components

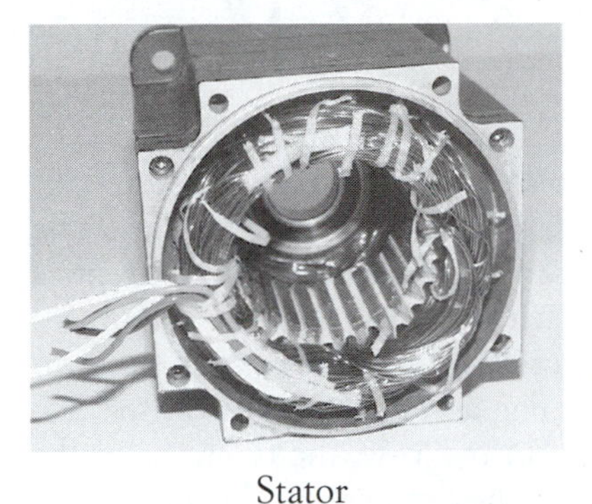

Stator

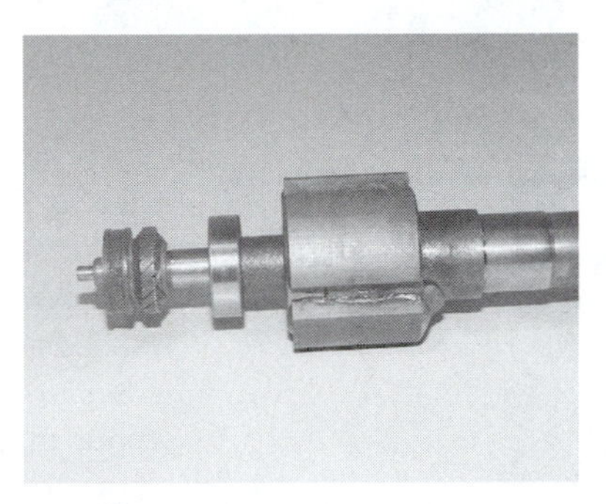

Permanent magnet rotor

Electric magnet rotor with slip rings

FIGURE 5.34
Conceptual representation of a synchronous machine

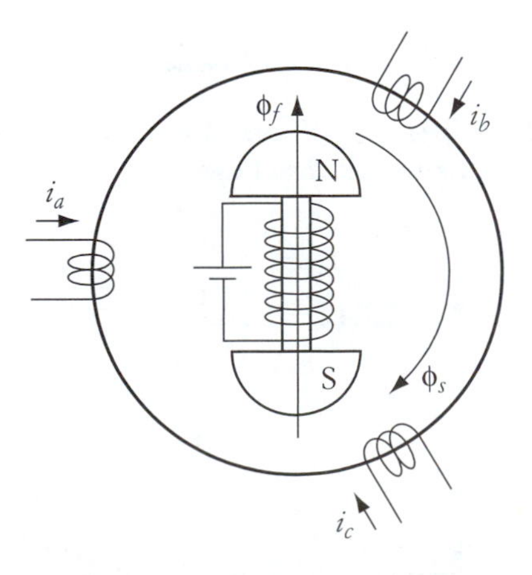

FIGURE 5.35
Simplified diagram of a singly-excited synchronous machine at no load

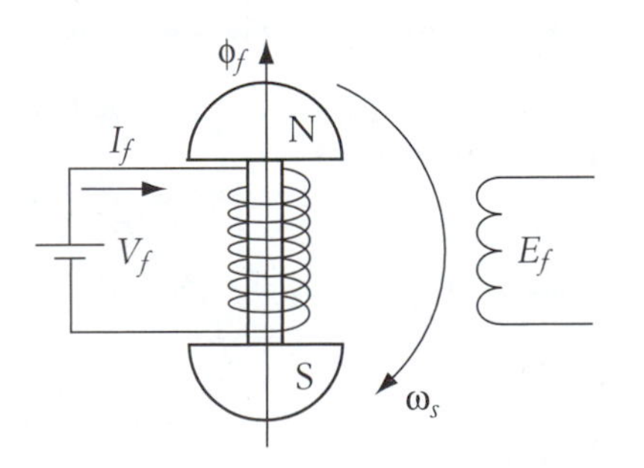

that we are rotating the machine externally at a synchronous speed ω_s. The field then cuts the stator windings and induces a voltage E_f.

$$E_f \sim \frac{d\phi_f}{dt} \tag{5.64}$$

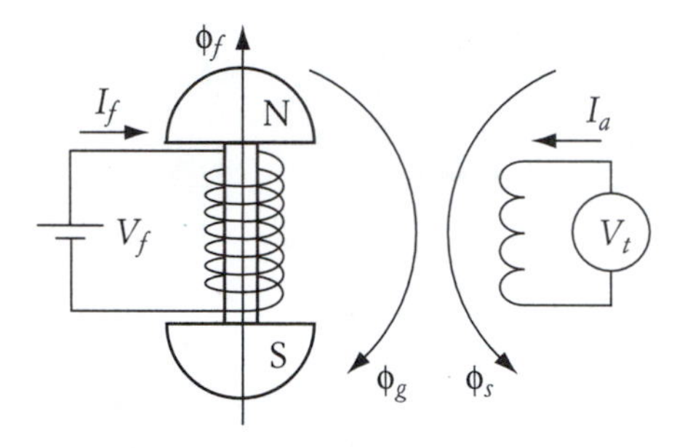

FIGURE 5.36
Simplified diagram of a synchronous machine at no load

E_f is known as the no-load equivalent excitation voltage. If the saturation of the rotor circuit is ignored, E_f is directly proportional to the excitation current I_f. The frequency of E_f is proportional to the synchronous speed ω_s given in Equation (5.32).

Now let us discuss the case in which the synchronous machine is running as a motor. Consider the diagram of Figure 5.36. In this case, the terminals of the stator are connected to an ac source V_t. The rotor is also connected to a dc source V_f. The rotor circuit produces a magnetic field ϕ_f. The current in the stator windings I_a (armature current) also produces a magnetic field ϕ_s that is rotating at the synchronous speed. The net magnetic field in the airgap ϕ_g is the phasor sum of both fields.

$$\bar{\phi}_g = \bar{\phi}_f + \bar{\phi}_s \tag{5.65}$$

Since the rotor field is generated by a dc circuit, we do not have to worry about the hysteresis and eddy current of the rotor.

We can simplify the equivalent circuit of the synchronous machine to that shown in Figure 5.37. The reactance X_s is known as the synchronous reactance. It is the reactance of the stator windings plus the equivalent reactance associated with the armature reaction. R is the resistance of the armature windings.

The equivalent circuit of Figure 5.37 can be simplified further by ignoring the resistance of the armature circuit. This is justified for large machines where the stator windings carry large current, and therefore the wire cross section is large. The simplified circuit is shown in Figure 5.38.

FIGURE 5.37
Equivalent circuit of a synchronous machine

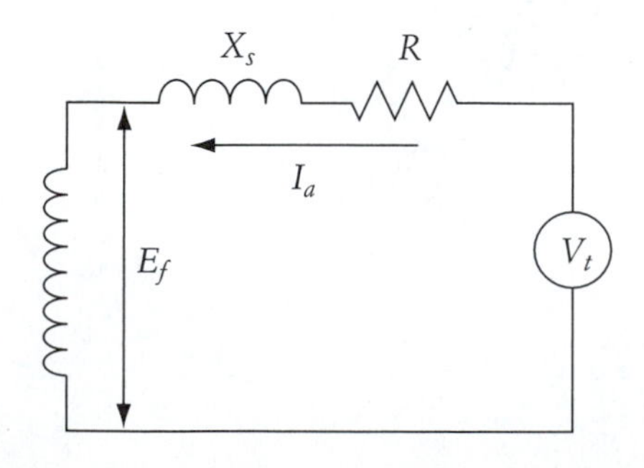

FIGURE 5.38
Simplified equivalent circuit of a synchronous machine

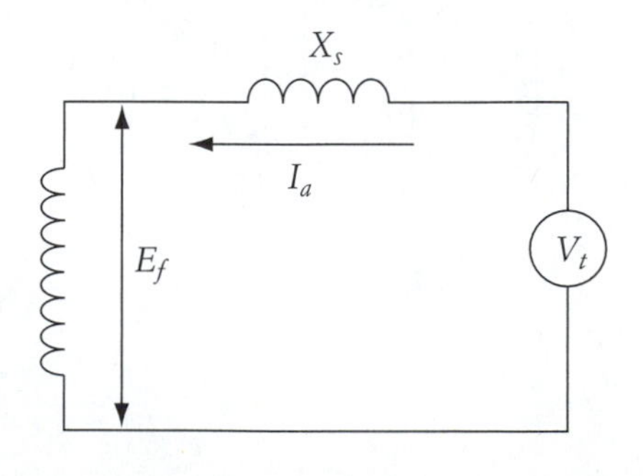

5.3.1 REACTIVE POWER

Equation (5.66) is the main equation for the synchronous motor. Both V_t and E_f are independent variables; V_t is adjusted by controlling the supply voltage and E_f is adjusted by controlling the magnitude of the dc current in the rotor circuit (field current).

$$\overline{V}_t = \overline{E}_f + \overline{I}_a \overline{X}_s \tag{5.66}$$

The armature current is then a dependent variable, with its magnitude and phase shift dependent on the adjustments of V_t and E_f. Moreover, the equivalent field voltage E_f always lags the terminal voltage V_t when the machine is running as a motor.

Three phasor diagrams of Equation (5.66) are shown in Figure 5.39. In Figure 5.39(a), E_f is adjusted so that $E_f \cos\delta > V_t$. In this case, the angle of the voltage drop $I_a X_s$ must be greater than 90°. Since I_a lags the voltage drop $I_a X_s$ by 90°, I_a leads V_t, and the power factor measured at the terminals of the motor ($\cos\theta$) is

FIGURE 5.39
Phasor diagram of a synchronous motor

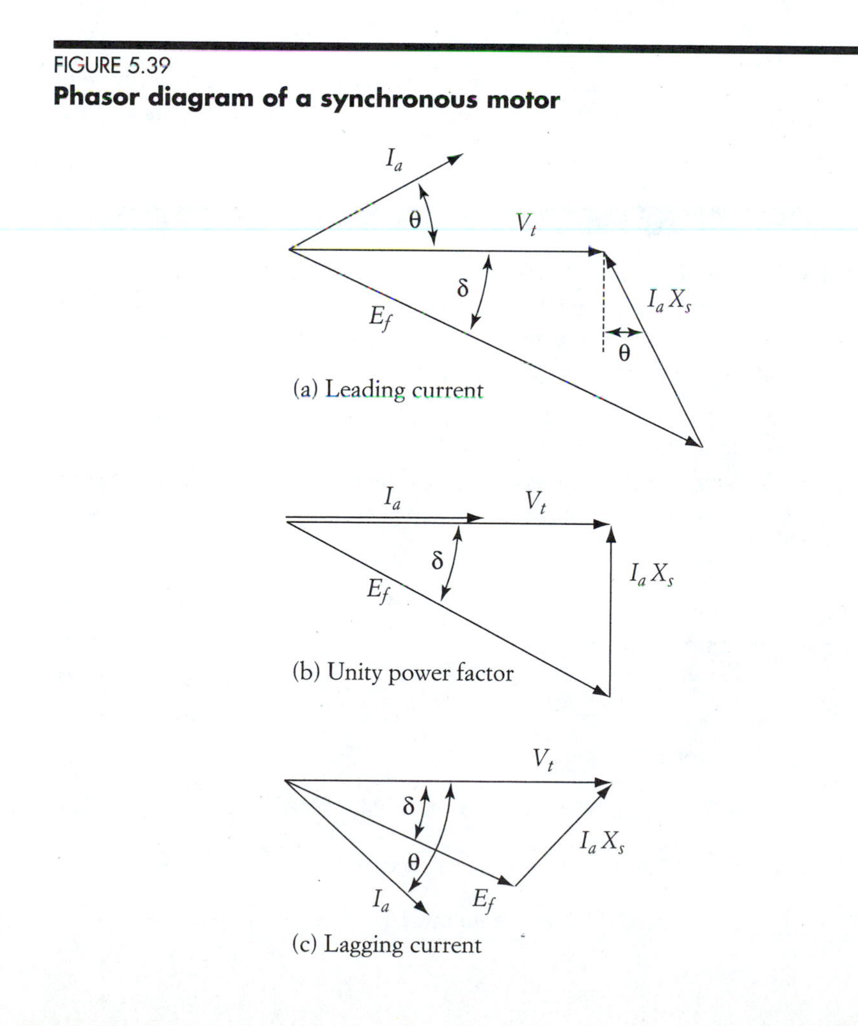

leading. In Figure 5.39(b), E_f is reduced so that $E_f \cos \delta = V_t$. In this case, the angle of the voltage drop $I_a X_s$ is exactly 90°. Hence, I_a is in phase with V_t, and the power factor measured at the terminals of the motor is unity. In Figure 5.39(c), E_f is further reduced so that $E_f \cos \delta < V_t$. In this case, the angle of the voltage drop $I_a X_s$ is less than 90°, I_a lags V_t, and the power factor measured at the terminals of the motor is lagging. The reactive power Q at the terminal of the motor can be computed:

$$Q = 3\, V_t I_a \sin \theta \tag{5.67}$$

V_t is a phase quantity. By examining the phasor diagrams of Figure 5.39(a), we can show that

$$I_a X_s \sin \theta = E_f \cos \delta - V_t \tag{5.68}$$

Substituting the current of Equation (5.68) into (5.67) yields

$$Q = \frac{3V_t}{X_s} (E_f \cos \delta - V_t) \tag{5.69}$$

The reactive power at the terminals of the motor is leading when the magnitude of Q in Equation (5.69) is positive. When Q is negative, the reactive power is lagging.

EXAMPLE 5.5

The load of an industrial plant is 40 MW at 0.85 power factor lagging. A 2 MW synchronous motor is used to improve the overall power factor of the plant. The motor is rated at 5 kV and has a synchronous reactance of 5 Ω. The equivalent phase value of the equivalent field voltage can be expressed by

$$E_f = 200\, I_f$$

where I_f is the dc excitation current. Assume that the motor is unloaded, and compute the excitation current to improve the overall power factor of the plant to 0.95 lagging.

SOLUTION

The power factor angle of the load is

$$\theta = \cos^{-1} 0.85 = 31.8°$$

The load reactive power is

$$Q_l = P \tan \theta = 40 \tan 31.8 = 24.8 \text{ kVAR}$$

The total reactive power for 0.95 power factor lagging is

$$Q_{tot} = P \tan(\cos^{-1} 0.95) = 40 \tan 18.2 = 13.15 \text{ kVAR}$$

FIGURE 5.40
Phasor diagram of a synchronous motor running at no load

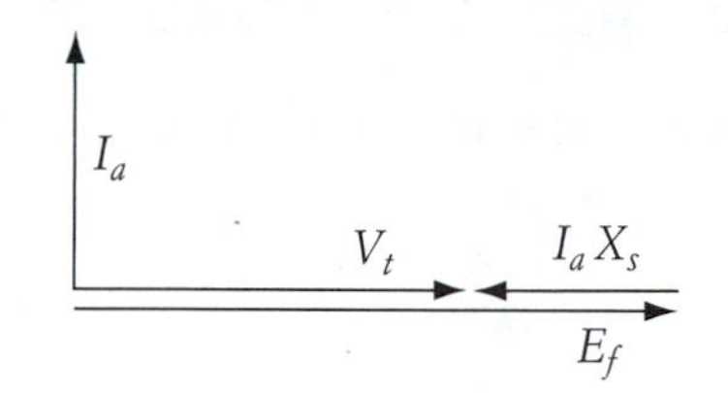

The reactive power to be generated by the synchronous motor is

$$Q_m = Q_{tot} - Q_l = (-13.15) - (-24.8) = 11.65 \text{ kVAR}$$

The negative sign implies a lagging reactive power. Since the motor is running at no load, the power factor angle at the terminals of the motor must be 90° (I_a leads V_t by 90°). In this case, E_f is in phase with V_t. The phasor diagram is shown in Figure 5.40. The excitation voltage E_f must be greater than V_t for leading current.

Using Equation (5.69), the magnitude of E_f can be computed.

$$Q = \frac{3V_t}{X_s}(E_f \cos \delta - V_t)$$

$$11650 = \frac{3\frac{5000}{\sqrt{3}}}{5}\left(E_f - \frac{5000}{\sqrt{3}}\right)$$

Hence,

$$E_f = 2.89 \text{ kV}$$

To achieve the desired level of reactive power, the excitation current must be adjusted to

$$I_f = \frac{E_f}{200} = 14.45 \text{ A}$$

5.3.2 POWER FLOW

The input power to the synchronous motor is from the armature circuit only. If we ignore losses in the rotor windings, there is no power consumed in the field circuit. Hence, the input power is

$$P = 3V_t I_a \cos \theta \tag{5.70}$$

V_t is a phase quantity. By examining the phasor diagrams of Figure 5.39, we can show that

$$I_a X_s \cos\theta = E_f \sin\delta \tag{5.71}$$

Substituting I_a of Equation (5.71) into Equation (5.70) yields

$$P = 3\frac{V_t E_f}{X_s}\sin\delta \tag{5.72}$$

Since the synchronous machine rotates at a synchronous speed, we can write the developed torque equation as

$$T = \frac{P}{\omega_s} = \frac{3}{\omega_s}\frac{V_t E_f}{X_s}\sin\delta \tag{5.73}$$

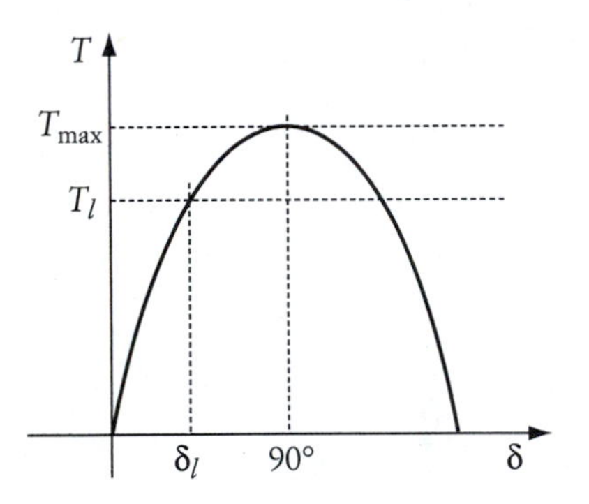

FIGURE 5.41
Torque curve of a synchronous motor

δ is known as the power angle. Figure 5.41 shows the torque curve representing Equation (5.73). If the excitation current, terminal voltage, and supply frequency are all maintained constant, changes in the load torque T_l result in changes in the power angles. As the figure shows, the load torque must always be limited to below the maximum torque T_{max} at $\delta = 90°$. If the load torque exceeds T_{max}, the motor cannot support the load torque and stops spinning.

EXAMPLE 5.6

A 2300 V, 60 Hz, six-pole synchronous motor is driving a constant-torque load of 5000 Nm. The synchronous reactance of the motor is 6 Ω. Compute the minimum excitation that the machine must maintain to provide the needed torque.

SOLUTION

First, let us compute the synchronous speed of the motor.

$$n_s = 120\frac{f}{p} = 120\frac{60}{6} = 1200 \text{ rpm}$$

The minimum excitation occurs when the load torque equals the maximum developed torque by the motor.

$$T_l = T_{max} = \frac{P}{\omega_s} = \frac{3}{\omega_s}\frac{V_t E_f}{X_s}$$

$$5000 = \frac{3}{2\pi\frac{1200}{60}}\frac{\frac{2300}{\sqrt{3}}E_f}{6}$$

Then

$$E_f = 946 \text{ V}$$

Note that E_f is a phase quantity. For any reduction of the excitation voltage below this value, the motor torque will be less than the load torque.

5.3.3 TORQUE CHARACTERISTICS

As mentioned earlier, the synchronous machine must spin at the synchronous speed of the rotating field generated by the stator windings. Hence, the speed of the motor at any loading condition is

$$n_s = 120 \frac{f}{p}$$

The speed of the machine is only changed when the number of poles or the supply frequency is changed. Figure 5.42 shows the speed–torque characteristics of a synchronous motor.

If the load torque increases to a level at which the fields in the airgap can no longer be aligned, the motor stops spinning. In this case, the load torque exceeds the maximum delivered torque of the motor, as explained in Example 5.6.

5.3.4 STARTING PROCEDURE

For heavily loaded motors with large inertia, the fields in the airgap at starting may not be strong enough to increase the rotor speed from standstill to synchronous. In this case, a starting circuit may be needed. The most common method is to install damper windings in the rotor circuit, similar to the rotor windings of a squirrel cage induction motor shown in Figure 5.43. At starting, the damper windings cause the synchronous motor to start as an induction machine. When the speed of the rotor

FIGURE 5.42
Speed–torque characteristics of a synchronous machine

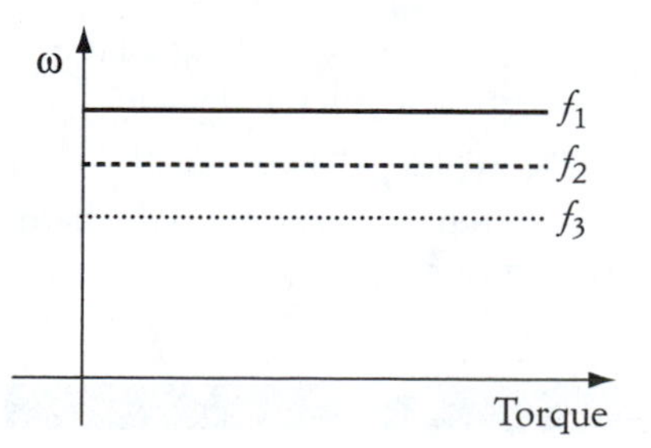

FIGURE 5.43
Damper windings

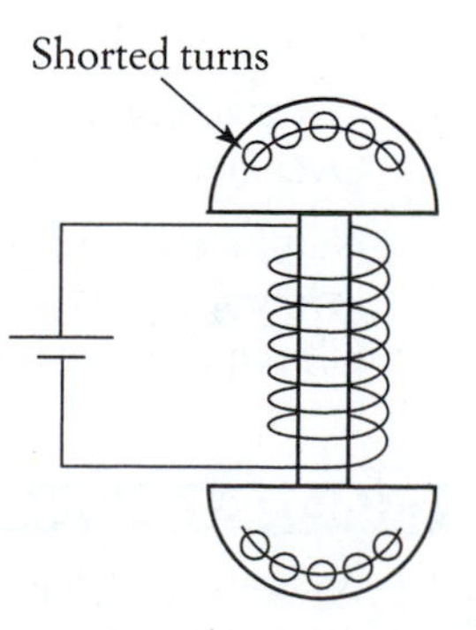

is close enough to the synchronous speed, the rotor field ϕ_f aligns with, and locks itself to, the synchronous field ϕ_s. Once the motor is running at the synchronous speed, the current inside the damper windings is zero (no relative speed between the damper windings and the rotating field). Remember that the rotor voltage of the induction motor when running at synchronous speed is zero.

5.4 DAMAGE TO ELECTRIC MACHINES

Overvoltage or overcurrent can damage electric motors. Excessive voltage can cause damage to the insulation of the windings that may lead to a permanent short circuit. Overcurrent produces excessive heat due to the energy dissipated in the windings' resistance and may result in melting down the winding's insulation, eventually causing a short circuit. For permanent magnet motors, large armature currents may also demagnetize the permanent magnet.

Damage due to overvoltage is usually rapid—so as a rule, motor voltage should not exceed the rated value by more than 10%. However, motors may tolerate high currents for a very short period of time.

In addition to the electrical constraints, one should keep in mind the mechanical limitation and integrity of the complete system. Excessive speed may result in a damage to bearings or to rotor windings, due to excessive centrifugal forces.

For most electric drive applications, several performance properties should be maintained to avoid *premature fatality* of the hardware, especially for large systems. Among these properties are the following:

1. The system should be able to achieve *soft transition;* for example, soft starting, soft speed change, and soft braking. Abrupt, large changes in speed may eventually result in ruinous effects on the mechanical integrity of the motor or load and unnecessary electrical stresses on the motor or converter. A soft transition, however, does not necessarily mean a slow transition.
2. The system should have a sufficient damping for speed oscillation at all times, including the equilibrium state (holding state).
3. Large, abrupt changes in the supply voltage should be avoided. Voltage must not exceed the tolerable limit of the system components.
4. The magnitude of the inrush current should be kept under control at all times. Overshoots of inrush current should be limited to some tolerated value.
5. Natural electromechanical oscillations should be avoided. These usually occur at low speeds when the electrical modes of the system correspond to the natural frequencies of the load and supporting structure.

CHAPTER 5 PROBLEMS

5.1 A 600 V, dc shunt motor has armature and field resistance of 1.5 Ω and 600 Ω, respectively. When the motor runs unloaded, the line current is 3 A, and the speed is 1000 rpm. Calculate the developed torque at a full load armature current of 50 A.

5.2 A dc, separately excited motor has the following parameters and ratings:

$$K\phi = 3 \text{ Vsec} \qquad R_a = 2\ \Omega$$

$$\text{Terminal voltage} = 600 \text{ V} \qquad \text{Full load torque} = 21 \text{ Nm}$$

a. Calculate the armature current at full load torque.
b. Calculate the starting current. Show how you can reduce the starting current by 80%.

5.3 A dc, separately excited motor has a load torque of 140 Nm and a frictional torque of 10 Nm. The motor is rated at 240 V. The armature resistance of the motor is 1 Ω. The motor speed at the given load is 600 rpm. Ignore the field losses and calculate the motor efficiency.

5.4 A dc series motor has an armature current of 10 A at full load. The motor terminal voltage is 300 V. The armature and field resistances are 2 Ω and 3 Ω, respectively. The motor speed at full load is 250 rpm. Calculate the starting torque of the motor.

5.5 A 1000 V, 50 hp compound motor runs at a speed of 750 rpm at full load. The armature, series, and shunt field resistances are 0.5 Ω, 1 Ω, and 200 Ω, respectively. The motor efficiency at this condition is 80%. Calculate the motor's starting current.

5.6 A 15 hp, 209 V, three-phase, six-pole, Y-connected induction motor has the following parameter values per phase:

$$R_1 = 0.128\ \Omega \qquad R'_2 = 0.0935\ \Omega \qquad X_{eq} = X_1 + X'_2 = 0.49\ \Omega$$

The motor slip at full load is 3%, and the efficiency is 90%.

a. Calculate the starting current. (Ignore the magnetizing current.)
b. Determine the starting torque.
c. Determine the maximum torque.
d. Calculate the value of the resistance that should be added to the rotor circuit to reduce the starting current by 50%.
e. What is the starting torque in part (d)?
f. Calculate the value of the resistance that should be added to the rotor circuit to increase the starting torque to maximum.
g. What is the starting current in part (f)?

5.7 Show how the starting current of the following machines can be reduced. Discuss the effect of your methods on the starting torque.

a. dc shunt motor (not a separately excited motor)
b. dc series motor
c. Induction motor

Use circuit diagrams and motor characteristics to explain your answer.

5.8 The shaft output of a three-phase, 60 Hz induction motor is 100 hp. The friction and windage losses are 900 W, the stator core loss is 4200 W, and the stator copper loss is 2700 W. If the slip is 3.75%, what is the efficiency of the motor?

5.9 A 500 hp, three-phase, 2200 V, 60 Hz, 12-pole, Y-connected, wound-rotor induction motor has the following parameters:

$$R_1 = 0.225\ \Omega \qquad R'_2 = 0.235\ \Omega \qquad X_{eq} = 1.43\ \Omega$$

$$X_m = 31.8\ \Omega \qquad R_m = 780\ \Omega$$

Calculate the following:

a. Slip at maximum torque
b. Input current and power factor at maximum torque
c. Maximum torque
d. Resistance that must be added to the rotor windings (per phase) to achieve maximum torque at starting

5.10 A four-pole, 60 Hz, Y-connected, squirrel cage induction motor has the following parameters:

$$R_1 = 0.2\ \Omega \qquad X_1 = 0.35\ \Omega \qquad R'_2 = 0.25\ \Omega \qquad X'_2 = 0.35\ \Omega$$

$$X_m = 12\ \Omega \qquad R_m >> X_m$$

The motor is connected to a 220 V supply through a cable of 1.30 Ω inductive reactance per phase. At a speed of 1710 rpm, calculate the following:

a. Motor current and input power
b. Terminal voltage
c. Developed torque

Also calculate the terminal voltage at starting. What is the percent change of the terminal voltage? Can you explain the change in the terminal voltage at starting?

5.11 A 15 hp (output power), 208 V, three-phase, six-pole, Y-connected induction motor has the following parameters:

$$R_1 = 0.1\ \Omega \qquad R'_2 = 0.1\ \Omega \qquad X_{eq} = 0.5\ \Omega$$

a. A fan-type load is connected to the motor. The slip of the motor in this case is 2%. If the terminal voltage of the motor is reduced by 20%, calculate the speed of the motor (you may use the small-slip approximation).
b. What is the percentage change of the maximum torque for case (a)?

5.12 A 500 hp, three-phase, 2200 V, 60 Hz, 12-pole, Y-connected, wound-rotor induction motor has the following parameters:

$$R_1 = 0.225\ \Omega \qquad R'_2 = 0.235\ \Omega \qquad X_1 + X'_2 = 1.43\ \Omega$$

$$X_m = 31.8\ \Omega \qquad R_m = 780\ \Omega$$

The motor is driving a constant-torque load at a speed of 570 rpm.

a. Calculate the load torque.
b. Calculate the motor speed when the source frequency is increased to 70 Hz.
c. Calculate the change in starting torque due to the frequency change.

5.13 A synchronous motor is rated at 100 kVA. The motor is connected to an infinite bus of 5 kV. The synchronous reactance of the motor is 0.1 Ω. The motor is running at a no-load condition (real power output is zero). All losses can be ignored. Calculate the equivalent field voltage E_f that operates the motor as a synchronous condenser delivering 100 kVAR to the infinite bus. Draw the phasor diagram. (A synchronous condenser produces reactive power and no real power.)

5.14 A three-phase synchronous motor is connected to an infinite bus of 416 V. The synchronous reactance of the motor is 1 Ω. The motor is driving a constant torque load. Ignore all losses. Calculate the change in power delivered to the load when the equivalent field voltage increases by 20%.

5.15 A four-pole synchronous motor is connected to an infinite bus of 5 kV through a cable. The synchronous reactance of the motor is 0.1 Ω, and the inductive reactance of the cable is 0.9 Ω. The reactive power at the motor terminals is zero when E_f is 4.8 kV (line-to-line). Calculate the following:

a. Terminal voltage of the motor
b. Developed torque
c. Output power

5.16 A six-pole synchronous motor is connected to an infinite bus of 480 V. The synchronous reactance of the motor is 0.5 Ω. The field current is adjusted so that the equivalent field voltage E_f is 500 V. Calculate the following:

a. Maximum torque
b. Power factor at maximum torque
c. Output power at maximum torque

6

Speed Control of Direct Current Motors

Direct current (dc) motors have several intrinsic properties, such as the ease by which they can be controlled, their ability to deliver high starting torque, and their near-linear performance. Direct current motors are widely used in applications such as actuation, manipulation, and traction.

Direct current motors have drawbacks that may restrict their use in some applications. For example, they are relatively high-maintenance machines due to their commutation mechanisms, and they are large and expensive compared to other motors, such as the induction. They may not be suitable for high-speed applications due to the presence of the commutator and brushes. Also, because of the electrical discharging between the commutator segments and brushes, dc machines cannot be used in clean or explosive environments unless they are encapsulated. Nevertheless, dc motors still hold a large share of the ASD (adjustable speed drive) market. Newer designs of dc motors have emerged that eliminate the mechanical commutator. The brushless motor, for example, is a dc motor that has the armature mounted on the stator and the field in the rotor. Like the conventional dc motor, the brushless motor switches the armature windings based on motor position. The switching, however, is done electronically, thereby eliminating the mechanical switching of the conventional dc motor.

6.1 SPEED CONTROL OF SHUNT OR SEPARATELY EXCITED MOTORS

As seen in Chapter 5, the speed–torque characteristics of a dc, separately excited (or shunt) motor can be expressed by the formula

$$\omega = \frac{V_t}{K\phi} - \frac{R_a}{(K\phi)^2} T_d = \omega_0 - \Delta\omega \tag{6.1}$$

or

$$\omega = \frac{V_t}{K\phi} - \frac{R_a}{K\phi} I_a = \omega_0 - \Delta\omega \tag{6.2}$$

where ω_0 is the no-load speed and $\Delta\omega$ is the speed drop. The no-load speed is computed when the torque and current are equal to zero. The speed drop is a function of the load torque. The load torque and rotational torques (such as friction) determine the magnitude of the motor's developed torque at steady state. For a given torque, the motor speed is a function of the following three quantities:

1. *Resistance in armature circuit.* When a resistance is inserted in the armature circuit, the speed drop $\Delta\omega$ increases and the motor speed decreases.
2. *Terminal voltage (armature voltage).* Reducing the armature voltage V_t of the motor reduces the motor speed.
3. *Field flux (or field voltage).* Reducing the field voltage reduces the flux ϕ, and the motor speed increases.

As explained in Chapter 5, we cannot operate electric motors with voltages higher than the rated value. Therefore, we cannot control the motor speed by increasing the armature or field voltages beyond the rated values. Only voltage reduction can be implemented. Hence, the second method of speed control (armature voltage) is only suitable for speed reduction, whereas the third method (field voltage) is suitable for speed increase. For a full range of speed control, more than one of the three methods must be employed.

6.1.1 CONTROLLING SPEED BY ADDING RESISTANCE

Figure 6.1 shows a dc motor setup with resistance added in the armature circuit. Figure 6.2 shows the corresponding speed–torque characteristics. Let us assume that the load torque is unidirectional and constant. A good example of this type of

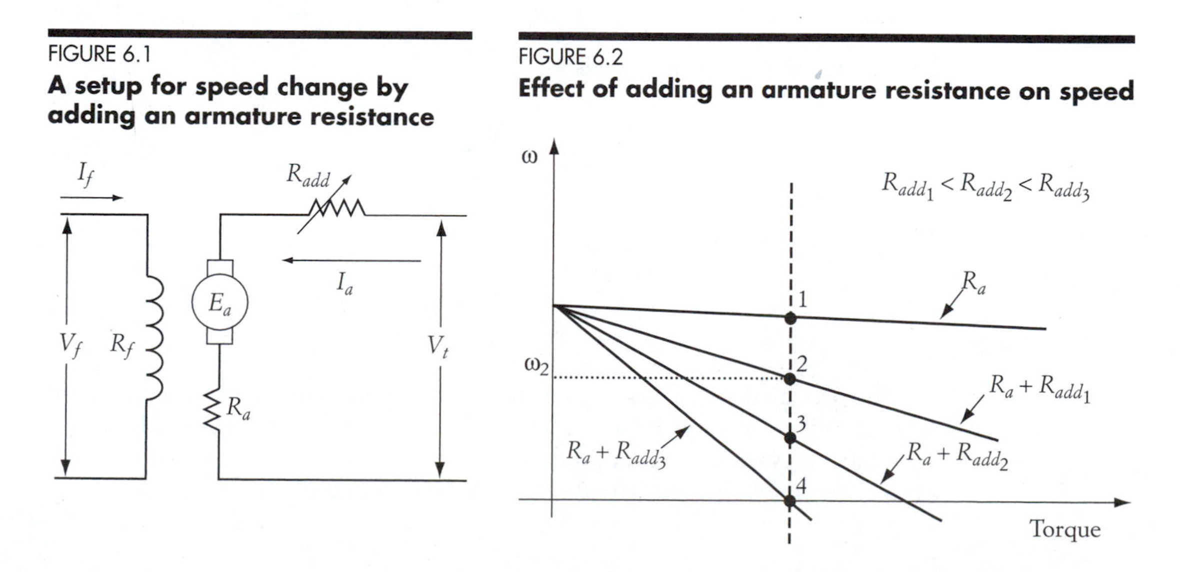

FIGURE 6.1
A setup for speed change by adding an armature resistance

FIGURE 6.2
Effect of adding an armature resistance on speed

torque is an elevator. Also assume that the field and armature voltages are constant. At point 1, no external resistance is in the armature circuit. If a resistance R_{add_1} is added to the armature circuit, the motor operates at point 2, where the motor speed ω_2 is

$$\omega_2 = \frac{V_t}{K\phi} - \frac{R_a + R_{add_1}}{(K\phi)^2} T_d = \omega_0 - \Delta\omega_2 \tag{6.3}$$

or

$$\omega_2 = \frac{V_t}{K\phi} - \frac{R_a + R_{add_1}}{K\phi} I_a = \omega_0 - \Delta\omega_2 \tag{6.4}$$

Note that the no-load speed ω_0 is unchanged regardless of the value of resistance in the armature circuit. The second term of the speed equation is the speed drop $\Delta\omega$, which increases in magnitude when R_{add} increases. Consequently, the motor speed is reduced.

If the added resistance keeps increasing, the motor speed decreases until the system operates at point 4, where the speed of the motor is zero. The operation of the drive system at point 4 is known as "holding." It is quite common to operate the motor under electrical holding conditions in applications such as robotics and actuation. An electrical drive system under holding may jiggle unless a feedback control circuit is used to stabilize the system.

When the motor is operating under a holding condition, the speed drop $\Delta\omega_4$ is equal in magnitude to the no-load speed ω_0.

$$\omega_4 = \omega_0 - \Delta\omega_4 = \frac{V_t}{K\phi} - \frac{R_a + R_{add_3}}{(K\phi)^2} T_d = 0 \tag{6.5}$$

The resistance R_{add_3} in this case is

$$R_{add_3} = \frac{K\phi V_t}{T_d} - R_a \tag{6.6}$$

or

$$R_{add_3} = \frac{V_t}{I_a} - R_a \tag{6.7}$$

Keep in mind that operating a dc motor for a period of time with a resistance inserted in the armature circuit is a very inefficient method. The use of resistance is acceptable only when the heat produced by the resistance is utilized as a by-product or when the resistance is used for a very short period of time.

EXAMPLE 6.1

A 150 V, dc shunt motor drives a constant-torque load at a speed of 1200 rpm. The armature and field resistances are 1 Ω and 150 Ω, respectively. The motor draws a line current of 10 A at the given load.

a. Calculate the resistance that should be added to the armature circuit to reduce the speed by 50%.

b. Assume the rotational losses to be 100 W. Calculate the efficiency of the motor without and with the added resistance.

c. Calculate the resistance that must be added to the armature circuit to operate the motor at the holding condition.

SOLUTION

a. Let us use Figure 6.2 to help in solving this problem. Assume that operating point 1 represents the motor without any added resistance, and point 2 is for the operating point at 50% speed reduction. Since the motor is a shunt machine, the line current is equal to the armature current plus the field current.

$$I_{a_1} = I - I_f = 10 - \frac{150}{150} = 9 \text{ A}$$

Also, the speed equations at these two operating points are

$$E_{a_1} = K\phi\omega_1 = V - I_{a_1}R_a$$

$$E_{a_2} = K\phi\omega_2 = V - I_{a_2}(R_a + R_{add_1})$$

The armature current is constant regardless of the value of the added resistance, because $I_a = \frac{T_d}{K\phi}$ and T_d and ϕ are constants. Hence, $I_{a_1} = I_{a_2}$.

$$\frac{E_{a_1}}{E_{a_2}} = \frac{\omega_1}{\omega_2} = \frac{n_1}{n_2} = \frac{V - I_a R_a}{V - I_a(R_a + R_{add_1})}$$

$$\frac{1200}{0.5 \times 1200} = \frac{150 - 9 \times 1}{150 - 9 \times (1 + R_{add_1})}$$

$$R_{add_1} = 7.83 \ \Omega$$

b. To calculate the motor efficiency, first calculate the input power

$$P_{in} = VI = 150 \times 10 = 1500 \text{ W}$$

Next, calculate the motor losses.

$$\text{Losses} = \text{field losses} + \text{armature losses} + \text{rotational losses}$$
$$= I_f^2 R_f + I_a^2 R_a + \text{rotational losses}$$

$$\text{Losses before adding armature resistance} = 150 + 81 + 100 = 331 \text{ W}$$

$$\text{Losses after adding armature resistance} = 150 + 81\,(1 + 7.83) + 100 = 965.23 \text{ W}$$

$$\text{Efficiency without resistance} = \frac{1500 - 331}{1500}\,100 = 77.93\%$$

$$\text{Efficiency after adding resistance} = \frac{1500 - 965.23}{1500}\,100 = 35.66\%$$

Note how low the motor efficiency is when a resistance is added to the armature circuit.

c. To calculate the resistance to be added to the armature for the holding operation, set the motor speed equal to zero.

$$K\phi\omega = V - I_a(R_a + R_{add}) = 0$$

$$R_{add} = \frac{V}{I_a} - R_a = \frac{150}{9} - 1 = 15.67\ \Omega$$

6.1.2 CONTROLLING SPEED BY ADJUSTING ARMATURE VOLTAGE

A common method of controlling speed is to adjust the armature voltage. This method is highly efficient and stable and is simple to implement. The circuit of Figure 6.3 shows the basic concept of this method. The only controlled variable is the armature voltage of the motor, which is depicted as an adjustable-voltage source. Based on Equation (6.1), when the armature voltage is reduced, the no-load speed ω_0 is also reduced. Moreover, for the same value of load torque and field flux, the armature voltage does not affect the speed drop $\Delta\omega$. The slope of the speed–torque characteristic is $R_a/(K\phi)^2$, which is independent of the armature voltage. Hence, the characteristics are parallel lines as shown in Figure 6.4. Note that we are assuming the field voltage is unchanged when the armature voltage varies.

FIGURE 6.3
A setup for changing speed by adjusting the armature voltage

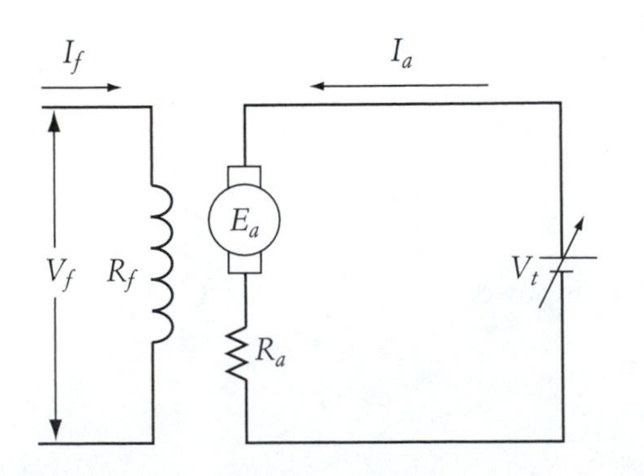

Electric holding can be done if the armature voltage is reduced until $\Delta\omega$ is equal to ω_0. This operating point is shown in Figure 6.4 at an armature voltage equal to V_4.

FIGURE 6.4
Motor characteristics when armature voltage changes

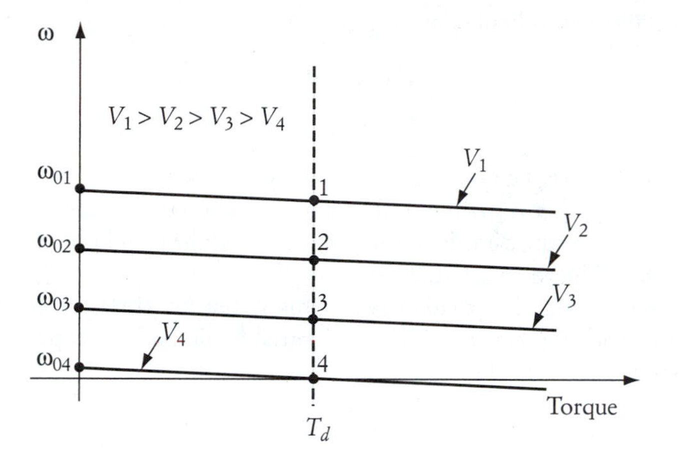

$$\omega_4 = \frac{V_4}{K\phi} - \frac{R_a}{(K\phi)^2} T_d = 0 \tag{6.8}$$

or

$$V_4 = \frac{R_a}{K\phi} T_d \tag{6.9}$$

6.1.3 CONTROLLING SPEED BY ADJUSTING FIELD VOLTAGE

Equations (6.1) and (6.2) show the dependency of motor speed on the field flux. The no-load speed is inversely proportional to the flux, and the slope of Equation (6.1) is inversely proportional to the square of the flux. Therefore, as explained next in Example 6.2, the speed is more sensitive to flux variations than to variations in the armature voltage.

EXAMPLE 6.2

For a 20% increase in the armature voltage, calculate the percentage change in the no-load speed. Assume that the load torque is unchanged. Repeat the calculations for a 20% reduction in the field flux.

SOLUTION

Let us assume that ω_{01}, ϕ_1, and V_1 are the variables for the initial condition. From Equation (6.1), the ratio of the no-load speeds, assuming that the armature voltage changes and the field is kept constant, is

$$\frac{\omega_{01}}{\omega_{02}} = \frac{V_1}{V_2}$$

where ω_{02} and V_2 are the new speed and voltage, respectively. Since $V_2 = 1.2\ V_1$, then $\omega_{02} = 1.2\ \omega_{01}$. Because the speed drop in Equation (6.1) is independent of the armature voltage for the same load torque, the speed drop will remain unchanged. Hence, the speed increase is also 20%.

Let us now repeat the calculation, assuming that the armature voltage is constant and the field is reduced by 20%. The variables ϕ_2 and ω_{02} represent the new field and speed, respectively.

$$\frac{\omega_{01}}{\omega_{02}} = \frac{\phi_2}{\phi_1}$$

If $\phi_2 = 0.8\ \phi_1$, then $\omega_{02} = 1.25\ \omega_{01}$. The increase in no-load speed is 25% as compared to the 20% of the previous case. Similarly, the ratio of the speed drop using Equation (6.1) is

$$\frac{\Delta\omega_1}{\Delta\omega_2} = \left(\frac{\phi_2}{\phi_1}\right)^2$$

Then for $\phi_2 = 0.8\ \phi_1$, $\Delta\omega_2 = 1.5625\ \Delta\omega_1$. The speed drop is increased by 56.25%.

FIGURE 6.5

Setup for controlling speed by adjusting field voltage

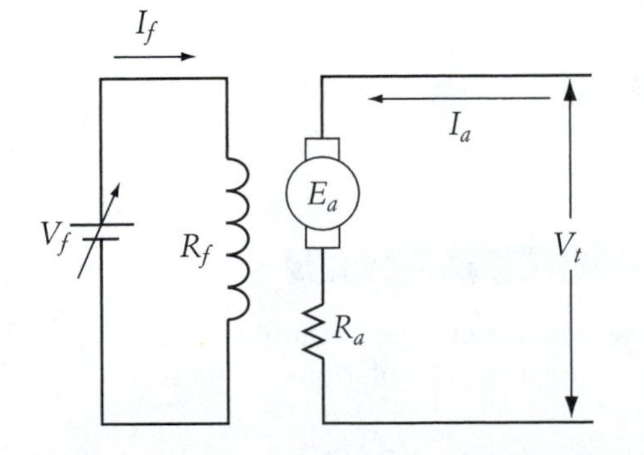

Figure 6.5 shows a setup for controlling speed by adjusting the field flux. If we reduce the field voltage, the field current and consequently the flux are reduced. Figure 6.6 shows a set of speed–torque characteristics for three values of field voltages. When the field flux is reduced, the no-load speed ω_0 is increased in inverse proportion to the flux, and the speed drop $\Delta\omega$ is also increased. The characteristics show that because of the change in speed drops, the lines are not parallel. Unless the motor is excessively loaded, the motor speed increases when the field is reduced. When motor speed is controlled by adjusting the field current, the following considerations should be kept in mind:

FIGURE 6.6
Effect of field voltage on motor speed

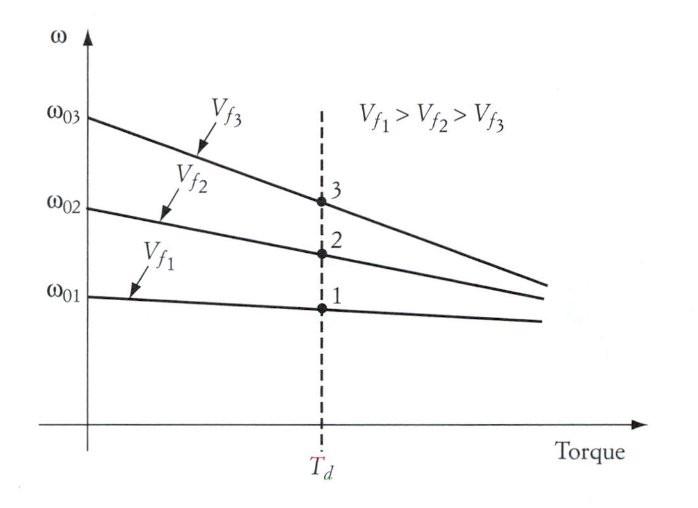

1. The field voltage must not exceed the absolute maximum rating.
2. Since dc motors are relatively sensitive to variations in field voltage, large reductions in field current may result in excessive speed.
3. Because the armature current is inversely proportional to the field flux ($I_a = T_d/K\phi$), reducing the field results in an increase in the armature current (assuming that the load torque is unchanged).

Because of (2) and (3), field voltage control should be done with special care to prevent mechanical and electrical damage to the motor. Furthermore, the field current should not be interrupted while the motor is running. If an interruption occurs, the residual magnetism will maintain a small amount of flux in the airgap. Consequently, the motor current will be excessively large, and the motor will accelerate to unsafe speeds. Although the system may have overcurrent breakers, special care should be given to this type of control to avoid an unpleasant experience!

EXAMPLE 6.3

A 150 V, dc shunt motor drives a constant-torque load at a speed of 1200 rpm. The armature and field resistances are 2 Ω and 150 Ω, respectively. The motor draws a line current of 10 A. Assume that a resistance is added in the field circuit to reduce the field current by 20%. Calculate the armature current, motor speed, value of the added resistance, and extra field losses.

SOLUTION

The armature current before inserting a resistance in the field circuit is

$$I_{a_1} = I - I_{f_1} = 10 - \frac{150}{150} = 9 \text{ A}$$

Since the load torque is constant,

$$T_d = K\phi_1 I_{a_1} = K\phi_2 I_{a_2}$$

$$I_{a_2} = \frac{\phi_1}{\phi_2} I_{a_1}$$

Assume that the flux is linearly proportional to the field current.

$$I_{a_2} = \frac{I_{f_1}}{I_{f_2}} I_{a_1} = \frac{1}{0.8} 9 = 11.25 \text{ A}$$

Notice that the armature current is increased by 25%. To calculate the speed, consider the two equations

$$E_{a_1} = K\phi_1\omega_1 = V - I_{a_1}R_a$$

$$E_{a_2} = K\phi_2\omega_2 = V - I_{a_2}R_a$$

or

$$\frac{\phi_1}{\phi_2}\frac{n_1}{n_2} = \frac{V - I_{a_1}R_a}{V - I_{a_2}R_a}$$

$$\frac{1}{0.8}\frac{1200}{n_2} = \frac{150 - 9 \times 2}{150 - 11.25 \times 2}$$

$$n_2 = 1448.86 \text{ rpm}$$

The result is a 20.73% increase in speed.

The value of the resistance that should be inserted in the field circuit can be calculated using Ohm's law:

$$V_f = V_t = I_{f_1}R_f = I_{f_2}(R_f + R_{add})$$

$$R_{add} = 37.5\ \Omega$$

The losses due to R_{add} are

$$P = I_{f_2}^2 R_{add} = (0.8)^2 \times 37.5 = 24 \text{ W}$$

Note that the additional losses are small when a resistance is added to the field circuit. This is why the technique is acceptable in industry even when solid-state field control devices are available. Compare the losses in this case with the losses in Example 6.1.

6.1.4 SOLID-STATE CONTROL

Solid-state control is used for enhanced efficiency and for versatile operation of electric drive systems. For dc machines, converters are often used in the armature circuit to control the terminal voltage of the motor. In some cases, the converter is also used to control the field voltage. When a converter is used, the power source can be either dc or ac, which makes the selection of the machine independent of the available power source at the site.

In this section, we will analyze the dc separately excited (or shunt) motors when energized by two types of power sources: ac (single- or multi-phase) and dc.

6.1.4.1 SINGLE-PHASE, HALF-WAVE DRIVES

The circuit in Figure 6.7 shows an example of a dc motor with converter. The armature circuit of the motor is connected to the converter, which is fed from an ac source. The field circuit of the motor is excited from the ac source through a full-wave rectifier circuit, which may contain filters.

The circuit of Figure 6.8 shows the armature loop. The converter in this case is a simple SCR triggered by a control circuit not shown in the figure. The waveforms of the circuit are shown in Figure 6.9. Before the triggering of the SCR at α, the instantaneous voltage across the motor terminals v_t is equal to E_a. During the SCR conduction, v_t is equal to the instantaneous source voltage v_s. The voltage

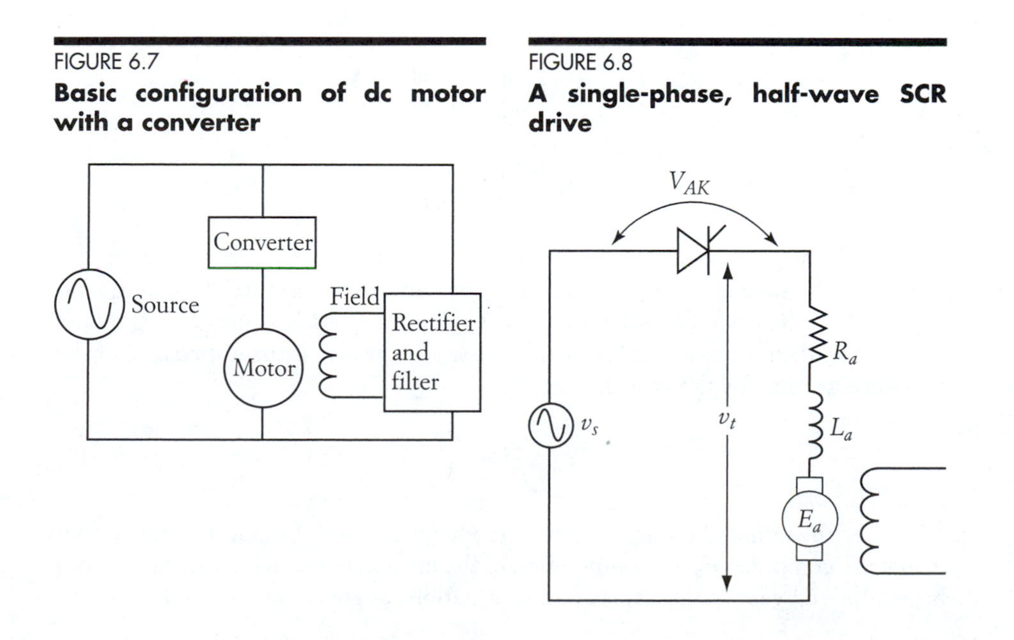

FIGURE 6.7
Basic configuration of dc motor with a converter

FIGURE 6.8
A single-phase, half-wave SCR drive

FIGURE 6.9
Waveforms of circuit in Figure 6.8

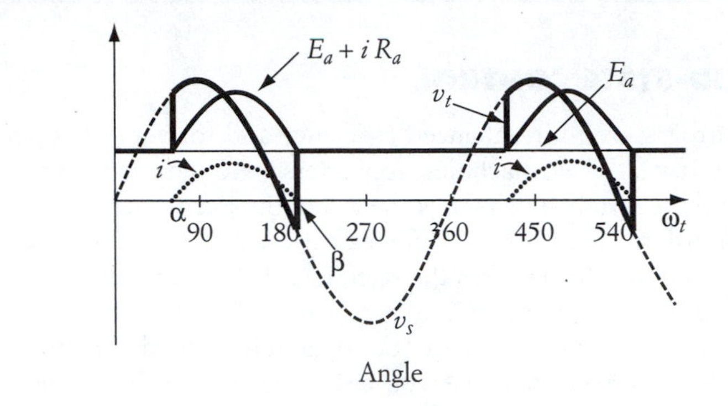

across the resistive component of the armature winding is identical in shape to the instantaneous armature current. The instantaneous voltage across the inductive element of the armature impedance v_{la} is

$$v_{la} = v_t - E_a - i_a R_a \tag{6.10}$$

The instantaneous terminal voltage can be expressed mathematically by

$$v_t = v_s(u_\alpha - u_\beta) + E_a[1 - (u_\alpha - u_\beta)]$$

$$v_t = V_{\max}\sin(\omega t)(u_\alpha - u_\beta) + E_a[1 - (u_\alpha - u_\beta)] \tag{6.11}$$

where u_α and u_β are step functions

$$u_\alpha = u(\omega t - \alpha)\begin{cases} u_\alpha = 1; \ \omega t \geq \alpha \\ u_\alpha = 0; \ \omega t < \alpha \end{cases}$$

$$u_\beta = u(\omega t - \beta)\begin{cases} u_\beta = 1; \ \omega t \geq \beta \\ u_\beta = 0; \ \omega t < \beta \end{cases}$$

and β is the angle at which the instantaneous current reaches its zero crossing. Assume that the speed of the motor is fairly constant during the steady-state operation, and the field voltage is kept constant. Hence, E_a is also constant. The load current can then be computed by dividing the voltage across the impedance of the armature winding by the impedance itself.

$$i = \frac{v_t - E_a}{R_a + jX_a} \tag{6.12}$$

Since the terminal voltage contains step functions and the load impedance has imaginary components, the computation of the instantaneous armature current can be simplified by using the Laplace transformations as given in Chapter 3. The current equation during the conduction period (between α and β) can be expressed by Equation (6.13).

$$i(t) = \mathcal{L}^{-1} I(s)$$

$$i(t) = \frac{V_{max}}{Z} \sin(\omega t - \phi) - \frac{E_a}{R_a} + \left(\frac{E_a}{R_a} - \frac{V_{max}}{Z} \sin(\alpha - \phi)\right) e^{-\frac{(\omega t - \alpha)}{\omega \tau}} \quad (6.13)$$

where Z is the impedance of the armature winding, ϕ is the phase angle of Z, and τ is the time constant of Z.

$$\phi = \tan^{-1} \frac{\omega L_a}{R_a}$$

$$\tau = \frac{L_a}{R_a}$$

$$Z = \sqrt{R_a^2 + (\omega L_a)^2}$$

$$\omega = 2\pi f$$

f is the frequency of the ac source.

For a given α, the conduction period can be determined by using Equation (6.13). At β, the instantaneous armature current $i(t_\beta)$ is equal to zero.

$$i(t_\beta) = \frac{V_{max}}{Z} \sin(\beta - \phi) - \frac{E_a}{R_a} + \left[\frac{E_a}{R_a} - \frac{V_{max}}{Z} \sin(\alpha - \phi)\right] e^{-\frac{(\beta - \alpha)}{\omega \tau}} = 0 \quad (6.14)$$

where $t_\beta = \beta/\omega$ is the time at β. Solving Equation (6.14) yields the value of β. Note that the equation is nonlinear in terms of β and that iterative methods may be used.

The average voltage across the terminals of the motor $V_{t\ ave}$ can be computed using the armature loop. First write the instantaneous-voltage equation of the armature loop as

$$v_t = E_a + v_L + iR_a \quad (6.15)$$

Then compute the average values

$$V_{t\ ave} = E_a + V_{L\ ave} + I_{ave} R_a \quad (6.16)$$

where $V_{L\ ave}$ is the average voltage across the inductive element of the armature impedance. As we discussed earlier in Chapter 3 (Section 3.3.1), $V_{L\ ave}$ must be equal to zero. Hence,

$$V_{t\ ave} = E_a + I_{ave} R_a$$

or

$$\frac{1}{2\pi}\int_0^{2\pi} v_t\,d\omega t = E_a + \frac{R_a}{2\pi}\int_0^{2\pi} i\,d\omega t$$

$$\frac{1}{2\pi}\left[\int_\alpha^\beta v_t\,d\omega t + \int_\beta^{\alpha+2\pi} v_t\,d\omega t\right] = E_a + \frac{R_a}{2\pi}\int_\alpha^{2\pi} i\,d\omega t$$

During the interval from α to β, v_t is equal to the source voltage, and between β and the triggering angle of the next cycle ($\alpha + 2\pi$), v_t is equal to E_a.

$$\frac{1}{2\pi}\left[\int_\alpha^\beta v_s\,d\omega t + E_a(2\pi + \alpha - \beta)\right] = E_a + R_a I_{\text{ave}}$$

Then

$$\frac{1}{2\pi}\int_\alpha^\beta V_{\max}\sin(\omega t)\,d\omega t = \frac{\gamma}{2\pi}E_a + R_a I_{\text{ave}} \tag{6.17}$$

where γ is the conduction period

$$\gamma = \beta - \alpha$$

Equation (6.17) can also be expressed as

$$\frac{V_{\max}}{2\pi}[\cos(\alpha) - \cos(\beta)] = \frac{\gamma}{2\pi}E_a + R_a I_{\text{ave}} \tag{6.18}$$

Replacing E_a with $K\phi\omega$ yields

$$\frac{V_{\max}}{2\pi}[\cos(\alpha) - \cos(\beta)] = \frac{\gamma}{2\pi}K\phi\omega + R_a I_{\text{ave}} \tag{6.19}$$

EXAMPLE 6.4

A 1 hp, dc shunt motor is loaded by a constant torque of 10 Nm. The armature resistance of the motor is 5 Ω, and the field constant $K\phi = 2.5$ V sec. The motor is driven by a half-wave SCR converter. The power source is 120 V, 60 Hz. The triggering angle of the converter is 60°, and the conduction period is 150°. Calculate the motor speed and the developed power.

SOLUTION

The average armature current is determined by the load torque and motor excitation.

$$I_{\text{ave}} = \frac{T}{K\phi} = \frac{10}{2.5} = 4\text{ A}$$

Direct substitution in Equation (6.19) yields

$$\frac{\sqrt{2} \times 120}{2\pi}[\cos(60) - \cos(60 + 150)] = \frac{150}{360} \times 2.5 \times \omega + 5 \times 4$$

$$\omega = 16.22 \text{ rad/sec}$$

$$n = 154.88 \text{ rpm}$$

The developed power is

$$P_d = E_a I_{\text{ave}} = K\phi\omega I_{\text{ave}} = 2.5 \times 16.22 \times 4 = 162 \text{ W}$$

which is about 21% of the motor rating.

6.1.4.2 SINGLE-PHASE, FULL-WAVE DRIVES

A full-wave drive can be realized by using one of the two circuits shown in Figures 6.10 and 6.11. The circuit in Figure 6.10 consists of four SCRs connected in a full-wave bridge. The switching of the SCRs is dependent on the polarity of the source voltage v_s. The current i_1 (solid lines) flows when the ac waveform of the source voltage is in the positive half-cycle, and SCRs S_1 and S_2 are triggered. Similarly, current i_2 (dashed lines) flows when the waveform of the source voltage is in the negative half, and S_3 and S_4 are triggered. In either half of the cycle, the current will flow in the same direction inside the motor.

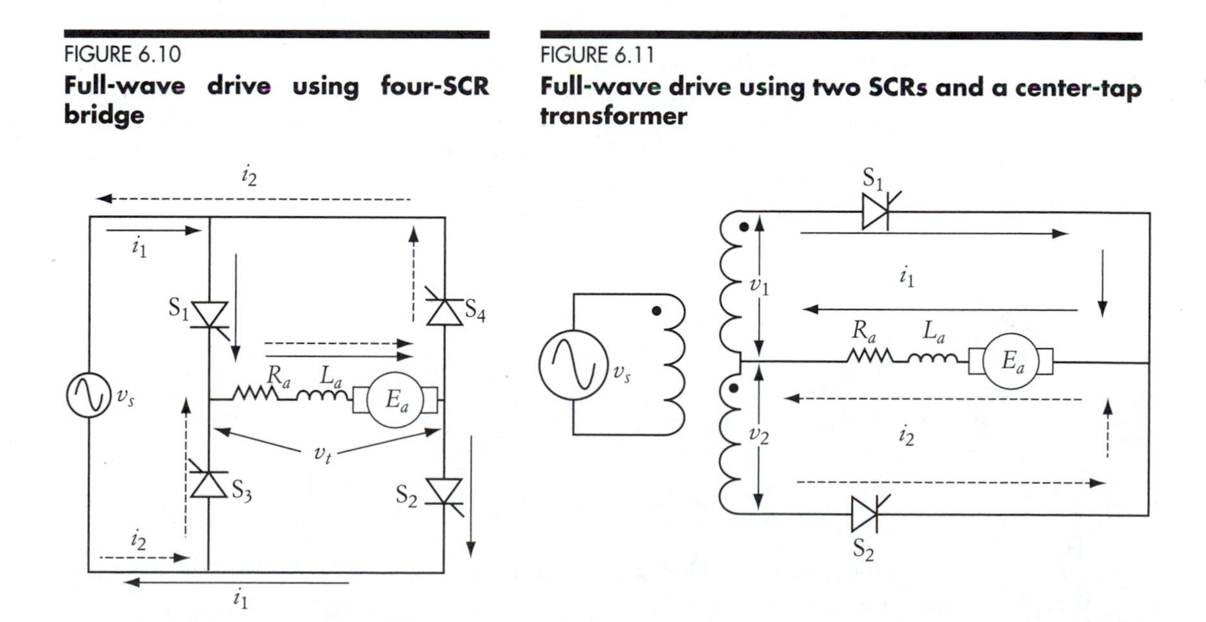

FIGURE 6.10
Full-wave drive using four-SCR bridge

FIGURE 6.11
Full-wave drive using two SCRs and a center-tap transformer

FIGURE 6.12
Waveforms of circuit in Figure 6.11

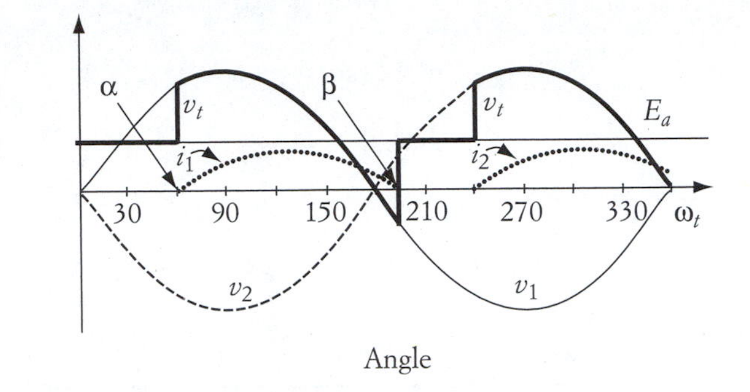

The circuit in Figure 6.11 shows another alternative where two SCRs and a center-tap transformer are used. The secondary of the transformer should have double the voltage rating of the motor; that is,

$$V_1 = V_2 = \text{rated armature voltage}$$

When the source voltage v_s is in the positive half of its cycle and S_1 is triggered, i_1 flows in the upper half of the transformer's secondary windings. When the source voltage is in the negative part and S_2 is closed, i_2 flows in the lower half of the secondary windings. Again, in either half of the source waveform, the armature current of the machine is unidirectional.

The waveforms of the circuit in Figure 6.11 are shown in Figure 6.12. The figure shows v_1 and v_2 in reference to the center point of the transformer. When v_1 is in the positive part of its cycle and S_1 is triggered at α, the terminal voltage of the motor v_t is equal to v_1 and the motor current is i_1. Because i_1 flows beyond 180°, the terminal voltage of the motor v_t becomes negative. When i_1 reaches zero at β, v_t is equal to E_a until S_2 is triggered. Similarly, during the positive half of v_2, i_2 flows and the terminal voltage of the motor equals v_2.

The average terminal voltage of the motor is calculated by

$$V_{t\,\text{ave}} = E_a + I_{\text{ave}} R_a$$

or

$$\frac{2}{2\pi}\int_0^{\pi} v_t \, d\omega t = E_a + \frac{2R_a}{2\pi}\int_0^{\pi} i \, d\omega t$$

$$\frac{1}{\pi}\left[\int_{\alpha}^{\beta} v_t \, d\omega t + \int_{\beta}^{\alpha+\pi} v_t \, d\omega t\right] = E_a + \frac{R_a}{\pi}\int_{\alpha}^{\beta} i \, d\omega t$$

Using the same procedure explained in Equations (6.16) to (6.18), the equation of the armature circuit is

$$\frac{V_{max}}{\pi}[\cos(\alpha) - \cos(\beta)] = \frac{\gamma}{\pi}E_a + R_a I_{ave} \tag{6.20}$$

Replacing E_a with $K\phi\omega$ yields

$$\frac{V_{max}}{\pi}[\cos(\alpha) - \cos(\beta)] = \frac{\gamma}{\pi}K\phi\omega + R_a I_{ave} \tag{6.21}$$

EXAMPLE 6.5

For the motor in Example 6.4, assume that the converter is a full-wave type. The triggering angle of the converter is 60°, and the conduction period is 150°. Calculate the motor speed and the developed power delivered to the load.

SOLUTION

The load torque and the excitation determine the average armature current of the motor. The average current is not affected by whether we are using a half-wave or a full-wave converter.

$$I_{ave} = \frac{T}{K\phi} = \frac{10}{2.5} = 4 \text{ A}$$

This average current is produced in a half-cycle for half-wave converters. For full-wave converters, the current is produced by the two halves of the cycle.

Direct substitution in Equation (6.21) yields

$$\frac{\sqrt{2} \times 120}{\pi}[\cos(60) - \cos(60 + 150)] = \frac{150}{180} \times 2.5 \times \omega + 5 \times 4$$

$$\omega = 25.82 \text{ rad/sec}$$

$$n = 246.56 \text{ rpm}$$

Note that the speed in this case is higher than that in Example 6.4. This is because full-wave converters allow more power to be transmitted to the motor.

The developed power can be calculated by

$$P_d = K\phi\omega I_{ave} = 2.5 \times 25.82 \times 4 = 258.2 \text{ W}$$

which is about a 56% increase in power as compared to the half-wave converter of Example 6.4.

6.1.4.3 CONTINUOUS ARMATURE CURRENT

For heavily loaded motors with high armature inductance, the conduction period may equal or exceed 180°. This may result in an overlap of i_1 and i_2, which results in continuous motor current. Figure 6.13 shows the armature current for several

FIGURE 6.13
Current waveforms of several loads

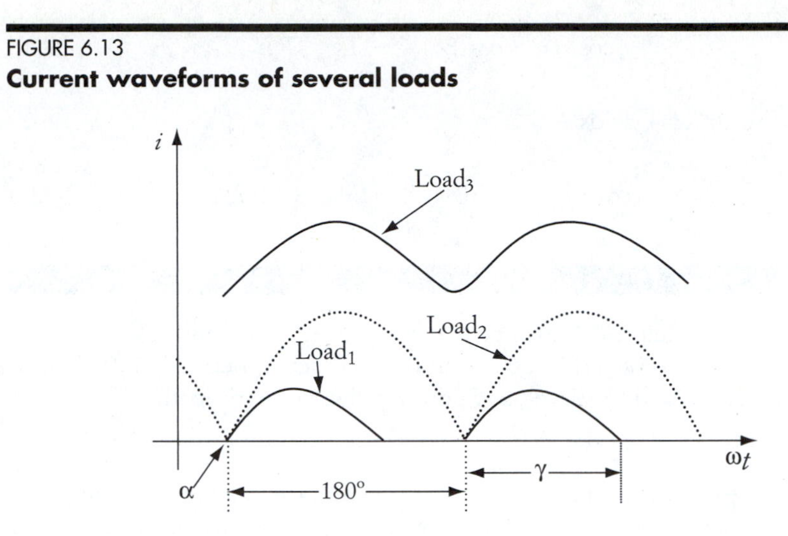

load cases, where $load_1 < load_2 < load_3$. In the figure, the triggering angle α is the same for all loads. When the load increases, the conduction period and the peak current increase. When the conduction period equals 180°, the current is said to be continuous.

For continuous current, Equations (6.20) and (6.21) can be simplified by replacing γ with 180°.

$$\frac{V_{max}}{\pi}[\cos(\alpha) - \cos(\alpha + 180)] = E_a + R_a I_{ave}$$

$$\frac{2V_{max}}{\pi}\cos(\alpha) = E_a + R_a I_{ave} \tag{6.22}$$

Replacing E_a with $K\phi\omega$ yields

$$\frac{2V_{max}}{\pi}\cos(\alpha) = K\phi\omega + R_a I_{ave} \tag{6.23}$$

or

$$\omega = \frac{\frac{2V_{max}}{\pi}\cos(\alpha)}{K\phi} - \frac{R_a}{K\phi} I_{ave} \tag{6.24}$$

Equation (6.23) is similar to (6.2)—the terminal voltage in Equation (6.2) is replaced by $(2V_{max}/\pi)\cos(\alpha)$ in Equation (6.24).

EXAMPLE 6.6

A dc, separately excited motor has a constant torque load of 60 Nm. The motor is driven by a full-wave converter connected to a 120 V, ac supply. The field constant of the motor $K\phi = 2.5$ and the armature resistance is 2 Ω. Calculate the triggering angle α for the motor to operate at 200 rpm. The motor current is continuous.

SOLUTION

From Equation (6.24),

$$\alpha = \cos^{-1}\left[\frac{\pi}{2V_{\max}}(R_a I_{ave} + K\phi\omega)\right]$$

$$\alpha = \cos^{-1}\left[\frac{\pi}{2V_{\max}}\left(R_a \frac{T}{K\phi} + K\phi\omega\right)\right]$$

Hence,

$$\alpha = \cos^{-1}\left[\frac{\pi}{2\sqrt{2} \times 120}\left(2\,\frac{60}{2.5} + 2.5 \times 2\pi\,\frac{200}{60}\right)\right] = 21.7°$$

The speed–torque characteristics of the dc motor under solid-state control are depicted in Figure 6.14. Since the armature current can be either discontinuous or continuous, the speed–torque characteristics are dependent on the magnitude of the load torque. Because of the nonlinearity in Equation (6.13) in terms of β, the speed–torque characteristics for discontinuous armature current are nonlinear.

FIGURE 6.14
Speed–torque characteristics of a dc motor driven by solid-state converter

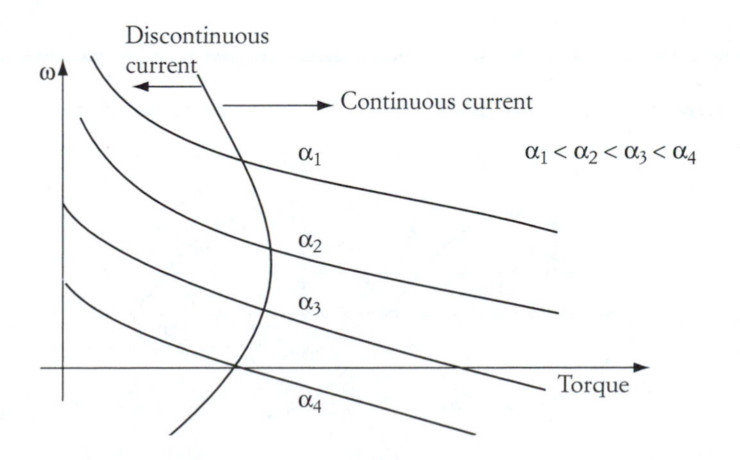

Equation (6.24), however, is not a function of $\beta(\beta = 180°)$. Then the speed–torque characteristics for continuous current are linear and similar to those represented by Equation (6.2).

6.1.4.4 EFFECT OF FREEWHEELING DIODE

As seen in Figure 6.12, the terminal voltage of the motor may become negative due to the energy stored in the inductance of the armature winding. As we explained in Chapter 3, a freewheeling diode can be used to dissipate this energy in the load itself, thus preventing the terminal voltage from becoming negative. The circuit with the freewheeling diode is shown in Figure 6.15. The polarities of the diodes are opposite to those of the SCRs.

The operation of the circuit can be explained by the waveforms in Figure 6.16. During the period from α to 180°, S_1 is closed and the current i_1 flows in the

FIGURE 6.15
Full-wave converter with freewheeling diode

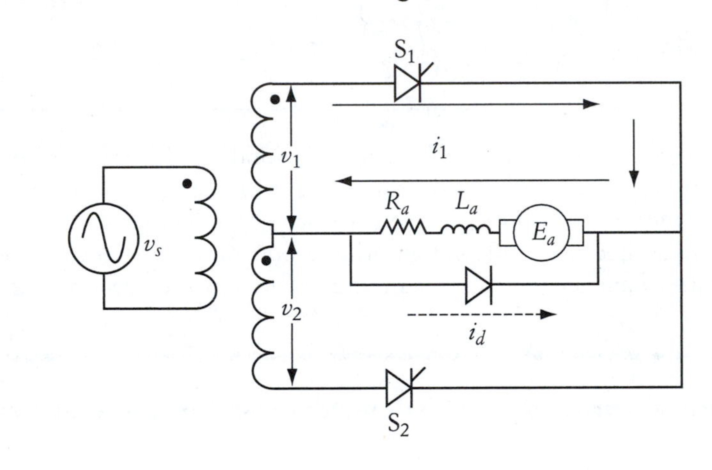

FIGURE 6.16
Waveforms for full-wave converter with freewheeling diode

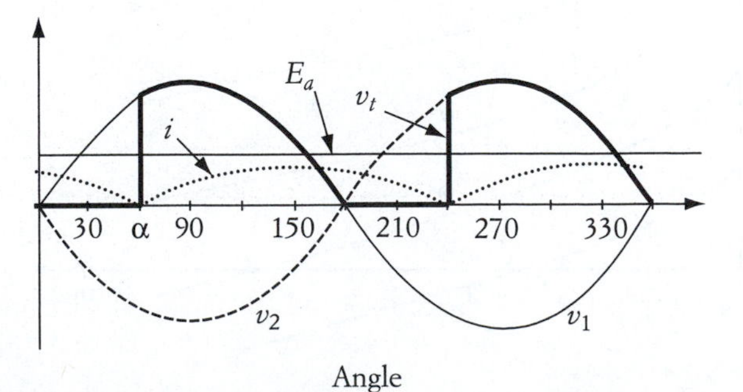

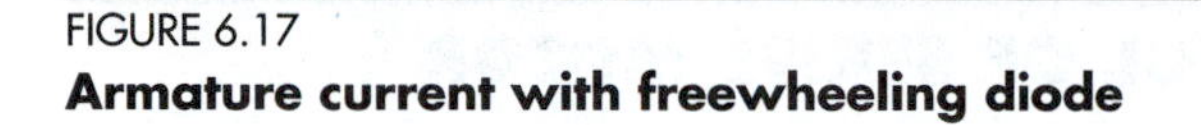

FIGURE 6.17
Armature current with freewheeling diode

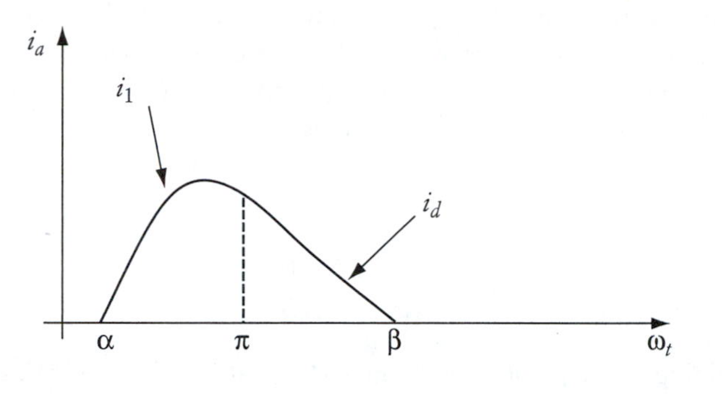

motor. Just after 180°, v_1 reverses its polarity and S_1 opens. Consequently, the freewheeling diode conducts, and the motor current is i_d, which flows from 180° until S_2 is turned on in the second half of the cycle. This process is repeated every half-cycle. The voltage across the motor is equal to the source voltage while the SCRs are conducting, and is equal to zero when the freewheeling diode is conducting.

Figure 6.17 shows the armature current of the motor i_a and the freewheeling current i_d. The armature current can be divided into two regions. The first is from α to π, in which the freewheeling diode has no effect, and $i_a = i_1$. The second is from π to β, where the diode is conducting ($i_a = i_d$), and no current is flowing from the source.

For the first region, the instantaneous current is given by Equation (6.13). During the diode conduction, the armature current i_d flows in the armature–diode loop, which is composed of the armature impedance and the back emf E_a. We can ignore the impedance of the diode during conduction. The current i_d can then be expressed by

$$E_a = R_a i_d + L_a \frac{di_d}{dt} \tag{6.25}$$

The solution of the differential equation (6.25) for the period from π to β is

$$i_d = I(\pi)\, e^{-(\omega t - \pi)/\omega\tau} - \frac{E_a}{R_a}\left[1 - e^{-(\omega t - \pi)/\omega\tau}\right] \tag{6.26}$$

where $I(\pi)$ is the initial condition at $\omega t = \pi$, which can be computed using Equation (6.13). Equation (6.26) shows that, mathematically, i_d is zero when $\omega t = \infty$. Practically, we can assume that i_d is almost zero when it reaches about 5% of its maximum value. In fact, because of the freewheeling diode, the armature current is likely to be continuous.

6.2 SPEED CONTROL OF SERIES MOTOR

The concept of speed control of series machines is almost identical to that for the shunt machines. The basic types of control used for shunt machines can also be implemented for series machines. The implementation, however, requires special consideration of the fact that the field and armature currents are directly correlated.

Equation (5.18) describes the speed–torque characteristics of the series motor. The equation is repeated here.

$$\omega = \frac{V_t}{K\phi} - \frac{R_a + R_f}{(K\phi)^2} T_d \tag{6.27}$$

After examining this equation, one concludes that three methods can be used to control the motor speed:

1. Adding a resistance in the armature circuit
2. Adjusting the armature voltage
3. Adjusting the field current

6.2.1 CONTROLLING SPEED BY ADDING RESISTANCE TO THE ARMATURE CIRCUIT

Consider Equation (6.28) for the series motor. In this equation, we are assuming that the flux is linearly proportional to the armature current ($\phi \approx CI_a$).

$$T = K\phi I_a \approx KC(I_a)^2 \tag{6.28}$$

The equation shows that, for a given load torque, the armature current of the motor is constant. The change in armature current is only proportional to the change in the load torque. This is also true even if the flux–current characteristic is in the nonlinear region.

Let us approximate Equation (6.27) by assuming that $\phi \approx CI_a$.

$$\omega = \frac{V_t}{KCI_a} - \frac{R_a + R_f}{KC} = \omega_0 - \Delta\omega \tag{6.29}$$

Now let us assume that a resistance R_{add} is inserted in series with the armature circuit as shown in Figure 6.18. Equation (6.29) can then be modified to

$$\omega = \frac{V_t}{KCI_a} - \frac{R_a + R_f + R_{add}}{KC} = \omega_0 - \Delta\omega \tag{6.30}$$

If we assume that the load torque is unchanged, then the armature current is constant. In addition, if the supply voltage is unchanged, then the motor speed is reduced when a resistance is added to the armature circuit. This is due to the increase in the speed drop $\Delta\omega$.

FIGURE 6.18
Controlling speed of series motor by adding a resistance in the armature circuit

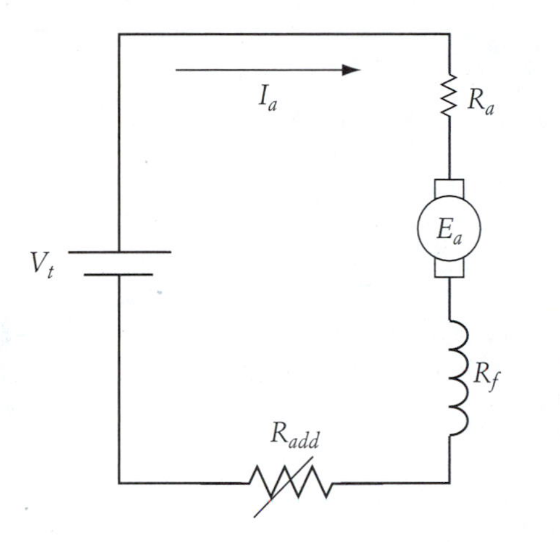

FIGURE 6.19
Speed–torque characteristics of series motor due to the insertion of a resistance in the armature circuit

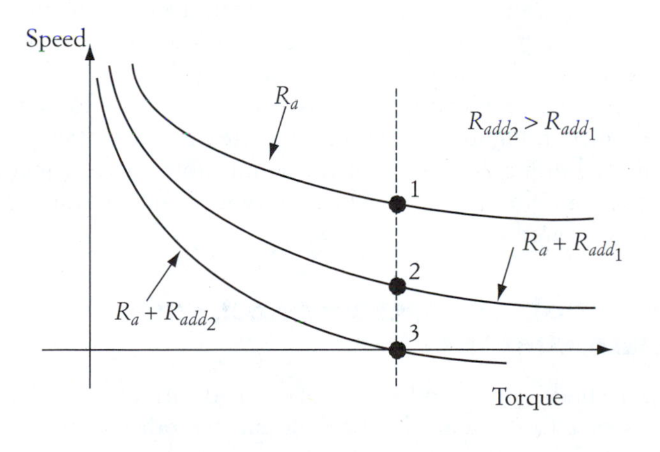

$$\Delta\omega = \frac{R_a + R_{add} + R_f}{KC}$$

The motor characteristics in this case are shown in Figure 6.19. The shapes of the characteristics are similar when a resistance is added, but the speed drop increases when the value of the added resistance increases.

FIGURE 6.20
Circuit for controlling speed of series motor by varying terminal voltage

FIGURE 6.21
Speed–torque characteristics of series motor under voltage control

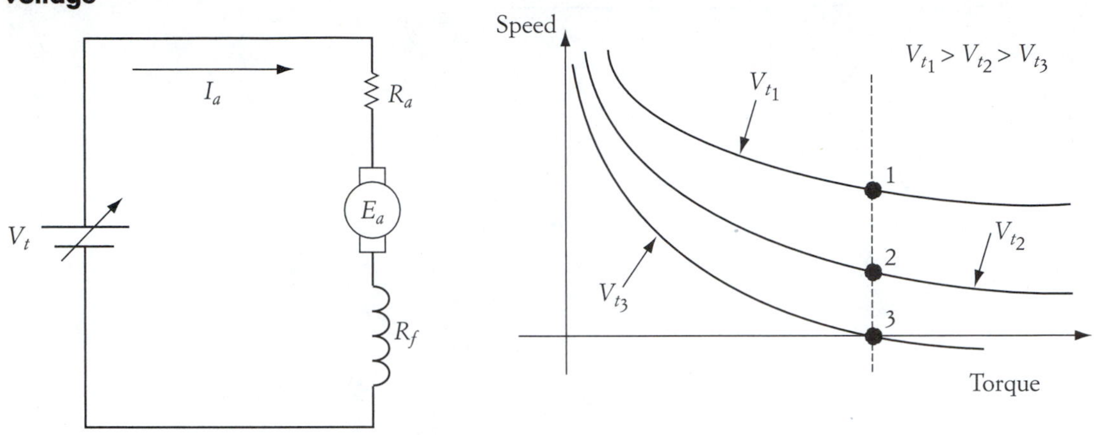

6.2.2 CONTROLLING SPEED BY ADJUSTING ARMATURE VOLTAGE

The change in armature voltage has a similar effect on the series motor as the insertion of an armature resistance. For a constant torque, the motor current is constant, and the first term, ω_0, in Equation (6.29) decreases with the decrease of voltage. The second term, $\Delta\omega$, remains unchanged. This results in a decrease in motor speed. The circuit and the characteristics of the motor with this type of control are shown in Figures 6.20 and 6.21. Note that the voltage control can be done by any technique described for the shunt motor drive. Since the voltage must be kept at or below the rated value, this type of control is suitable for speed reduction below rated speed.

6.2.3 CONTROLLING SPEED BY ADJUSTING FIELD CURRENT

Two simple methods can be used to control the field current. One of them is to add a shunt resistance to the series field circuit and the other is to use a solid-state switching device across the field windings to regulate the field current. These two methods provide similar performance.

If a resistance R_{fadd} is inserted in shunt with the series field winding, as shown in Figure 6.22, the field current is reduced by the following ratio:

$$I_f = \frac{R_{fadd}}{R_{fadd} + R_f} I_a = A_R I_a \tag{6.31}$$

FIGURE 6.22
Simple circuit for controlling speed of series motor by varying field current

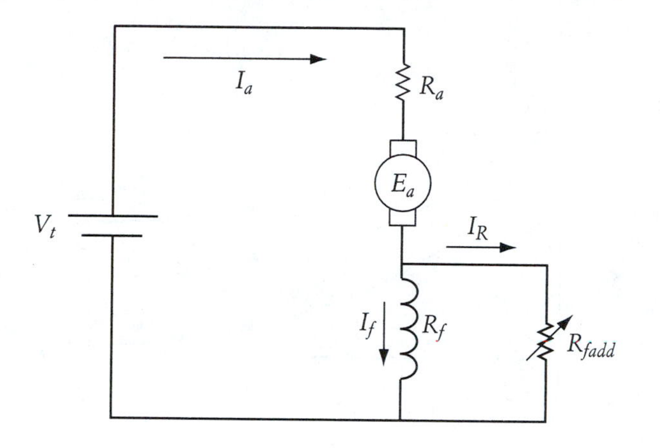

where A_R is a resistance ratio. If we assume that the field flux is proportional to the field current—that is, $\phi = CI_f$—then

$$K\phi = KCA_R I_a \tag{6.32}$$

and the load torque equation is

$$T_d = K\phi I_a = KCA_R I_a^2 \tag{6.33}$$

Modify Equation (6.27) to include the new added resistance

$$\omega = \frac{V_t}{K\phi} - \frac{R_a + \dfrac{R_f R_{fadd}}{R_f + R_{fadd}}}{K\phi} I_a = \frac{V_t}{K\phi} - \frac{R_a + A_R R_f}{K\phi} I_a \tag{6.34}$$

Substituting Equations (6.32) and (6.33) into (6.34) yields

$$\omega = \frac{V_t}{KCA_R I_a} - \frac{R_a + A_R R_f}{KCA_R} \tag{6.35}$$

A reduction of R_{fadd} results in current reduction of the field windings, which leads to an increase in motor speed. The characteristics of the series motor under this type of control are shown in Figure 6.23. Note that since the technique results in a field reduction, it is suitable for speed increase. This type of speed control must be performed with care so the current of the motor is not excessive. This can be seen by examining Equation (6.33); if you assume that the torque is constant, the armature current will increase if R_{fadd} is reduced.

Solid-state devices can also implement the field reduction. An example is shown in Figure 6.24. In this figure, a transistor and a diode shunt the field circuit.

FIGURE 6.23
Series motor characteristics by varying field current

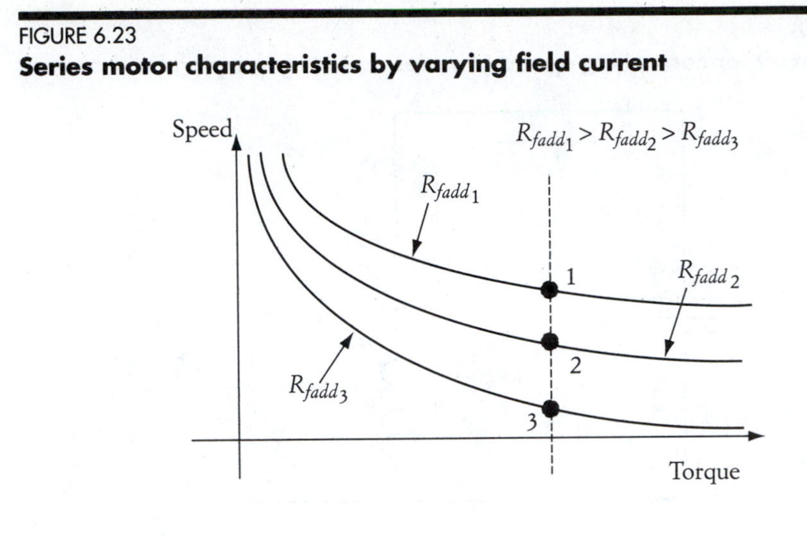

FIGURE 6.24
Solid-state circuit for controlling speed of a series motor by varying field current

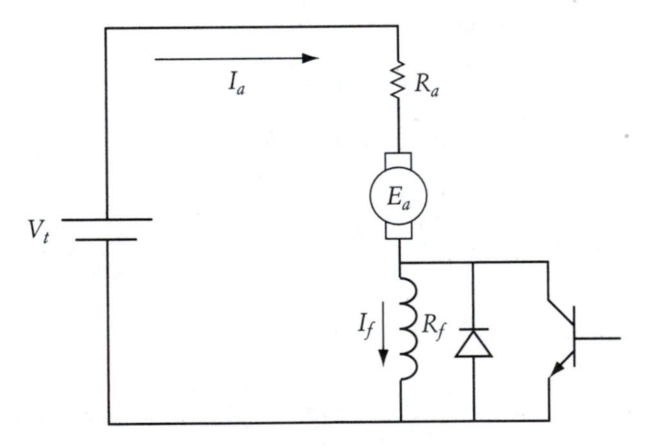

The diode is connected in the reverse direction of the armature current, and acts as a freewheeling diode that prevents the field circuit from being abruptly opened, which hence eliminates the surge in voltage across the transistor. When the transistor is closed, it carries the armature current. When it is open, the armature current flows through the field windings creating the flux of the motor. By controlling the switching of the transistor, the field current can be regulated: the higher the duty ratio of the transistor, the lower the field current.

EXAMPLE 6.7

A dc series motor has an armature resistance of 2 Ω and a series field resistance of 3 Ω. At a terminal voltage of 320 V and full load torque of 60 Nm, the motor speed is 600 rpm.

a. Calculate the field current.
b. Assuming that the load torque is constant, calculate the motor voltage required to reduce the speed to 400 rpm.
c. If at full voltage the field circuit is shunted by a 6 Ω resistance, calculate the motor speed.

SOLUTION

a. Assuming that the field flux is linearly proportional to the field current, then

$$V_t = I_a(R_a + R_f) + E_a = I_a(R_a + R_f) + K\phi\omega = I_a(R_a + R_f) + KCI_a\omega$$

$$V_t = I_a(R_a + R_f + KC\omega)$$

$$320 = I_a\left(2 + 3 + KC\left(2\pi\frac{600}{60}\right)\right)$$

and the torque equation is

$$T = KCI_a^2$$

$$60 = KCI_a^2$$

Solving these two equations results in two values for KC:

$$KC = 0.0256 \text{ or } 0.248$$

The armature currents for these two values of KC are

$$I_a = \sqrt{\frac{60}{0.0256}} = 48.44 \text{ A} \qquad \text{or} \qquad I_a = \sqrt{\frac{60}{0.248}} = 15.55 \text{ A}$$

Which of these answers is correct? Obviously, it is the one close to the expected value of the armature current for the given load. You can find the expected current by using the torque and speed to compute the output power. Then compute the expected current by dividing the power by the voltage. Of course, the actual current is larger than this value because of the presence of the armature and field resistance. However, we are using this calculation to get an idea of the neighborhood of the armature current. The expected current in this example is close to 12 A. Then the correct value of KC is the one that results in the next higher value. Hence, $KC = 0.248$.

b. For a constant load torque, the armature voltage is kept at 15.55 A. The armature voltage for 400 rpm is calculated by

$$V_t = 15.55\left(2 + 3 + 0.248\left(2\pi \frac{400}{60}\right)\right) = 239.28 \text{ V}$$

c. When a resistance is added in shunt with the series field, the armature current will increase if the load torque is unchanged. According to Equation (6.33), the new current can be computed from

$$T = KCA_R I_a^2$$

where

$$A_R = \frac{R_{fadd}}{R_f + R_{fadd}} = \frac{2}{3}$$

Hence,

$$I_a = \sqrt{\frac{60 \times 3}{0.248 \times 2}} = 19.05 \text{ A}$$

Now we can use Equation (6.35) to compute the new speed.

$$\omega = \frac{V_t}{KCA_R I_a} - \frac{R_a + A_R R_f}{KCA_R}$$

$$\omega = \frac{320}{0.248 \times \frac{2}{3} \times 19.05} - \frac{2 + \frac{2}{3} 3}{0.248 \times \frac{2}{3}} = 77.4 \text{ rad/sec}$$

$$n = 739.18 \text{ rpm}$$

CHAPTER 6 PROBLEMS

6.1 A 220 V, 1500 rpm, 11.6A (armature current) separately excited motor is controlled by a single-phase, full-wave SCR converter. The armature resistance of the motor is 2 Ω. The ac source voltage is 230 V (rms) 60 Hz. Enough filtering inductance is added to ensure continuous conduction for any torque greater than 25% of the rated value. You may ignore the rotational losses. What should be the value of the firing angle (triggering angle) to drive a mechanical load of rated torque at 1000 rpm?

6.2 A 600 V, dc shunt motor has armature and field resistances of 1.5 Ω and 600 Ω, respectively. When the motor runs unloaded, the line current is 3 A, and the speed is 1000 rpm.

a. Calculate motor speed when the load draws an armature current of 30 A.
b. If the load is constant-torque type, what is the motor speed when 3 Ω resistance is added to the armature circuit?
c. Calculate the motor speed if the field is reduced by 10%.

6.3 A dc shunt motor drives a centrifugal pump at a speed of 1000 rpm when the terminal voltage and line currents are 200 V and 50 A, respectively. The armature and field resistances are 0.1 Ω and 100 Ω, respectively.

a. Design a starting resistance for a maximum starting current of 120 A in the armature circuit.
b. What resistance should be added to the armature circuit to reduce the speed to 800 rpm?
c. If the terminal voltage is reduced by 25%, what is the speed of the motor?

6.4 A dc, separately excited motor has an armature resistance of 1 Ω. When a dc supply of 100 volts is applied to the motor, the armature current is 4 A and the motor speed is 300 rpm. A half-wave SCR converter is designed to control the motor speed. The supply voltage is 120 volts (rms), and the triggering angle of the converter is adjusted to 60°. When the motor is loaded with a constant-load torque of 10 Nm, the conduction angle is 175°. Assume that the field current is constant.

a. Calculate the average speed of the motor.
b. If the triggering angle is 35° and the average speed is 300 rpm, what is the conduction period of the motor?
c. A full-wave SCR converter is designed for the same motor. If the triggering angle is 50° and the conduction period is 120°, what is the speed of the motor?
d. A sufficiently large inductance is added in series with the armature circuit of the motor described in (c). Calculate the minimum triggering angle of the SCR in order to run the motor at a speed of 100 rpm with continuous armature current.

6.5 A dc shunt motor is driving a constant-torque load at the rated speed and rated terminal voltage. The motor has the following rated data:

Terminal voltage = 115 V

Speed = 312 rpm

Field constant ($K\phi$) = 3 V sec

If the terminal voltage of the motor is reduced by 10%, what is the motor speed? Assume that the field voltage is also reduced by the same ratio.

6.6 A dc, separately excited motor is connected to a fan-type load. The armature circuit of the motor is connected to a full-wave, ac/dc SCR converter. The input voltage to the converter is 200 V (rms). The triggering angle of the converter is adjusted for a motor speed of 500 rpm. The armature current in this case is 16 A. The armature resistance of the motor is 0.5 Ω, and the field constant ($K\phi$) is 2.5 V sec. Assume that the armature current is always continuous.

   a. Calculate the triggering angle to run the motor at 500 rpm.
   b. If the motor speed is to be reduced to 100 rpm, what is the triggering angle?

6.7 A 220 V, 1500 rpm, 11.6 A (armature current), separately excited motor is controlled by a single-phase, full-wave SCR converter. The armature resistance of the motor is 2 Ω. The ac source voltage is 230 V (rms) 60 Hz. Enough inductance is added to ensure continuous conduction for any torque greater than 25% of rated value. (Ignore the rotational losses.)

   a. Calculate the triggering angle for a speed of 1000 rpm at rated torque.
   b. Assuming that the torque is a fan-type, calculate the triggering angle for a motor speed of 900 rpm.
   c. Sketch the speed–torque characteristics showing the operating points of cases (a) and (b).

6.8 A dc, separately excited motor is used to drive a constant-torque load. The field circuit is excited by a full-wave, ac/dc SCR converter. The armature circuit of the motor is connected to a constant dc voltage source of 160 V. The inductance of the field circuit is large and the field current is continuous. The ac voltage (input to the converter) is 120 V (rms), and the field resistance is 100 Ω. The armature resistance is 2 Ω. When the triggering angle of the SCRs is adjusted to zero, the motor speed is 1200 rpm, and the armature current is 10 A.

   a. Calculate the average current and dc power of the field circuit when the triggering angle is equal to 20°.
   b. Calculate the rms voltage across the field windings for the condition given in (a). Explain how the rms voltage is dependent on the triggering angle.
   c. Calculate the no-load speed of the motor. Ignore the friction and windage losses.
   d. Calculate the triggering angle to operate the motor at a speed of 1400 rpm.
   e. Can you use the field converter to reduce the motor speed to 1000 rpm? How?

6.9 A dc, separately excited motor is driving a hoist. The motor has an armature resistance of 1.5 Ω, and a field constant ($K\phi$) equal to 3.5 V sec. The terminals of the armature circuit are connected directly across a 240 V, dc source. The field circuit is connected to an ac/dc, full-wave, solid-state converter. The control circuit of the converter is designed to maintain the speed of the motor constant. At full load, when the field current is at its rated value, the motor speed is 600 rpm. One day while the motor was operating at full-load conditions, a failure in the dc source caused the armature voltage to change suddenly. The control circuit of the field converter acted rapidly to maintain the motor speed at 600 rpm. Due to the action of the control circuit, the field flux was reduced by 25%. What was the percentage change in the armature voltage? Indicate whether the change was a voltage increase or decrease.

6.10 A dc, separately excited motor drives a constant-torque load.

a. If a resistance is added in series with the armature circuit, does the armature current change? Explain why.
b. How does the speed change in case (a)?

6.11 A separately excited, dc motor has the following name plate ratings:

Terminal voltage = 400 V

Speed = 1250 rpm

Developed torque at full load = 90 Nm

Full-load armature current = 30 A

A single-phase, full-wave, ac/dc converter is connected between a 480 V (rms) ac source and the armature terminals. A fan-type load is connected to the motor. When the triggering angle is adjusted to 40°, the motor speed is 1050 rpm. Calculate the triggering angle required to operate the motor at 1200 rpm. Assume that the armature current is always continuous.

6.12 A robot manipulator with a dc, separately excited motor on the driving end has ratings similar to those given in Problem 6.11. The motor is running at full load and is used to drill holes in solid material. If the field current is decreased to 80% of the rated value by using a solid-state converter in the field circuit, what is the percentage change in its speed?

6.13 A dc, separately excited motor has the following ratings:

Armature voltage = 200 V

Field constant ($K\phi$) = 3 V sec

Armature resistance = 1 Ω

The motor is used in a drilling operation. When the armature voltage of the loaded motor is 200 V, the motor speed is 500 rpm. Calculate the following:

a. Armature current when the motor speed is 500 rpm
b. Load torque when the motor speed is 500 rpm
c. Motor speed when the armature voltage is reduced by 10%
d. Armature current at the condition described in (c)

6.14 A dc, separately excited motor has the following data:

Rated field voltage = 300 V
Field constant ($K\phi$) = 3 V sec
Armature resistance = 2 Ω
Field resistance = 150 Ω

The motor is used to drive an assembly line consisting of a conveyor belt moving horizontally. The load on the belt varies depending on the amount of goods being moved. The load torque seen by the motor varies from a maximum of 24 Nm to a minimum of 3 Nm. At all loading conditions, the speed of the motor must be maintained constant and equal to 200 rpm.

a. To achieve the required operation, the armature voltage of the motor is adjusted by a single-phase, full-wave, ac/dc converter. The input to the converter is 240 V (rms). Calculate the range of the triggering angle (minimum and maximum) required to maintain the motor speed at the specified value at all loading conditions. Assume that the armature current is always continuous.
b. Another method to achieve the desired operation is to adjust the field voltage. In this case, the armature voltage can be kept constant at some value. Calculate the range of the field voltage (minimum and maximum). *Hints:* Assume that the maximum value of the field voltage is equal to its rated value (i.e., 300 V). Also assume that the field constant ($K\phi$) is linearly proportional to the field voltage.

6.15 A 1000 V, 50 hp, dc series motor is used as a hoist. The motor runs at a speed of 750 rpm at full load. The armature and field resistances are 0.5 Ω and 2.0 Ω, respectively.

a. Calculate the motor speed and line current when the load torque is reduced by 50%.
b. For the load condition in part (a), assume that a resistance of 5 Ω is added in series with the field windings. Calculate the motor speed and line current.

6.16 A dc series motor drives a fan-type load. At rated current, the motor speed is 600 rpm. If a resistance equal to 0.25 of the field resistance shunts the field winding, what is the approximate motor speed?

6.17 A dc series motor runs a constant-torque load. The terminal voltage of the motor is 200 V, the speed is 500 rpm, the armature current is 25 A, the armature resistance is 0.2 Ω, and the field resistance is 0.6 Ω. If the armature is shunted by a 10 Ω resistance, what is the speed of the motor?

6.18 A dc, separately excited motor has an armature resistance of 1 Ω and an inductive reactance of 2 Ω. The motor is powered by an ac/dc converter. The average torque of the motor is 12 Nm. The field constant $K\phi$ is 3 V sec, and the average terminal voltage is 100 V. Calculate the motor speed.

6.19 Explain the basic methods for speed control of a dc shunt motor. Use circuit diagrams and motor characteristics in your answer. Comment on the following issues:
   a. Suitability of the method for speed increase or speed reduction relative to the no-load speed
   b. Effect of the method on the overall efficiency of the system

6.20 A dc, separately excited motor is used to hoist a constant-weight load. The motor is driven by a full-wave, ac/dc converter. The voltage on the ac side is 110 V (rms). The field constant of the motor $K\phi$ is 3 V sec, and the armature resistance is 1 Ω. The armature current is continuous under loaded conditions. When the triggering angle is 30°, the motor speed is 60 rpm. Calculate the following:
   a. Load torque
   b. Load power
   c. Armature current when the triggering angle is adjusted to 45°
   d. Motor speed when the triggering angle is reduced to 30° and the field current is reduced by 10%

6.21 A dc, separately excited motor is driving a load torque composed of two components as given in the equation.

$$T = 25 + 0.1\,\omega^2$$

The armature circuit of the motor is connected to a full-wave, ac/dc SCR converter. The input voltage to the converter is 300 V (rms). The armature resistance of the motor is 0.5 Ω, and the field constant ($K\phi$) is 2.5 V sec. Assume that the armature current is always continuous. Calculate the range of the triggering angle to operate the motor at a speed range of 0 to 600 rpm.

6.22 A dc, separately excited motor is driven by a full-wave, ac/dc SCR converter. The voltage on the ac side is 240 V (rms). The armature resistance of the motor is 5 Ω. The armature current is continuous when a full-load torque of 400 Nm is applied. The motor speed under full-load torque is 100 rad/sec, the motor efficiency is 95%, and the rotational losses are 105 W. Ignore the field losses. Calculate the following:
   a. Output power of the motor
   b. rms voltage across the motor terminals
   c. Input power to the motor
   d. Losses in the armature resistance
   e. rms armature current

6.23 A dc series motor is driving a fan-type load. The armature and field resistances of the motor are 2 Ω and 3 Ω, respectively. When the terminal voltage of the motor is 200 V, the motor speed is 250 rpm and the armature current is 10 A. Assume that the motor operates at the linear region of the field–current characteristic. Calculate the terminal voltage needed to reduce the motor speed to 100 rpm. Also sketch the speed–torque characteristics and show all operating conditions.

6.24 A dc, separately excited motor drives a conveyor belt (constant torque). The terminal voltage of the motor is 120 V. When the conveyor belt is fully loaded, the armature current of the motor is 15 A and the speed of the motor is 180 rpm. The armature resistance of the motor is 2 Ω.

a. Calculate the steady-state speed of the motor if the field voltage is reversed.
b. Calculate the motor speed if after the field voltage is reversed, the field voltage is reduced by 10%.
c. Sketch the speed–torque characteristics and show all operating points.

Hint: Assume that the field *MMF* is linearly proportional to the field voltage.

6.25 A 300 V, dc, separately excited motor drives a conveyor belt (constant torque). The armature resistance of the motor is 1 Ω. When the conveyor belt is loaded at 150 Nm, the motor speed is 800 rpm. The field constant $K\phi$ of the motor is always greater than 2 V sec. The motor is controlled by a full-wave, ac/dc converter. At 150 Nm, the triggering angle is adjusted so that the speed of the motor is 400 rpm. The armature current of the motor is continuous. Calculate the triggering angle and the average terminal voltage of the motor.

6.26 A dc, separately excited motor drives a conveyor belt (constant torque). The terminal voltage of the motor is controlled by a full-wave, ac/dc converter. When the conveyor belt is fully loaded, the triggering angle is adjusted so that the average armature voltage is 150 V, the average armature current is 15 A, and the speed of the motor is 400 rpm. The armature resistance of the motor is 1 Ω. The armature current of the motor is continuous. Calculate the following:

a. Load torque
b. Triggering angle if the voltage on the ac side is 240 V (rms).
c. Steady-state speed of the motor if the triggering angle is changed to 60°.

7

Speed Control of Induction Motors

Until recently, induction machines were used in applications for which adjustable speed is not required. Compared to dc motors, changing the speed of an induction motor demands elaborate and complex schemes. Before the power electronics era, and the pulse width modulation in particular, the speed control of induction machines was limited to highly inefficient methods with a narrow range of speed.

With the advances in solid-state devices and variable-frequency power converters, different approaches to induction motor drive systems have emerged and developed that result in more sophisticated operations. Induction machines can now be used in high-performance applications where precise movement is required. Several models of robots, actuators, and guided manipulators are now equipped with induction machines that operate under precise control techniques.

The efficiency of the induction machine can also be improved when a proper solid-state converter is used. Depending on loading conditions, the efficiency at the rated voltage ranges from about 75% to 90%. High efficiency can be achieved when the sum of the copper losses of the windings and the core losses is minimized. This is achievable at a particular loading condition. However, when the load deviates, the efficiency is reduced. Since most industrial loads are varying, high-efficiency operation is not always achievable. However, with solid-state converters, efficiency can be improved at all loading conditions. For example, a 40% reduction in losses can be achieved at 25% load throughout the speed range. This reduction in losses translates into substantial annual savings that justify the use of solid-state converters even when speed control is not needed.

In this chapter, we will discuss several fundamental methods for speed control and efficiency enhancement. The reader should be familiar with the material in Chapters 2, 3, and 5 before reading this chapter.

7.1 BASIC PRINCIPLES OF SPEED CONTROL

The speed control of an induction motor requires more elaborate techniques than the speed control of dc machines. First, however, let us analyze the basic relationship for the speed–torque characteristics of an induction motor given in Equation (5.57).

$$T_d = \frac{P_d}{\omega} = \frac{V^2 R'_2}{s\omega_s\left[\left(R_1 + \frac{R'_2}{s}\right)^2 + X_{eq}^2\right]} \tag{7.1}$$

By examining this equation, one can conclude that the speed ω (or slip *s*) can be controlled if at least one of the following variables or parameters is altered:

1. armature or rotor resistance
2. armature or rotor inductance
3. magnitude of terminal voltage
4. frequency of terminal voltage

As discussed later in this chapter, each of the above techniques by itself is not sufficient. However, when more than one are combined, the control of the induction motor becomes more effective.

Although it is not evident by examining Equation (7.1), there are other useful and effective techniques for speed control. Among them are:

5. rotor voltage injection
6. slip energy recovery
7. voltage/frequency control

These seven techniques are described in this chapter, although in a different order so the information flow from one method to the other is logical.

7.2 CONTROLLING SPEED USING ROTOR RESISTANCE

Due to the complexity of equation (7.1), it is difficult to show the impact of rotor resistance on motor speed. However, if we are to study steady-state operation, we can use the small-slip approximation described in Equation (5.60). This is justifiable since at steady state, the speed of the motor is near the synchronous speed.

$$T_d \approx \frac{V^2 s}{\omega_s R'_2} \tag{7.2}$$

Keep in mind that *V* is a line-to-line quantity. If the voltage, frequency, and torque are kept constant, the increase in R'_2 results in an increase in the slip. Hence, the motor speed is reduced.

Figure 7.1 shows the motor characteristics for the case when a resistance R_{add} is added to the rotor circuit. As we explained in Chapter 5, the increase in rotor resistance does not change the synchronous speed or the magnitude of the maxi-

FIGURE 7.1
Effect of rotor resistance on motor speed

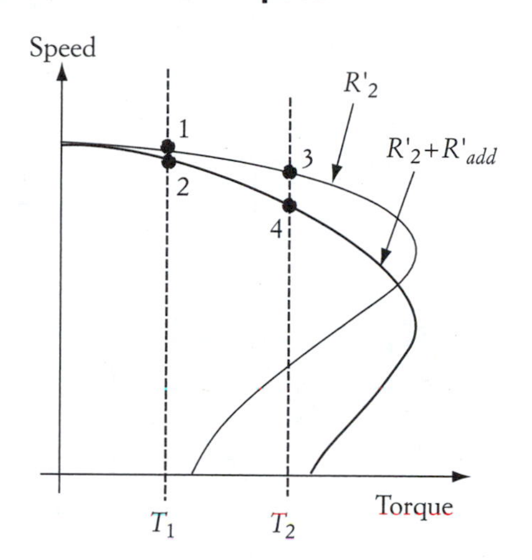

mum torque; it only skews the characteristics so the maximum torque occurs at a lower speed.

Adding a resistance to the rotor circuit does not cause the motor speed to change by any appreciable value at light loading conditions. The difference in speed between points 1 and 2 in Figure 7.1 is rather small. Although at heavy loading conditions, T_2, the motor speed may change by a wider range—from point 3 to point 4—the speed range is still narrow. Therefore, controlling the motor speed by changing the rotor (or stator) resistance is not considered a realistic option. In addition, this method increases the motor losses substantially as illustrated in the next example.

EXAMPLE 7.1

A three-phase, Y-connected, 30 hp (rated output), 480 V, six-pole, 60 Hz, slip ring induction motor has a stator resistance $R_1 = 0.5\ \Omega$ and a rotor resistance referred to stator $R'_2 = 0.5\ \Omega$. The rotational losses are 500 W and the core losses are 600 W. Assume that the change in the rotational losses due to the change in speed is minor. The motor load is a constant-torque type. At full-load torque, calculate the speed of the motor. Calculate the added resistance to the rotor circuit needed to reduce the speed by 20%. Calculate the motor efficiency without and with the added resistance. If the cost of energy is \$0.05/kWh, compute the annual cost of operating the motor continuously with the added resistance. Assume that the motor operates 100 hours a week.

SOLUTION

Consider the power flow of the induction machine given in Chapter 5, which is also shown in Figure 7.2. First, let us compute the rated developed power.

FIGURE 7.2
Power flow of induction motors

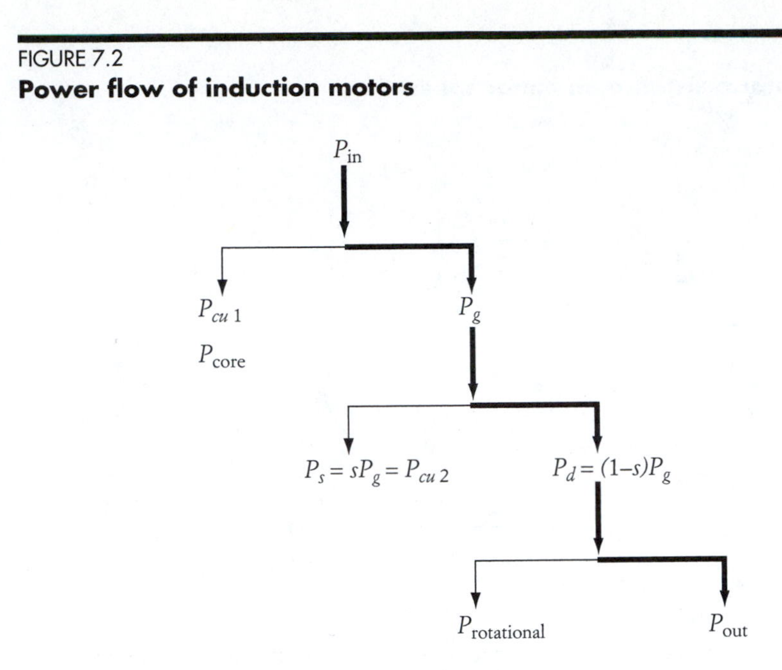

$$\text{Developed power} = \text{output power} + \text{rotational losses}$$

$$P_d = P_{out} + P_{rotational} = 30(746) + (500) = 22.88 \text{ kW}$$

To compute the motor speed, we can use Equation (7.1) or the small-slip approximation of Equation (7.2).

$$T_d \approx \frac{V^2 s}{\omega_s R'_2}$$

$$P_d = T_d \omega \approx \frac{V^2 s}{\omega_s R'_2}\,\omega = \frac{V^2 s(1-s)}{R'_2}$$

$$22{,}880 = \frac{480^2 s(1-s)}{0.4}$$

This equation has two solutions; one of them yields a large slip and should be ignored since the motor speed at full load is always near synchronous.

$$s = 0.0417$$

$$n = n_s(1-s) = \left(120\frac{60}{6}\right)(1 - 0.0417) = 1150 \text{ rpm}$$

To compute the winding losses, we first need to calculate the motor current. The easiest way is to use the developed power equation.

$$P_d = 3(I'_2)^2 \frac{R'_2}{s}(1 - s)$$

$$22{,}880 = 3(I'_2)^2 \frac{0.4}{0.0417}(1 - 0.0417)$$

$$I'_2 = 28.8 \text{ A}$$

Losses of motor winding = losses of rotor resistance + losses of stator resistance

These losses can be approximated by using the equivalent circuit in Figure 5.26(b).

$$P_{\text{winding}} = P_{cu\,1} + P_{cu\,2} \approx 3(I'_2)^2(R_1 + R'_2) = 2488 \text{ W}$$

Input power = developed power + winding losses + core losses

$$P_{\text{in}} = 22{,}880 + 2488 + 600 = 25.97 \text{ kW}$$

The motor efficiency without added resistance η is

$$\eta = \frac{P_{\text{out}}}{P_{\text{in}}} = \frac{30 \times 746}{25{,}970} = 86\%$$

Now let us calculate the speed and efficiency after a rotor resistance is added. The resistance is added to reduce the rotor speed by 20%, so the new rotor speed is

$$n_{\text{new}} = 0.8(1150) = 920 \text{ rpm}$$

$$s_{\text{new}} = \frac{n_s - n}{n_s} = 0.233$$

$$\frac{P_{d\,\text{new}}}{P_d} = \frac{T_{d\,\text{new}}\,\omega_{d\,\text{new}}}{T_d\,\omega}$$

since the load torque is constant.

$$\frac{P_{d\,\text{new}}}{P_d} = \frac{T_{d\,\text{new}}\,\omega_{d\,\text{new}}}{T_d\,\omega} = \frac{\omega_{d\,\text{new}}}{\omega}$$

$$P_{d\,\text{new}} = 18.3 \text{ kW}$$

Now we need to calculate the size of the added resistance. Let us use the small-slip approximation given in Equation (7.2) to create a ratio between the original and new torque.

$$\frac{T_d}{T_{d\,\text{new}}} = 1 = \frac{s(R'_2 + R_{add})}{s_{\text{new}}\,R'_2}$$

$$R_{add} = 2.3\ \Omega$$

The new current can be calculated by using the developed power equation

$$P_{d\,\text{new}} = 3(I'_{2\,\text{new}})^2 \frac{(R'_2 + R_{add})}{s_{\text{new}}} (1 - s_{\text{new}})$$

$$I'_{2\,\text{new}} = 25.73 \text{ A}$$

The new winding losses are

$$P_{\text{winding new}} \approx 3(I'_{2\,\text{new}})^2 (R_1 + R'_2 + R_{add}) = 6.552 \text{ kW}$$

$$\text{Input power} = 18{,}300 + 6552 + 600 = 25.452 \text{ kW}$$

$$\text{Output power} = \text{new developed power} - \text{rotational losses} = 18{,}300 - 500 = 17.8 \text{ kW}$$

The new efficiency with added resistance η_{new} is

$$\eta_{\text{new}} = \frac{P_{\text{out}}}{P_{\text{in}}} = \frac{17.8}{25.452} = 70\%$$

Note that the new efficiency is much lower when a resistance is added to the rotor circuit. In this example, a 20% reduction in motor speed resulted in about a 20% reduction in efficiency. If the motor operates with this added resistance for an extended time, the energy loss will be costly. This is the main drawback of this type of control.

Now let us calculate the cost of energy when an added resistance controls the speed of the motor. The losses due to the added resistance are

$$P_{add} \approx 3(I'_{2\,\text{new}})^2 R_{add} = 4.568 \text{ kW}$$

The total hours of operation t in one year are

$$t = 100(52) = 5200 \text{ hr}$$

The cost of energy C is

$$C = P_{add}\, t(0.05) \approx \$1{,}188$$

Keep in mind that this is the cost of one machine. If the plant has more machines operating by this method, the cost of speed control accumulates to an unacceptable level.

7.3 ROTOR VOLTAGE INJECTION

If the induction machine is a slip-ring type, we can access its rotor circuit, which would allow us to insert a resistance or connect the rotor to an external source. The latter is a more efficient method of speed control.

Consider the equivalent circuit of the induction motor in Figure 7.3(a). Ignore the magnetizing branch and concentrate on the windings' impedances. In this circuit, instead of shorting the terminals of the slip rings, we are connecting the slip rings to an external voltage source V_i. The magnitude of this voltage source is adjustable, and its frequency f_r always tracks the frequency of the rotor-induced voltage ($s\,E_2$). Keep in mind that E_2 is the standstill voltage across the rotor windings.

FIGURE 7.3
Equivalent circuit of induction motor with voltage injection

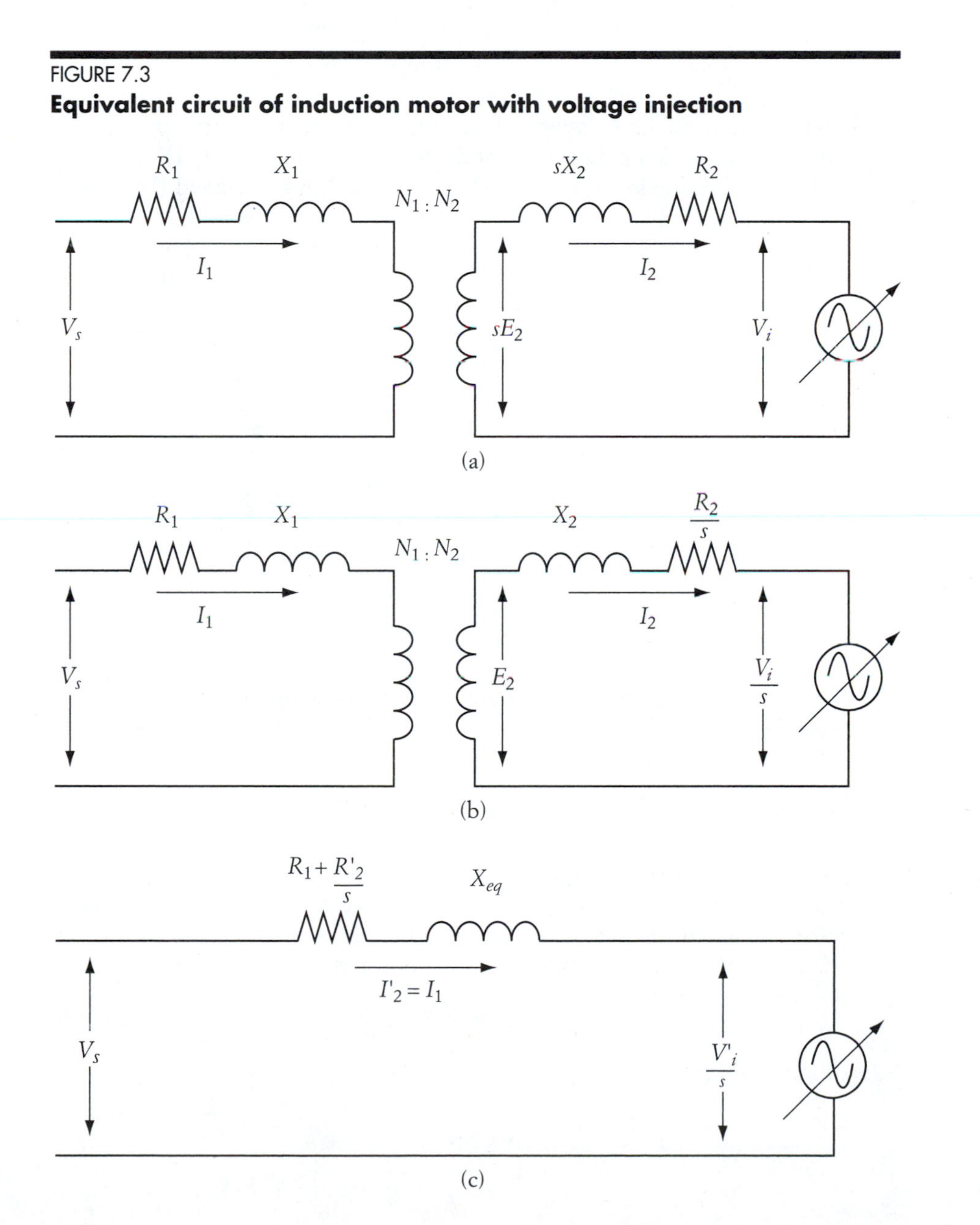

The frequency of E_2 is equal to the frequency of the supply voltage V_s. The frequency f_r is dependent on the motor speed n and the stator frequency f_s.

$$f_r = sf_s = \frac{n_s - n}{n_s} f_s = \left(1 - \frac{n}{\frac{120 f_s}{p}}\right) f_s \tag{7.3}$$

$$f_r = f_s - \frac{np}{120}$$

As shown in Figure 5.24, we can modify the equivalent circuit of the rotor to that in Figure 7.3(b) by dividing the voltage and impedances by the slip s. The new representation of the induction motor keeps the rotor current unchanged. The model is merely a more convenient representation for induction machines.

The equivalent circuit in Figure 7.3(c) is a modification of that in Figure 7.3(b). All variables and parameters are referred to the stator side using the windings ratio N_1/N_2. Assume that the motor is Y-connected, and $\overline{V}_s$ and $\overline{V}_i$ are phase-to-neutral quantities. The rotor current referred to stator I'_2 can be computed by

$$I'_2 = \frac{\overline{V}_s - \frac{\overline{V}'_i}{s}}{\left(R_1 + \frac{R'_2}{s}\right) + jX_{eq}} \tag{7.4}$$

As shown in Equation (5.52), the equation of the developed torque T_d is

$$T_d = \frac{P_g}{\omega_s} \tag{7.5}$$

where P_g is the airgap power (three-phase power). Using Figure 7.3(b), the airgap power can be computed as

$$P_g = 3(I'_2)^2 \frac{R'_2}{s} + 3 \frac{V'_i}{s} I'_2 \cos(\theta_r)$$

where θ_r is the angle between V_i and I'_2. P_g is divided into three components: one is converted to mechanical power driving the load, the second is losses in the rotor resistance, and the third is power delivered to the source connected across the slip rings. The sum of the last two components is known as slip power or sP_g.

$$sP_g = 3(I'_2)^2 R'_2 + 3\, V'_i I'_2 \cos(\theta_r) \tag{7.6}$$

Substituting P_g of Equation (7.6) into (7.5) yields

$$T_d = \frac{3}{s\omega_s} [(I'_2)^2 R'_2 + V'_i I'_2 \cos(\theta_r)] \tag{7.7}$$

Equations (7.4) and (7.7) are the foundations for speed control of the induction motor using voltage injection. Normally, the load torque T (or the range of the load torque) is known. Thus, you need to compute the voltage that must be injected in the rotor circuit to drive the machine at a certain speed. With this scenario, you will have three unknown variables in Equations (7.4) and (7.7): the rotor current, the magnitude of the injected voltage, and the angle of the injected voltage with respect to the source voltage. We can simplify the calculations if the injected voltage is in phase with the source voltage. In this case, the magnitude of the rotor current can be expressed by

$$I'_2 = \frac{V_s - \dfrac{V'_i}{s}}{\sqrt{\left(R_1 + \dfrac{R'_2}{s}\right)^2 + X_{eq}^2}} \tag{7.8}$$

which leaves us with two nonlinear equations and two unknowns. The magnitude of the injected voltage can then be computed. Substituting Equation (7.8) into Equation (7.7) leads to the equation of the speed–torque characteristics

$$T_d = \frac{3}{s\omega_s}\left[\frac{\left(V_s - \dfrac{V'_i}{s}\right)^2 R'_2}{\left(R_1 + \dfrac{R'_2}{s}\right)^2 + X_{eq}^2} + \frac{\left(V_s - \dfrac{V'_i}{s}\right) V'_i \cos(\theta_r)}{\sqrt{\left(R_1 + \dfrac{R'_2}{s}\right)^2 + X_{eq}^2}}\right] \tag{7.9}$$

Note that the no-load speed of the motor with injected voltage is no longer equal to the synchronous speed.

Figure 7.4 is a graphical representation of Equation (7.9). The figure shows a family of characteristics for various values of injected voltage $v_{i_1} < v_{i_2} < v_{i_3}$. As seen

FIGURE 7.4
Speed–torque characteristics of induction motor with rotor-injected voltage

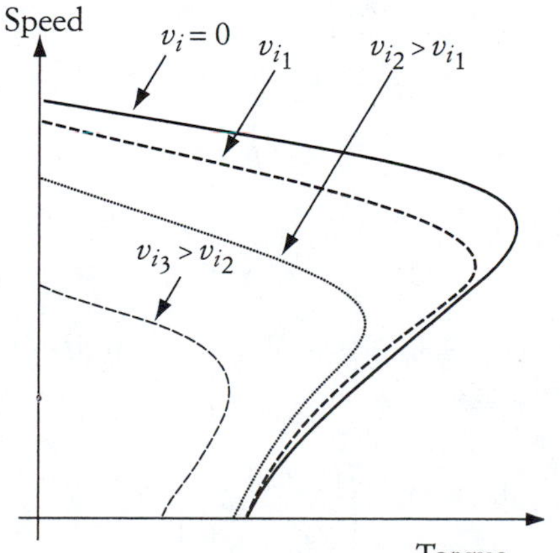

in the figure, the injected voltage tends to reduce the maximum torque of the motor and the speed at maximum torque. The figure also shows that a wide range of speed control can be achieved by this method.

EXAMPLE 7.2

A three-phase, 480 V, four-pole, 60 Hz induction motor is driving a constant-torque load of 60 Nm. The parameters of the motor are

$$R_1 = 0.4\ \Omega \qquad R_2 = 0.1\ \Omega \qquad X_{eq} = 4\ \Omega \qquad \frac{N_1}{N_2} = 2$$

Calculate the magnitude of the injected voltage that would reduce the motor speed to 1000 rpm. Also calculate the power received by the source of the injected voltage.

SOLUTION

The synchronous speed of the motor is

$$n_s = 120\frac{f_s}{p} = 1800 \text{ rpm}$$

The slip at the new speed is

$$s = \frac{n_s - n}{n_s} = \frac{1800 - 1000}{1800} = 0.44$$

We are assuming that the injected voltage is in phase with the supply voltage. Hence, the motor current is

$$\bar{I}'_2 = \frac{\dfrac{480}{\sqrt{3}} - \dfrac{V'_i}{0.44}}{\sqrt{\left(0.4 + \dfrac{0.1(2^2)}{0.44}\right)^2 + (4^2)}} \angle -\theta_r \text{ A} \tag{7.10}$$

$$= \frac{122 - V'_i}{1.85} \angle -\theta_r \text{ A}$$

Since the injected voltage V_i is in phase with V_s, then the phase angle θ_r is the angle of the windings impedance.

$$\theta_r = \tan^{-1}\left(\frac{X_{eq}}{R_1 + \dfrac{R'_2}{s}}\right) = 71.9°$$

Using Equation (7.7) for the torque yields

$$60 = \frac{3}{0.44\left(2\pi \frac{1800}{60}\right)} [(I'_2)^2 0.4 + V'_i I'_2 \cos(71.9)] \qquad (7.11)$$

$$0.4(I'_2)^2 + 0.31\, V'_i I'_2 = 1658.7$$

By substituting the magnitude of I'_2 of Equation (7.10) into (7.11), we get

$$0.4\left(\frac{122 - V'_i}{1.85}\right)^2 + 0.31\, V'_i \frac{122 - V'_i}{1.85} = 1658.7$$

$$0.05(V'_i)^2 + 8\, V'_i - 80.74 = 0$$

which yields

$$V'_i = 9.5 \text{ V} \quad \text{or} \quad V'_i = -169.5 \text{ V}$$

The negative value of the voltage is not applicable for normal motor operation. Hence, the injected voltage is

$$V_i = V'_i \frac{N_2}{N_1} = 9.5 \frac{1}{2} = 4.75 \text{ V}$$

The line-to-line injected voltage is

$$\sqrt{3}\, 4.75 = 8.22 \text{ V}$$

Note that the speed of the induction motor is changed by 44% when only 8.22 V is injected in the rotor circuit. This change in speed is a very desirable feature. The drawback, however, is that this method requires the tracking of the frequency of the rotor circuit and the phase angle of the supply voltage. Such requirements make this technique more involved.

The power received by the injected source P_r is

$$P_r = 3V'_i I'_2 \cos(\theta_r)$$

$$P_r = 3(9.5)\frac{122 - 9.5}{1.85} \cos(71.9) = 538.4 \text{ W}$$

EXAMPLE 7.3

For the motor in Example 7.2, compute the starting current and the starting torque when no voltage is injected in the rotor circuit. Repeat the solution for the injected voltage computed in Example 7.2.

SOLUTION

Without injected voltage. The starting current can be obtained using Equation (7.8) when $V'_i = 0$ and $s = 1$.

$$I'_{2\,st} = \frac{V_s}{\sqrt{(R_1 + R'_2)^2 + X_{eq}^2}}$$

$$= \frac{\dfrac{480}{\sqrt{3}}}{\sqrt{[0.4 + 0.1(2^2)]^2 + (4^2)}} = 68 \text{ A}$$

Similarly, the starting torque can be obtained from Equation (7.7).

$$T_{st} = \frac{3(I'_{2\,st})^2 R'_2}{\omega_s}$$

$$T_{st} = \frac{3}{\left(2\pi \dfrac{1800}{60}\right)} (68)^2\, 0.4 = 29.4 \text{ Nm}$$

With injected voltage

$$I'_{2\,st} = \frac{V_s - V'_i}{\sqrt{(R_1 + R'_2)^2 + X_{eq}^2}}$$

$$I'_{2\,st} = \frac{\dfrac{480}{\sqrt{3}} - 9.5}{\sqrt{[0.4 + 0.1(2^2)]^2 + (4^2)}} = 65.6 \text{ A}$$

$$T_{st} = \frac{3}{\omega_s}[(I'_{2\,st})^2 R'_2 + V'_i I'_{2\,st} \cos(\theta_r)]$$

At starting, $\theta_r = \tan^{-1}\left(\dfrac{X_{eq}}{R_1 + R'_2}\right) = 78.7°$

$$T_{st} = \frac{3}{\left(2\pi \dfrac{1800}{60}\right)}[(65.6)^2\, 0.4 + 65.6(9.5)\cos(78.7)] = 29.34 \text{ Nm}$$

Note that for a small voltage injection, the speed of the motor changes substantially. However, the starting current and torque do not significantly change.

7.4 SLIP ENERGY RECOVERY

Consider the power flow of the induction motor given in Chapter 5 and also shown in Figure 7.2. Most of the input electric power P_{in} is converted to mechanical power P_{out} to support the load. However, part of P_{in} is lost in the resistive element of the stator circuit $P_{cu\,1}$. The rest is power transmitted to the rotor via the airgap P_g. At high speeds, most of P_g is converted to mechanical developed power $P_d = (1-s)P_g$. The rest is known as the slip power $P_s = sP_g$. Slip power is an electrical power dissipated in the rotor resistance in the form of rotor copper losses $P_{cu\,2}$.

Slip power P_s can be substantial at low speeds. Example 7.1 shows that when a resistance in the rotor circuit is used to reduce the motor speed, the efficiency of the motor is substantially reduced. The speed reduction is due to the extra power dissipated in the rotor circuit, which results in less mechanical power for the load. We can still use this principle to reduce the motor speed, but instead of dissipating the extra power in the rotor resistance, we send it back to the source.

A slip energy recovery (SER) circuit, also known as the static Scherbius circuit, is shown in Figure 7.5. The rotor in this circuit is linked back to the stator windings via two converters: three-phase ac/dc and three-phase dc/ac. The ac/dc converter is often a simple full-wave diode rectifier circuit. The output of the converter is connected to a dc/ac converter through an inductive element. The output of the dc/ac converter is a three-phase system connected to the same source feeding the induction motor. This SER circuit divides the slip power into two parts: the losses in the rotor resistance and the power returned back to the source.

To simplify the analysis, let us assume that the copper losses in the rotor resistance are small compared to the energy that returns back to the source. This assumption is depicted in Figure 7.5. The entire slip power is flowing through the converters back to the source.

Let us describe the flow of power at a given moment. First we assume that the voltage e_{ab} of the rotor terminals is in phase with v_{ab} of the supply. Moreover, we assume that e_{ab} is positive and large enough to allow diodes D_1 and D_6 to conduct. Point 1 will have the potential of phase a, and point 2 the potential of phase b. Hence, $v_2 = e_{ab}$. The airgap power and the current can flow from D_1 to the inductor L. If we trigger S_3 and S_4, point 3 will have the potential of phase b and point 2 the potential

FIGURE 7.5
Slip energy recovery circuit

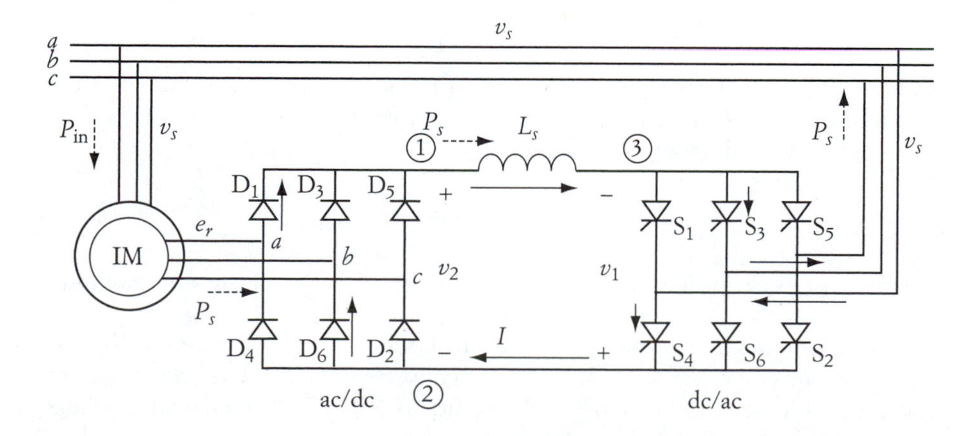

of phase a. In this case, $v_1 + v_2 + v_{L_s} = 0$, where v_{L_s} is the voltage drop across the inductance L_s. Hence, the current loop will be closed through D_1, S_3, S_4, and D_6, and the current (and power) will flow back to the source.

7.4.1 CONTROLLING SPEED BY THE SLIP ENERGY RECOVERY METHOD

Since the three-phase supply is a constant voltage source, v_s is sinusoidal with fixed peak value. Hence, a change in the triggering angle of the SCRs changes the average value of v_1. Because the balance between v_1 and v_2 is always maintained in the loop of the dc link, v_2 must also change. When v_2 changes, the rotor voltage e_r on the input side of the diode will change accordingly. e_r is a function of the motor speed.

$$e_r = sE_2$$

where E_2 is the rotor voltage at standstill, which is constant. If we ignore the voltage drop of the stator windings, E_2 is constant when the stator voltage is maintained constant.

$$E_2 \approx \frac{N_2}{N_1} V_s \tag{7.12}$$

where N_1 and N_2 are the number of turns of the stator and rotor windings, respectively. If the rms voltage v_s is maintained constant, any change in e_r changes the motor slip in a linear relation.

$$s = \frac{e_r}{E_2} \tag{7.13}$$

The new speed of the motor is dependent on the value of e_r, which is a function of the triggering angle of the SCRs as shown in the following analyses. We will assume that the motor is Y-connected and that all voltages on the ac sides are phase quantities.

The output voltage of the ac/dc converter is v_2. Its average value at standstill can be computed as shown in Chapter 3, Equation (3.55), assuming full conduction.

$$V_{2\,\text{ave}} = \frac{3\sqrt{3}\, E_{2\,\text{max}}}{\pi} \tag{7.14}$$

where $E_{2\,\text{max}}$ is the peak value of E_2. This is the rotor voltage at standstill. We can rewrite Equation (7.14) to a more general form for a rotating machine using e_r instead of E_2. Accordingly, the average voltage of V_2 is

$$V_{2\,\text{ave}} = \frac{3\sqrt{3}\, E_{r\,\text{max}}}{\pi} = \frac{3\sqrt{3}}{\pi} s\, E_{2\,\text{max}} \tag{7.15}$$

where $E_{r\,\text{max}}$ is the peak value of the rotor voltage when the machine is rotating at slip s.

Now let us analyze the second half of the circuit shown in Figure 7.5. Keep in mind that the equations used for the dc/ac converter discussed in Section 3.9 are not applicable here. In Section 3.9, the ac side is a load with rectangular voltage waveforms. However, in the circuit of Figure 7.5, the ac side is a source voltage with sinusoidal waveforms. The more accurate relationship between the input and output of the dc/ac converter can be represented by Equation (3.55).

It is justifiable to assume that the voltage drop across the inductor in the dc link is small. Note that the orientations of the diodes and SCRs make $v_2 = -v_1$. Hence, the average voltage at the input of the dc/ac converter $V_{2\ ave} = -V_{1\ ave}$.

The average voltage across the SCR circuit can be computed using the ac/dc conversion formula similar to Equation (3.54), but modified for the ac voltage v_s and including the triggering angle as shown in Section 3.7.

$$V_{1\ ave} = \frac{3\sqrt{3}\ V_{s\ max}}{\pi} \cos(\alpha) \tag{7.16}$$

where α is the triggering angle of the dc/ac converter, measured from the zero crossing of the line-to-line voltage. In the dc link, we can write the loop voltage as

$$V_{1\ ave} + V_{2\ ave} = 0$$

Substituting the values of $V_{1\ ave}$ and $V_{2\ ave}$ of Equations (7.15) and (7.16), we get

$$\begin{aligned} s &= -\frac{V_{s\ max}}{E_{2\ max}} \cos(\alpha) = -\frac{V_s}{E_2} \cos(\alpha) \\ &= -\frac{V_s}{\frac{N_2}{N_1} V_s} \cos(\alpha) = -\frac{N_1}{N_2} \cos(\alpha) \end{aligned} \tag{7.17}$$

Since

$$n = n_s(1 - s)$$

then

$$n = n_s\left[1 + \frac{N_1}{N_2} \cos(\alpha)\right] \tag{7.18}$$

Equation (7.18) shows that adjusting the triggering angle of the dc/ac converter can control the speed of the machine. The range of α is from $\pi/2$ to π. In this range, the induction machine operates as a motor where the speed is less than the synchronous speed. The motor cannot operate at $\alpha < 90°$, because the motor speed exceeds the synchronous speed. This is not possible for the circuit in Figure 7.5, since it requires the current to flow in the opposite direction in the dc link—the diodes do not allow it.

7.4.2 TORQUE–CURRENT RELATIONSHIP

According to Equation (7.18), the speed of the motor is independent of the load torque, because we assumed that the motor is ideal and the impedance of the stator windings is negligible. Nevertheless, the speed computed this way is very close to the actual speed under loading conditions.

In this section, we shall study the effect of the load torque on the motor current. It is intuitive that an increase in load torque is expected to result in an increase in current everywhere in the circuit. This can be shown by first substituting Equation (7.12) into (7.15).

$$V_{2\,ave} = \frac{3\sqrt{3}}{\pi} s \frac{N_2}{N_1} V_{s\,max} \tag{7.19}$$

where $V_{s\,max}$ is the peak value of the phase voltage of the source. To use the rms line-to-line voltage V_s, we can rewrite Equation (7.19) as

$$V_{2\,ave} = \frac{3\sqrt{2}}{\pi} s \frac{N_2}{N_1} V_s = KsV_s \tag{7.20}$$

where
$$K = \frac{3\sqrt{2}}{\pi} \frac{N_2}{N_1}$$

The slip power sP_g is also equal to the power at the input of the dc/ac converter. If we ignore the ripples in the current i and voltage v_2, we can write the slip power as

$$sP_g = IV_{2\,ave} = KsV_s I \tag{7.21}$$

Hence,

$$P_g = KV_s I \tag{7.22}$$

The developed mechanical power P_d is

$$P_d = T_d \omega = (1 - s)p_g \tag{7.23}$$

where ω is the shaft speed of the motor defined by

$$\omega = \omega_s(1 - s)$$

Substituting Equation (7.22) into (7.23) yields

$$I = \frac{T_d \omega_s}{KV_s} \tag{7.24}$$

Note that the current here is dependent on the load torque, but independent of the motor speed, because of the assumption we made that the induction machine is an ideal motor.

7.4.3 EFFICIENCY

To understand the improvements made to motor efficiency, let us develop the power flow chart for the induction motor with slip-energy recovery. This flow chart

FIGURE 7.6
Power flow chart of induction motor under energy recovery

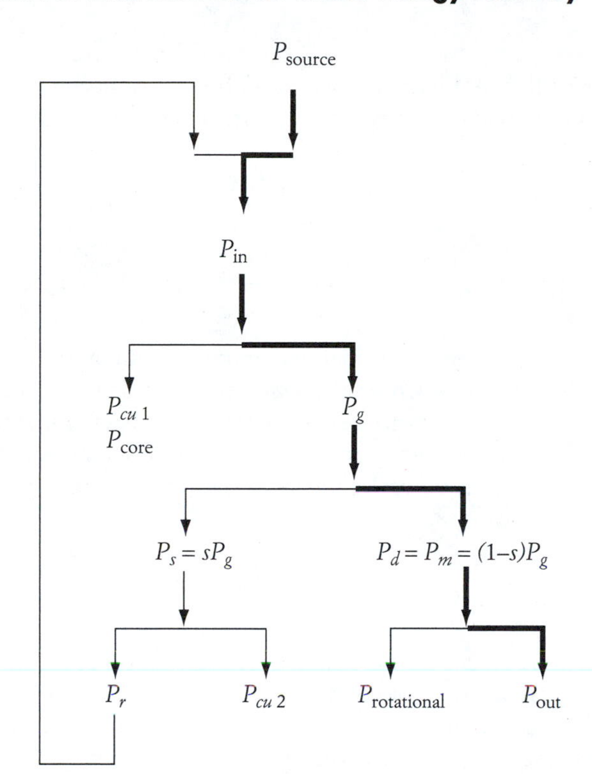

is given in Figure 7.6. Compare it to the chart in Figure 7.2. With SER, the slip power is divided into the copper losses of the rotor and the recovery power P_r. The recovery power is injected back to the source. Thus, the actual power delivered by the source is the input power required by the motor minus P_r.

The output power is the shaft torque multiplied by the shaft speed. If we add the rotational losses to the output power, we get the developed mechanical power P_d. The rotor copper losses $P_{cu\,2}$ in the rotor resistance can be computed as

$$P_{cu\,2} = 3\, I_2^2\, R_2 \tag{7.25}$$

Similarly, the stator losses of the motor can be expressed by

$$P_{cu\,1} = 3\, I_1^2\, R_1 \tag{7.26}$$

where I_2 and I_1 are the rotor and stator currents, respectively, including all harmonics. R_2 and R_1 are the rotor and stator resistances, respectively. Remember that all harmonic components of the current produce losses in the resistance.

Let us assume that the inductor in the dc link is large enough to allow the current to be continuous without ripples. Although v_1 contains harmonics, the recovery power, P_r, is due to current and voltage of the same frequency as explained in Section 3.1.2.3. Hence, the recovery power is due to the dc components.

$$P_r = V_{2\,\mathrm{ave}} I = V_{1\,\mathrm{ave}} I \tag{7.27}$$

Using the power flow of Figure 7.6, we can define two efficiencies: the efficiency of the motor without SER system η_m and the efficiency of the system with SER η_{SER}.

$$\eta_m = \frac{P_{out}}{P_{in}}$$

$$\eta_{SER} = \frac{P_{out}}{P_{source}} \tag{7.28}$$

Let us assume that we have two identical motors running at the same speed and driving equal load torques. Under this assumption, it is fair to assume that the rotational losses of both motors are equal. Now assume that the first motor has its rotor windings shorted, and the second motor has an SER system. Since both motors provide identical output power at identical speeds, the airgap powers of the machines are equal, because

$$P_g = \frac{P_d}{1 - s}$$

Also, the slip powers (sP_g) of both machines are equal. For the first machine, the slip power is equal to the rotor copper losses. However, for the second machine it is equal to the rotor copper losses plus the recovery power returned back to the source. Hence, the rotor copper losses of the machine with SER are less than those for the machine without SER.

The increase in system efficiency due to SER is illustrated by the following example.

EXAMPLE 7.4

A three-phase, six-pole, Y-connected, 480 V induction motor is driving a 300 Nm constant-torque load. The motor has the following parameters:

$$\frac{N_1}{N_2} = 1 \qquad P_{rotational} = 1\ \mathrm{kW}$$

The motor is driven by a slip energy recovery system. The triggering angle of the dc/ac converter is adjusted to 120°. Ignore all core and copper losses and calculate the following:

a. Motor speed
b. Current in the dc link

c. Rotor rms current

d. Stator rms current

e. Power returned back to the source

f. Assume that the motor is not driven by an SER system. If a resistance is added in the rotor circuit to reduce the speed to that calculated in part a., compute the additional losses.

SOLUTION

a. The speed of the motor can be computed using equation (7.18)

$$n_s = 120\frac{60}{6} = 1200 \text{ rpm}$$

$$n = n_s\left(1 + \frac{N_1}{N_2}\cos\alpha\right) = 1200\ [1 + \cos(120)] = 600 \text{ rpm}$$

$$s = \frac{1200 - 600}{1200} = 0.5$$

b. To compute the current in the dc link, you need to compute the output power P_{out} and developed power P_d

$$P_{out} = T_{out}\,\omega = 300\left(2\pi\frac{n}{60}\right) = 18.85 \text{ kW}$$

where T_{out} is the shaft torque.

$$P_d = P_{out} + P_{rotational} = 18.85 + 1.0 = 19.85 \text{ kW}$$

The developed power P_d is also equal to

$$P_d = T_d\,\omega$$

where T_d is the developed torque and ω is the speed in rad/sec. The current in the dc link is given in equation (7.24)

$$I = \frac{T_d\,\omega_s}{KV_s} = \frac{P_g}{KV_s} = \frac{\dfrac{P_d}{1-s}}{KV_s} = \frac{\dfrac{19850}{1-0.5}}{\dfrac{3\sqrt{2}}{\pi}(480)} = 61.2 \text{ A}$$

c. An approximate value for the rms current of the rotor I_2 can be computed by assuming that the current in the dc link is free from harmonics. Taking into account that the dc/ac converter is a three-phase, full-wave type, each diode is conducting for 120° only.

$$I_2 = \sqrt{\frac{1}{\pi}\int_0^{2\pi/3} I^2 \, d\omega t} = \sqrt{\frac{2}{3}}\, I \approx 0.82(61.2) = 50 \text{ A}$$

d. If we ignore the core losses, the rms current of the stator I_1 can be computed as

$$I_1 \approx I_2 \frac{N_2}{N_1} = 50 \text{ A}$$

e. The power returned back to the source P_r is

$$P_r = P_d = 19.85 \text{ kW}$$

Note that $P_r = P_d$ since the slip is equal to 0.5 and the copper losses of the rotor are ignored.

f. Let us first compute the inserted resistance R'_{add}. This can be simply done by using the small-slip approximation of Equation (7.2). But first, we need to compute the developed torque.

$$T_d = \frac{P_d}{\omega} = \frac{19850}{2\pi \frac{600}{60}} = 316 \text{ Nm}$$

Now, let us use Equation (7.2)

$$T_d = \frac{V_s^2 \, s}{\omega_s \, R'_{add}}$$

$$316 = \frac{480^2 \times 0.5}{2\pi \frac{1200}{60} R'_{add}}$$

$$R'_{add} = 2.9 \ \Omega$$

The rotor current is

$$I'_2 = \sqrt{\frac{\left(\frac{s\,P_g}{3}\right)}{R'_{add}}} = \sqrt{\frac{\left(\frac{s}{1-s}\right)\frac{P_d}{3}}{R'_{add}}}$$

$$I'_2 = \sqrt{\frac{\frac{19850}{3}}{2.9}} = 47.77 \text{ A}$$

Note that this current is almost the same as that computed for the system with SER. The difference is due to the assumption that the current in the dc link is harmonic-free.

The additional losses are

$$P_{add} = 3(I'_2)^2 R_{add} = 3(47.77)^2\, 2.9 = 19.85 \text{ kW}$$

These losses are very high, and equal to the developed power since the slip is 0.5. Note that the losses here are equal to the power returned back to the source P_r when the SER technique is used. It is now obvious that with SER technique, the drive system is highly efficient.

7.5 CONTROLLING SPEED USING INDUCTANCE

Adding inductance to the motor windings is an unrealistic option for the following reasons:

1. The physical size of the inductance required to make a sizable change in speed is likely to be larger than the motor itself.
2. Unlike variable resistance, variable inductance requires expensive and elaborate design.
3. The insertion of inductance reduces the starting torque.
4. The insertion of inductance consumes reactive power that further lowers the already low power factor of induction motors.

7.6 CONTROLLING SPEED BY ADJUSTING THE STATOR VOLTAGE

Several techniques can be used to change the stator voltage of the motor. Among them are fixed pulse modulation (FPM), explained in Chapter 3, or the phase control shown in Figure 7.7. The circuit configuration of phase control is a full-wave, three-phase SCR converter similar to the ones discussed in Chapter 3. In this circuit, the induction motor is connected to a three-phase supply voltage via back-to-back SCR pairs. For each phase, one SCR conducts the current in one direction (from the source to the motor), and the other SCR conducts the current in the second half of the cycle (from motor to source). If the triggering of these SCRs is controlled, the voltage across the stator terminals can change from zero to almost full voltage.

As seen in Equation (7.1), the torque of the motor is proportional to the square of its stator voltage. For the same slip and frequency, a small change in

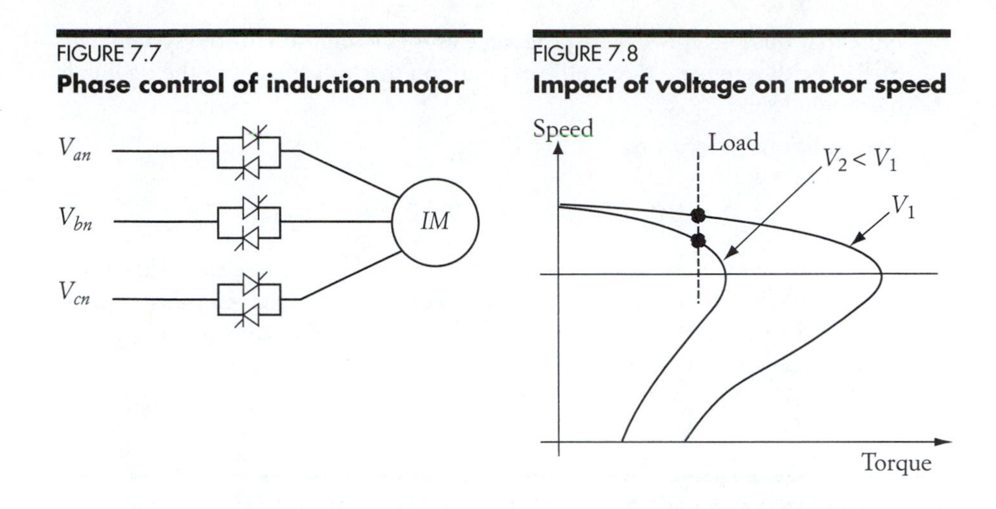

FIGURE 7.7
Phase control of induction motor

FIGURE 7.8
Impact of voltage on motor speed

motor voltage results in a relatively large change in torque. A 10% reduction in voltage causes a 19% reduction in developed torque as well as the starting and maximum torques.

The characteristics of the motor under voltage control are shown in Figure 7.8. The figure is based on Equation (7.1) and shows two curves for two different values of the stator voltage. Note that the slip at the maximum torque remains unchanged since it is not a function of voltage. For normal operation in the linear region, the figure shows that the motor speed can be modestly changed when the voltage is altered. However, a wide range of speed control cannot be accomplished by this technique. Nevertheless, it is an excellent method for reducing starting current and increasing efficiency during light loading conditions. The starting current is reduced since it is directly proportional to the stator voltage. The losses are reduced, particularly core losses, which are proportional to the square of the voltage.

Keep in mind that the terminal voltage cannot exceed the rated value to prevent the damage of the windings' insulation. Thus, this technique is only suitable for speed reduction below the rated speed.

EXAMPLE 7.5

For the motor given in Example 7.1, assume that the load torque is constant and equal to 120 Nm. Ignore the rotational losses and calculate the motor speed at full voltage. Repeat the computation if the voltage is reduced by 20%.

SOLUTION

First calculate the motor speed at the given load torque. The small-slip approximation can be used.

$$T_d \approx \frac{V^2 s}{\omega_s R'_2}$$

$$120 = \frac{480^2 s}{2\pi \frac{1200}{60} 0.5}$$

$$s = 0.0327$$

Thus, the speed at full voltage is

$$n = n_s(1 - s) = 1161 \text{ rpm}$$

Now calculate the new motor speed when the voltage is reduced.

$$\frac{T_{d\,\text{new}}}{T_d} = 1 = \frac{V^2_{\text{new}}}{V^2} \frac{s_{\text{new}}}{s}$$

$$s_{\text{new}} = 0.0511$$

The new speed of the motor is

$$n_{\text{new}} = 1200(1 - 0.0511) = 1139 \text{ rpm}$$

In this example, note that a 20% reduction in voltage yields about 5% reduction in speed.

7.7 CONTROLLING SPEED BY ADJUSTING THE SUPPLY FREQUENCY

In steady state, the induction motor operates in the small-slip region, where the speed of the motor is always close to the synchronous speed of the rotating flux.

$$n_s = 120 \frac{f}{p} \tag{7.29}$$

where f is the frequency of the stator voltage and p is the number of poles. Since the synchronous speed is directly proportional to the frequency of the stator voltage, any change in frequency results in an equivalent change in motor speed.

If you plot the motor characteristics of Equations (7.1) for different values of supply frequencies, you can obtain a family of characteristics similar to the ones shown in Figure 7.9. The effect of frequency on motor current is given by Equation (7.30), which is the same as Equation (5.55). The current characteristics of the motor are shown in Figure 7.10.

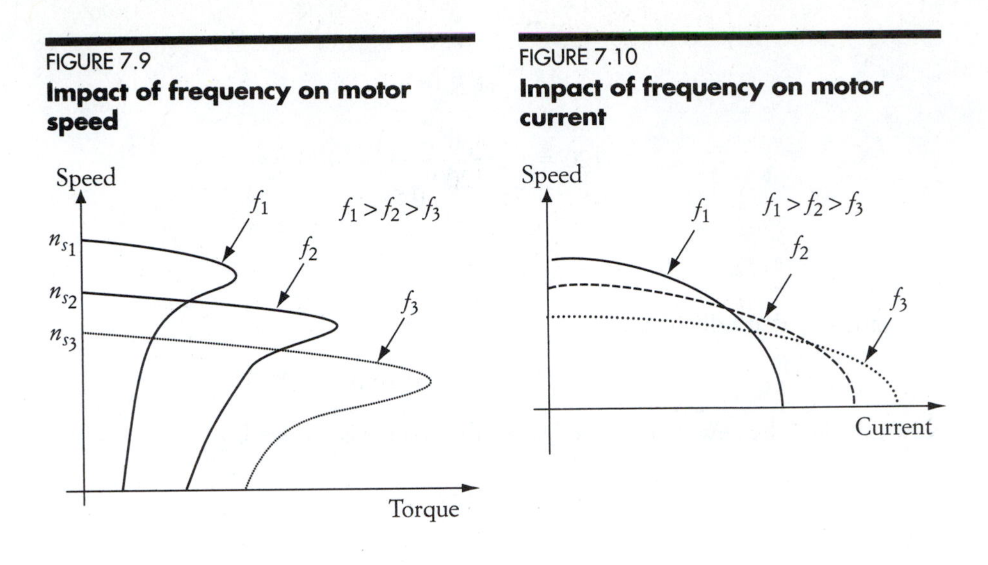

FIGURE 7.9
Impact of frequency on motor speed

FIGURE 7.10
Impact of frequency on motor current

$$I'_2 = \frac{V}{\sqrt{\left(R_1 + \frac{R'_2}{s}\right)^2 + X_{eq}^2}} \tag{7.30}$$

Frequency manipulation appears to be an effective method for speed control that requires a simple dc/ac converter with variable switching intervals similar to the ones shown in Figure 3.29. However, there are severe limitations to this method: very low frequencies may cause motor damage due to excessive currents, and large frequencies may stall the motor. These limitations are discussed in the following sections.

7.7.1 EFFECT OF EXCESSIVELY HIGH FREQUENCY

As shown in Figures 7.9 and 7.10, the increase in supply frequency results in the following five changes:

1. *An increase in the no-load speed (synchronous speed).* This increase is due to the increase in frequency as given by Equation (7.29).
2. *A decrease in the maximum torque.* The maximum torque is described in Chapter 5 and its equation is given by (5.62). The maximum torque equation is rewritten for a single phase in (7.31). The voltage V is a phase-to-neutral value. It shows that the maximum torque is inversely proportional to both the synchronous speed ω_s and the equivalent reactance of the windings X_{eq}. Each of these quantities increases by increasing the frequency. Hence, the maximum torque decreases when the frequency of the supply voltage increases.

$$T_{\max} = \frac{V^2}{2\omega_s \left[R_1 + \sqrt{R_1^2 + X_{eq}^2}\right]} \tag{7.31}$$

3. *A decrease in the starting torque.* The starting torque of the induction motor, T_{st}, is computed by Equation (5.59) for a three-phase system or Equation (7.32) for a single phase system.

$$T_{st} \approx \frac{V^2 R'_2}{\omega_s X_{eq}^2} \tag{7.32}$$

   As seen in this equation, the starting torque decreases when the synchronous speed and equivalent reactance increase. This is due to the increase in frequency.

4. *An increase in speed at the maximum torque.* Due to the increase in frequency, the slip at maximum torque s_{max} decreases when the equivalent reactance increases, as shown in Equation (5.61). Also, the speed at maximum torque n_{max} given by Equation (7.33) increases.

$$n_{max} = n_s(1 - s_{max}) \tag{7.33}$$

5. *A decrease in the starting current.* This can be seen from Equation (5.63).

$$I'_{2\,st} = \frac{V}{\sqrt{(R_1 + R'_2)^2 + X_{eq}^2}} \tag{7.34}$$

   When the frequency increases, the equivalent reactance increases and the starting current decreases. At high frequencies, the resistance of the motor windings may also increase due to the skin effect.

Now let us examine the case when the increase in frequency is excessive. Figure 7.11 shows two characteristics for two different values of stator frequency. Assume that the load torque is constant, and the motor operates initially at frequency f_1. The steady-state operation is represented by point 1. Now assume that the frequency of the stator voltage increases to a higher value, f_2, where the new maximum torque of the motor is less than the load torque. In this case, no steady-state operating point can be achieved, and the motor eventually stalls or even operates under braking. One solution to this problem is to increase the supply voltage when the frequency increases. This will be discussed in a later section.

FIGURE 7.11
Effect of excessively high frequency

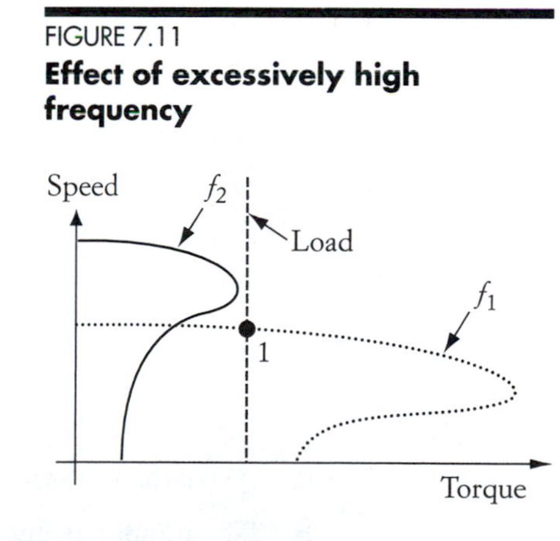

EXAMPLE 7.6

A 480 V, two-pole, 60 Hz, Y-connected induction motor has an inductive reactance of 4 Ω and a stator resistance of 0.2 Ω. The rotor resistance referred to the stator is 0.3 Ω. The motor is driving a constant-torque load of 60 Nm at a speed of 3500 rpm. Assume that this torque includes the rotational components.

a. Compute the maximum frequency of the supply voltage that would not result in stalling the motor.

b. Calculate the motor current at 60 Hz, and at the maximum frequency.

c. Calculate the power delivered to the load at 60 Hz, and at the maximum frequency.

SOLUTION

a. Let us look at the maximum torque equation

$$T_{max} = \frac{V^2}{2\omega_s(R_1 + \sqrt{R_1^2 + X_{eq}^2})}$$

If V is the line-to-line value, the maximum torque is due to the three phases. Based on the values of R_1 and X_{eq}, $R_1^2 << X_{eq}^2$, the maximum torque equation can be approximated by

$$T_{max} \approx \frac{V^2}{2\omega_s X_{eq}} \tag{7.35}$$

The upper limit of the supply frequency is determined by the maximum torque; the developed torque, at most, must be equal to the maximum torque—so let us modify the maximum torque of Equation (7.35), and make it more general for any frequency.

$$T_{max} = \frac{V^2}{2\left(\frac{f}{60}\omega_s\right)\left(\frac{f}{60}X_{eq}\right)}$$

Now set this equation equal to the developed torque and solve for the frequency.

$$T_{max} = T_d = 60 = \frac{480^2}{2\left(\frac{f}{60}2\pi\frac{3600}{60}\right)\left(\frac{f}{60}4\right)}$$

$$f \approx 67.7 \text{ Hz}$$

Thus, the increase in frequency should not exceed 67.7 Hz.

b. The motor current can be calculated using the current represented by Equation (7.30).

$$I'_2 = \frac{V}{\sqrt{\left(R_1 + \frac{R'_2}{s}\right)^2 + X_{eq}^2}}$$

At 60 Hz, the slip is

$$s = \frac{n_s - n}{n_s} = \frac{3600 - 3500}{3600} = 0.0277$$

and the current is

$$I'_2 = \frac{\frac{480}{\sqrt{3}}}{\sqrt{\left(0.2 + \frac{0.3}{0.0277}\right)^2 + 4^2}} = 23.62 \text{ A}$$

Now let us compute the current at the new frequency. Since the load torque is equal to the maximum torque, the slip can be computed using Equation (5.61).

$$s_{\max} = \frac{R'_2}{\sqrt{R_1^2 + X_{eq}^2}}$$

Note that the equivalent reactance of the motor increases due to the increase in frequency. The new equivalent reactance is

$$X_{eq} = \frac{67.7}{60} 4 = 4.51$$

$$s_{\max} = \frac{R'_2}{\sqrt{R_1^2 + X_{eq}^2}} = \frac{0.3}{\sqrt{0.2^2 + 4.51^2}} = 0.0665$$

The new motor speed is

$$n = 120 \frac{67.7}{2} (1 - 0.0665) = 3792 \text{ rpm}$$

The current at the new frequency is

$$I'_2 = \frac{\frac{480}{\sqrt{3}}}{\sqrt{\left(0.2 + \frac{0.3}{0.0665}\right)^2 + 4.51^2}} = 42.5 \text{ A}$$

The current at 67.7 Hz is higher than the current at 60 Hz—about an 80% increase. The load torque is constant and the speed increases, so the load power increases. In this current equation, you can also attribute the increase in current to the increase in slip.

c. The developed power at 60 Hz is

$$P_d = T_d \omega = 60 \times 2\pi \frac{3500}{60} \approx 22 \text{ kW}$$

At 67.7 Hz,

$$P_d = T_d \omega = 60 \times 2\pi \frac{3792}{60} \approx 23.83 \text{ kW}$$

This is an increase of 8.3%; it is imperative that the drive system be able to handle this increase in current and power demands.

7.7.2 EFFECT OF EXCESSIVELY LOW FREQUENCY

Reducing the supply frequency reduces the speed of the motor. However, frequency reduction may result in an increase in motor current as given in Equation (7.30) and Figure 7.10. At very low frequencies, the equivalent reactance of the motor X_{eq} is very low. Since X_{eq} is the limiting parameter for motor current at starting, its large reduction could lead to an excessive current beyond the ratings of the machine. The following example explains this effect.

EXAMPLE 7.7

For the motor described in Example 7.6, compute the motor speed and starting current if the frequency is decreased to 50 Hz.

SOLUTION

Let us first compute the new synchronous speed.

$$n_s = 120 \frac{50}{2} = 3000 \text{ rpm}$$

$$\omega_s = 2\pi \frac{3000}{60} = 314.16 \text{ rad/sec}$$

To compute the slip, let us use the small-slip approximation of Equation (7.2).

$$T_d \approx \frac{V^2 s}{\omega_s R'_2}$$

$$60 \approx \frac{480^2 s}{314.16(0.3)}$$

$$s = 0.0245$$

The new speed at 50 Hz is

$$n = 3000(1 - 0.0245) = 2926.5 \text{ rpm}$$

which is about a 19% reduction in speed. The starting current is given in Equation (7.34).

$$I'_{2\,st} = \frac{V}{\sqrt{(R_1 + R'_2)^2 + X_{eq}^2}}$$

The starting current at 60 Hz is

$$I'_{2\,st} = \frac{\frac{480}{\sqrt{3}}}{\sqrt{(0.5)^2 + 4^2}} = 68.75 \text{ A}$$

At 50 Hz,

$$I'_{2\,st} = \frac{\frac{480}{\sqrt{3}}}{\sqrt{(0.5)^2 + \left(\frac{50}{60}4\right)^2}} = 82.21 \text{ A}$$

which is about a 20% increase in the starting current. Note that the frequency reduction leads to a reduction in speed, and an increase in starting currents.

7.8 VOLTAGE/FREQUENCY CONTROL

As seen in Figure 7.11, the increase in the supply frequency increases the motor speed and also reduces the maximum torque of the motor. Furthermore, in Figure 7.8, we see that the increase in voltage results in an increase in the maximum torque of the motor. If we combine these two features, we can achieve a control design by which the speed increases and the torque is kept the same. This is known as voltage/frequency control, v/f.

Figure 7.12 shows three characteristics; one is used as our reference at voltage V_1 and frequency f_1. For the arbitrary fan-type load in the figure, the reference operating point is 1. If we increase the frequency of the supply to f_2 while keeping the voltage V_1 unchanged, the speed of the motor increases and the maximum torque decreases. The load torque in this case is higher than the maximum torque provided by the motor. Thus, no steady-state operating point can be achieved and the motor eventually stalls.

Now let us keep the supply frequency to the new value at f_2, but increase the magnitude of the voltage to V_2. The motor characteristics in this case stretch and the maximum torque increases. The motor operates at point 2, and a new steady-state point is achieved.

FIGURE 7.12
Impact of change in frequency and voltage

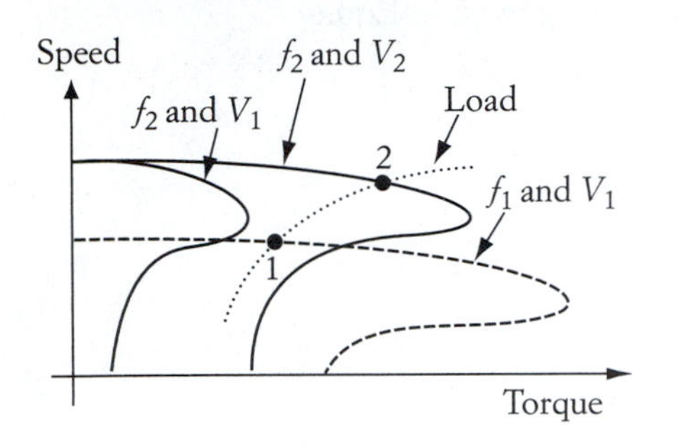

The change in voltage and frequency is a powerful method for speed control. Note that both frequency and voltage can change simultaneously by the pulse-width modulation technique described in Section 3.9.4. This type of control is common for induction motors. There are several variations where the *v/f* ratio is also adjusted to provide a special operating performance. The most common method, though, is the fixed *v/f* ratio.

An induction motor operating under constant *v/f* control exhibits the characteristics shown in Figures 7.13 and 7.14. Note that the changes in the maximum torque are not substantial. This can be explained by examining Equation (7.36), which is the same as equation (5.62). Keep in mind that V in Equation (7.36) is a line-to-line quantity. If we assume that the equivalent inductive reactance X_{eq}, at frequencies near the rated value, is much larger than the armature resistance, then Equation (7.36) can be approximated by Equation (7.37).

$$T_{\max} = \frac{V^2}{2\omega_s[R_1 + \sqrt{R_1^2 + X_{eq}^2}\,]} \tag{7.36}$$

$$T_{\max} \approx \frac{V^2}{2\omega_s X_{eq}} = \frac{V^2}{2\left(\frac{4\pi}{p}f\right)(2\pi f L_{eq})} \sim \left(\frac{V}{f}\right)^2 \tag{7.37}$$

where p is the number of poles and L_{eq} is the equivalent inductance of the motor windings. It is clear that when the *v/f* ratio is constant, the maximum torque is unchanged. Keep in mind that this approximation may not be valid at very low frequencies when X_{eq} is not much larger than R_1.

Another feature of the constant *v/f* control is that the magnitude of the starting current is almost constant. Examine Equation (7.34), and assume that $X_{eq}^2 >> (R_1 + R'_2)^2$. This assumption is valid for frequencies close to the rated frequency. The starting current can then be approximated by Equation (7.38).

$$I'_{2\,\mathrm{st}} = \frac{V}{X_{eq}} = \frac{1}{2\pi L_{eq}}\frac{V}{f} \tag{7.38}$$

Equation (7.38) shows that when *v/f* is kept constant, the starting current remains unchanged; this is another advantage of *v/f* control.

When the change in voltage is used to control the induction machine, whether it is a voltage control or *v/f* control, one must be careful not to increase the voltage magnitude beyond the ratings of the motor. Excessive voltage can cause instant damage to the insulation of the motor's windings, leading to shorts and internal faults. Usually the voltage should be kept below 110% of the rated value.

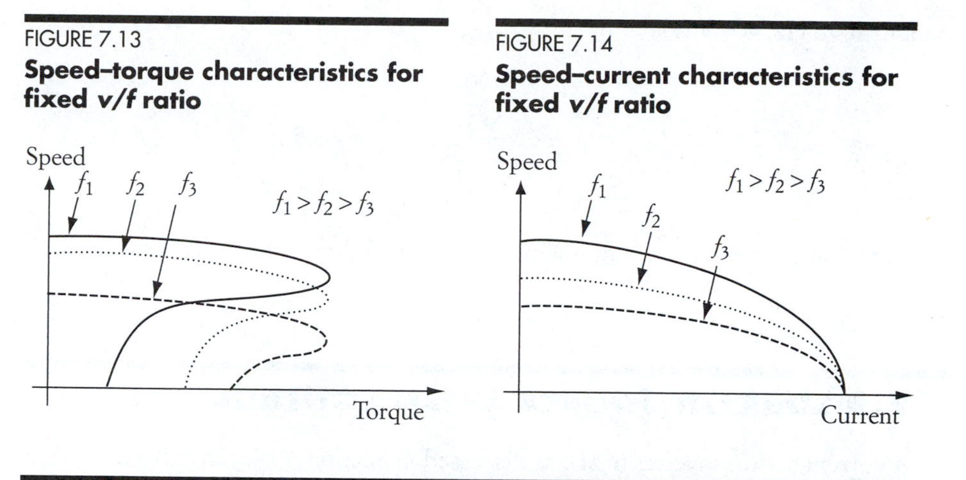

FIGURE 7.13
Speed–torque characteristics for fixed *v/f* ratio

FIGURE 7.14
Speed–current characteristics for fixed *v/f* ratio

EXAMPLE 7.8

Repeat Example 7.7 for a constant v/f control.

SOLUTION

The voltage frequency ratio is 480/60 = 8. When the frequency of the supply is reduced to 50 Hz, the supply voltage should also be reduced to

$$V_{new} = 50(8) = 400 \text{ V}$$

Since it depends on the supply frequency alone, the synchronous speed at 50 Hz is the same as that calculated in Example 7.7. However, the slip is dependent on the supply voltage as given in the small-slip approximation of Equation (7.2).

$$T_d \approx \frac{V^2 s}{\omega_s R'_2}$$

$$60 \approx \frac{400^2 s}{314.16(0.3)}$$

$$s = 0.0353$$

This slip is higher than the one calculated in Example 7.7. The new speed at 50 Hz is

$$n = 3000(1 - 0.0353) = 2894 \text{ rpm}$$

$$\omega = 303 \text{ rad/sec}$$

The starting current at 60 Hz, 480 V is

$$I'_{2\,st} = \frac{\frac{480}{\sqrt{3}}}{\sqrt{(0.5)^2 + 4^2}} = 68.75 \text{ A}$$

and at 50 Hz, 400 V it is

$$I'_{2\text{ st}} = \frac{\frac{400}{\sqrt{3}}}{\sqrt{(0.5)^2 + \left(\frac{50}{60}4\right)^2}} = 68.5 \text{ A}$$

Note that the starting current is almost unchanged due to the v/f control.

7.9 CURRENT SOURCE SPEED CONTROL

A current source inverter (CSI), as discussed in Section 3.12, can drive the induction machine. The model of Figure 7.15(a) represents the induction motor with CSI. In this circuit, we include the magnetizing branch in the stator circuit because it has a greater impact in the CSI drive. Figure 7.15(b) shows another equivalent circuit, discussed in Chapter 5. In this circuit, the effect of speed is transferred from the rotor-induced voltage sE_2 and the rotor reactance to the rotor resistance. The rotor current in both circuits is the same. Figure 7.15(c) shows the equivalent circuit referred to the stator by using the turns ratio.

When the induction motor is driven by a CSI, the stator current I_1 is equal to the current source I_s, which is constant.

$$\bar{I}_s = \bar{I}_1 = \bar{I}'_2 + \bar{I}_m \tag{7.39}$$

Since I_1 is constant, changes in I'_2 due to changes in mechanical load will result in changes in the magnetizing current I_m. Changes in I_m should be analyzed carefully. The magnetizing circuit of the induction machine has an iron alloy core that saturates with flux at large magnetizing currents. When the core saturates, the flux does not noticeably increase when the magnetizing current increases. This phenomenon changes the magnitude of the magnetizing inductance X_m in a nonlinear manner. A good approximation for X_m can be obtained using Figure 7.16. The figure shows the relationship between the flux density and flux intensity. The voltage across the magnetizing reactance E_1 is directly related to the flux density, while the current I_m represents the flux intensity. The slope of the solid curve is the magnetizing reactance X_m. The value of X_m changes according to the relationship between the flux intensity and density. With voltage source drive VSD, when the terminal voltage of the motor is maintained constant, the magnetizing current is also constant, and accordingly X_m is constant. However, with CSI, the magnetizing current is changing and so is the magnetizing reactance. Analyzing the machine under this condition is very involved and may require numerical computations. An approximate method can be used where the magnetizing curve is divided into regions, as shown in Figure 7.16. Based on two operating regions, two values for the magnetizing reactance are assumed: the unsaturated reactance X_{m_1} and the saturated reactance X_{m_2}. Of

FIGURE 7.15
Equivalent circuits of an induction motor with current source inverter

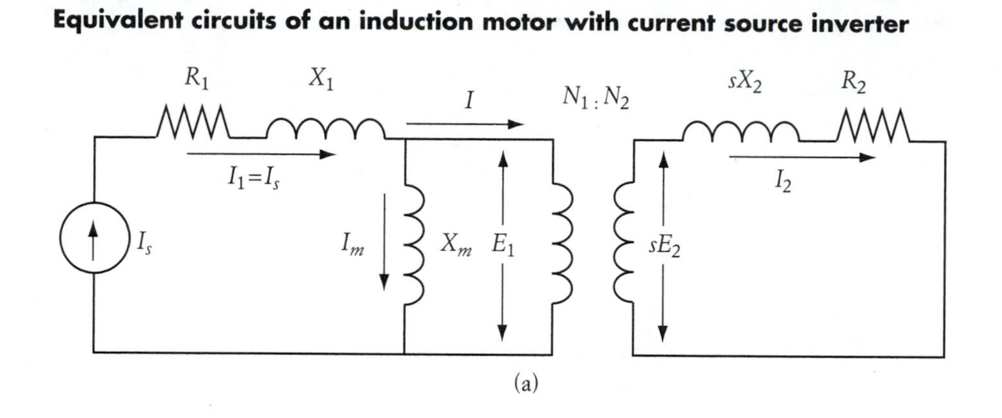

(a)

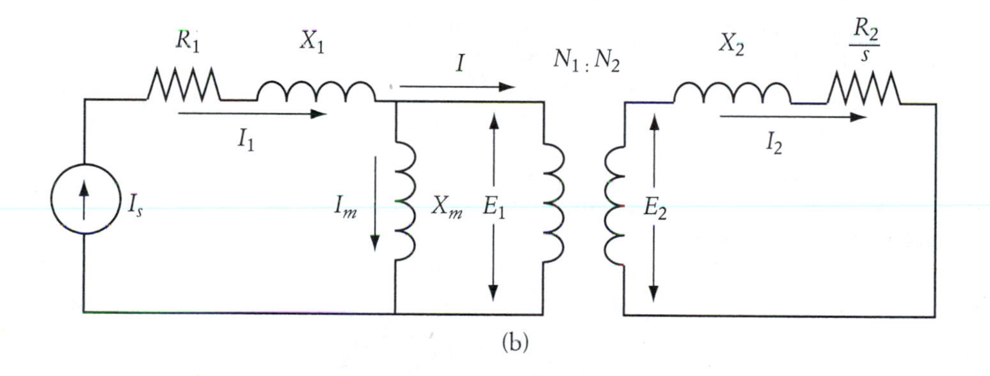

(b)

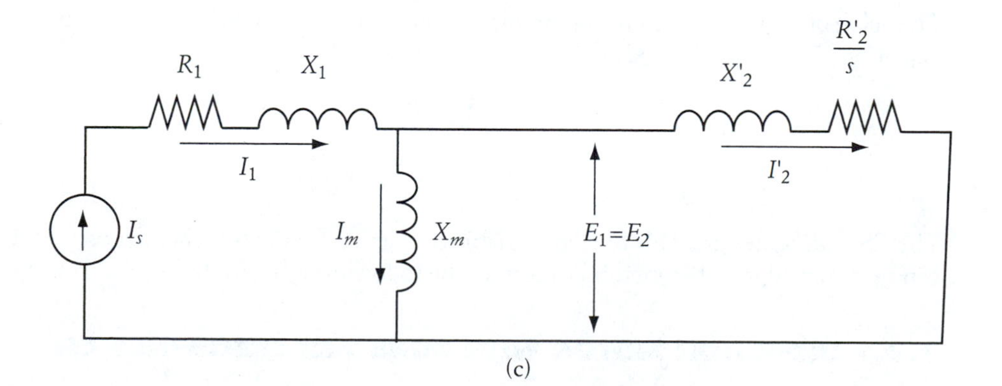

(c)

FIGURE 7.16
Effect of saturation on the magnetizing inductive reactance

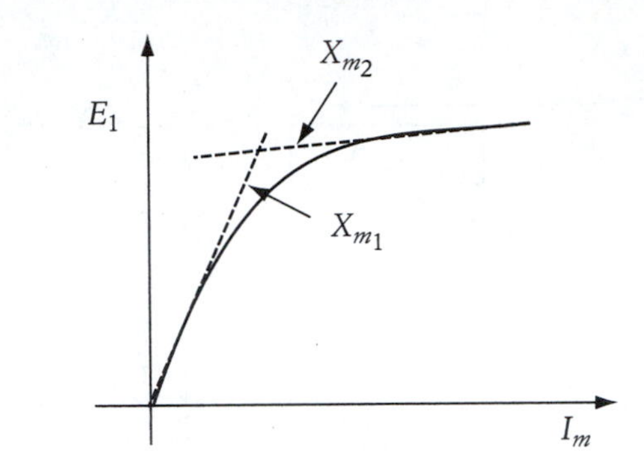

course, one could divide the curve into more regions to improve the accuracy, but in most cases, two divisions are quite adequate.

To obtain the speed–torque characteristics of the motor operating under CSI, let us write the basic equations for airgap power P_g:

$$P_g = T_d \omega_s = 3\, I'_2 \frac{R'_2}{s} \tag{7.40}$$

The rotor current I'_2 can be computed by using the impedance ratio of the parallel branches shown in Figure 7.15(c).

$$I'_2 = I_s \frac{X_m}{\sqrt{\left(\frac{R'_2}{s}\right)^2 + (X'_2 + X_m)^2}} \tag{7.41}$$

The value of X_m depends on the operating region in Figure 7.16. Substituting I'_2 of Equation (7.41) into (7.40) yields

$$T_d = \frac{3\, I_s^2\, X_m^2\, R'_2}{s\omega_s\left[\left(\frac{R'_2}{s}\right)^2 + (X'_2 + X_m)^2\right]} \tag{7.42}$$

The CSI can be designed for fixed or variable frequency. Distinctive characteristics can be obtained in either method as seen in the following subsections.

7.9.1 INDUCTION MOTOR WITH CONSTANT-FREQUENCY CSI

The speed–torque characteristics of the induction motor operating under CSI with fixed-supply frequency are shown in Figure 7.17. Equation (7.42) is used to construct these characteristics for two values of I_s. In addition, the figure shows the characteristics of the induction motor operating by a voltage source inverter (VSI) drive.

When a CSI is used, the induction motor exhibits different characteristics as compared to the VSI. The most noticeable one is the low starting torque. This is primarily due to the high rotor current I'_2 at starting, which reduces the magnetizing current I_m, as shown in Equation (7.39). Remember that the current I_1 is constant! The low magnetizing current at starting reduces the flux of the motor; hence, it reduces the starting torque.

The other noticeable difference is that the speed of the motor in the normal operating region with CSI is stiffer (has a flatter slope) than that of a VSI motor. This is because the core of the motor is saturated with flux in this region. If the core saturates, small changes in the magnetizing current tend to have little or no effect on the flux. When the load torque increases, the rotor current tends to increase, which reduces the magnetizing current (I_s is constant). When reductions of the magnetizing current do not reduce the flux (in the saturation region), the speed of the motor remains almost unchanged.

FIGURE 7.17
Speed–torque characteristics of an induction motor with CSI and VSI

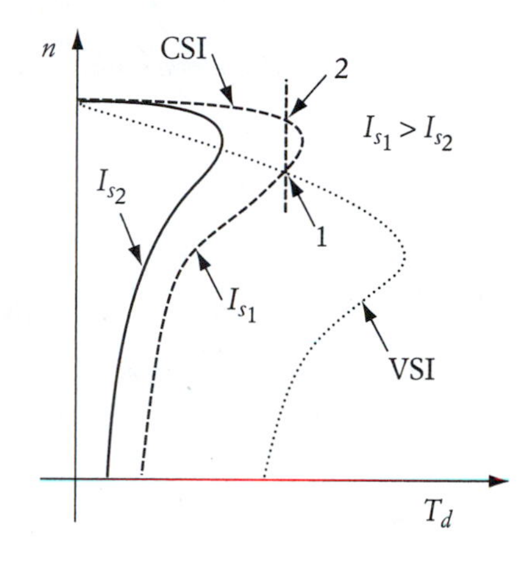

Let us assume that the voltage and flux of the VSI motor are at their rated values. Hence, the intersection point of the VSI and CSI (point 1), represents the operation of the motor at the rated flux and voltage. This point is in the unstable region of the induction machine. Now we assume that we want to operate the motor at point 2, where the torque is the same as that at point 1, and the speed is slightly higher. At point 2, the slip is smaller and the rotor current is accordingly smaller. This tends to increase the magnetizing current I_m. Normally, I_m at point 1 is at the rated value, which is normally close to the saturation region. Hence, I_m at point 2 is higher than the rated value, and the core of the motor is saturated.

In Figure 7.15, we represented the core of the machine as an ideal inductor. In reality, it includes a resistive component representing the core losses. Since at point 2 the magnetizing current increases, the core losses also increase. Hence, the overall efficiency of the motor is decreased. Therefore, point 1 is a preferable operating point because of the higher efficiency and the unsaturated core. However, because it is in the unstable region, a feedback control mechanism is needed to ensure the stability of the drive system.

EXAMPLE 7.9

A 480 V, six pole, 60 Hz, Y-connected induction motor has a stator inductive reactance of 3 Ω and a stator resistance of 0.2 Ω. The rotor inductive reactance and resistance referred to the stator are 2 Ω and 0.1 Ω, respectively. The magnetizing reactance is 120 Ω in the linear region and 40 Ω in the saturation region. The motor is driven by CSI, and its load is a constant torque of 100 Nm. The input current of the CSI is adjusted to run the machine at 900 rpm. Compute the input current.

SOLUTION

Let us first compute the slip.

$$s = \frac{n_s - n}{n_s} = \frac{1200 - 900}{1200} = 0.25$$

If the machine is in the linear region

$$T_d = \frac{3\, I_s^2\, X_m^2\, R'_2}{s\omega_s\left[\left(\frac{R'_2}{s}\right)^2 + (X'_2 + X_m)^2\right]}$$

$$100 = \frac{3\, I_s^2 (120^2)(0.1)}{0.25(125.66)\left[\left(\frac{0.1}{0.25}\right)^2 + (2 + 120)^2\right]}$$

Then

$$I_s = 104 \text{ A}$$

Let us compare this current to the rated current of the machine. Since the power rating of the machine is not given, we can assume that the following equation applies:

$$\sqrt{3}\, V_s I_{1 \text{ rated}} \cos\theta \approx T_d \omega_s$$

In this equation, we are ignoring the stator losses. Nevertheless, it is a good approximation for the current. We can further assume that the power factor $\cos\theta \approx 0.7$, which is a typical value. Based on these assumptions,

$$I_{1 \text{ rated}} \approx \frac{T_d \omega_s}{\sqrt{3}\, V_s \cos\theta} = \frac{100(125.66)}{\sqrt{3}\, 480\,(0.7)} = 21.6 \text{ A}$$

Note that the source current, I_s, is about 4.8 times larger than the rated current, $I_{1 \text{ rated}}$. This excessive current will be damaging to the machine. Hence, the machine cannot operate under this condition.

If the machine is in the saturation region. Let us examine whether the machine can operate in the saturation region of the magnetizing curve.

$$T_d = \frac{3\, I_s^2\, X_m^2\, R'_2}{s\omega_s\left[\left(\frac{R'_2}{s}\right)^2 + (X'_2 + X_m)^2\right]}$$

$$100 = \frac{3\, I_s^2 (120^2)(0.1)}{0.25(125.66)\left[\left(\frac{0.1}{0.25}\right)^2 + (2 + 42)^2\right]}$$

Then

$$I_s = 37.5 \text{ A}$$

Even in the saturation region, the current I_s is about 1.73 times larger than the rated current $I_{1 \text{ rated}}$. This excessive current will also be damaging to the machine. To reduce the current, the frequency of the source must be adjusted as seen in the next section.

7.9.2 INDUCTION MOTOR WITH ADJUSTABLE FREQUENCY CSI

As we have seen in the previous example, the induction motor operating under constant frequency cannot provide a good range of speed control. To correct this drawback, the frequency of the supply must change. The change of the supply frequency results in the following changes:

1. The synchronous speed is changed by f_2/f_1, where the subscript 1 indicated the original value and 2 the new value.

$$\omega_{s_2} = \omega_{s_1} \frac{f_2}{f_1}$$

2. The slip is changed according to the form

$$s_2 = s_1 \frac{n_{s_2} - n}{n_{s_1} - n} \frac{n_{s_1}}{n_{s_2}}$$

3. All reactances are changed by f_2/f_1.

$$X_{m_2} = X_{m_1} \frac{f_2}{f_1}$$

Substituting the new parameters and variables into Equation (7.42) yields

$$T_d = \frac{3\, I_s^2 \left(X_m \frac{f_2}{f_1} \right)^2 R'_2}{(\omega_{s_2} - \omega) \left[\left(\frac{R'_2}{s_2} \right)^2 + \left((X'_2 + X_m) \frac{f_2}{f_1} \right)^2 \right]} \tag{7.43}$$

Equation (7.43) is represented in Figure 7.18. Note that the reduction of the supply frequency reduces the motor speed by a wide range and also increases the starting torque—both desirable features.

FIGURE 7.18
Induction motor with variable-frequency CSI

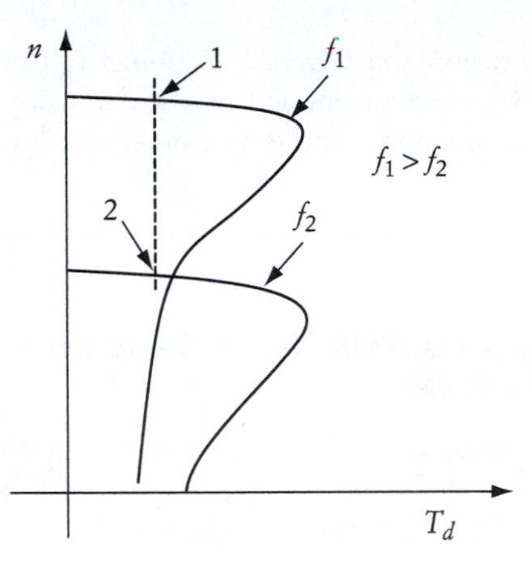

EXAMPLE 7.10

For the machine in Example 7.9, compute the frequency of the CSI to drive the machine at 900 rpm without exceeding the rated current.

SOLUTION

If the current is to be limited by the ratings of the motor, we can assume that the machine is operating in the linear region of the magnetizing curve. The torque equations can be modified for the new frequency as given next.

$$T_d = \frac{3\ I_s^2\ X_m^2\ R'_2}{s\omega_s\left[\left(\frac{R'_2}{s}\right)^2 + (X'_2 + X_m)^2\right]}$$

$$100 = \frac{3\ (21.6)^2\left(120\,\frac{f}{60}\right)^2\ 0.1}{\frac{2\pi}{60}\left(120\,\frac{f}{p} - 900\right)\left[\left(\frac{0.1}{(120f/p - 900)/(120f/p)}\right)^2 + \left[(2 + 120)\,\frac{f}{60}\right]^2\right]}$$

The solution of this equation could be very involved. However, with simple and valid assumptions, we can simplify the equation. For example, the magnetizing reactance X_m in the linear region is much larger than R'_2/s, even for small slips. Also, $X_m >> X'_2$. Hence,

$$T_d \approx \frac{3\, I_s^2\, R'_2}{s\omega_s}$$

$$100 \approx \frac{3(21.6)^2\, 0.1}{\frac{2\pi}{60}\left(120\frac{f}{p} - 900\right)}$$

The frequency of the CSI is then

$$f \approx 45.67 \text{ Hz}$$

Note that the torque equation given here is independent of the core reactance when the machine operates in the linear region.

CHAPTER 7 PROBLEMS

7.1 A 209 V, three-phase, six-pole, Y-connected induction motor has the following parameters:

$$R_1 = 0.128\ \Omega \qquad R'_2 = 0.0935\ \Omega \qquad X_{eq} = 0.49\ \Omega$$

The motor slip at full load is 2%. Calculate the following:
  a. Starting current (ignore the magnetizing current)
  b. Full load current
  c. Starting torque
  d. Maximum torque
  e. Motor efficiency (ignore rotational and core losses)

7.2 For the motor in Problem 7.1, assume that the motor load is fan-type. If an external resistance equal to the rotor resistance is added to the rotor circuit, calculate the following:
  a. Motor speed
  b. Starting torque
  c. Starting current
  d. Motor efficiency (ignore rotational and core losses)

7.3 For the motor in Problem 7.1 and for a fan-type load, calculate the value of the resistance that should be added to the rotor circuit to reduce the speed at full load by 20%. What is the motor efficiency in this case?

7.4 For the motor in Problem 7.1 and for a fan-type load, calculate the following if the voltage is reduced by 20%:
  a. Motor speed
  b. Starting torque
  c. Starting current
  d. Motor efficiency (ignore rotational and core losses)

7.5 For the motor in Problem 7.1 and for a fan-type load, calculate the following, assuming that the supply frequency is reduced by 20%:
   a. Motor speed
   b. Starting torque
   c. Starting current
   d. Motor efficiency (ignore rotational and core losses)

7.6 For the motor in Problem 7.1 and for a constant-load torque equal to half the full-load torque, calculate the minimum supply frequency that will not allow the motor current to exceed the full-load current. Calculate the motor speed.

7.7 For the motor in Problem 7.1 and for a fan-type load, calculate the following, assuming that the supply frequency is increased by 20%:
   a. Motor speed
   b. Starting torque
   c. Starting current
   d. Motor efficiency (ignore rotational and core losses)

7.8 For the motor in Problem 7.1 and for a constant-load torque, calculate the maximum increase in supply frequency and the motor speed.

7.9 For the motor in Problem 7.1 and for a fan-type load, calculate the following, assuming that the supply frequency is reduced by 20% and the v/f ratio is kept constant:
   a. Motor speed
   b. Starting torque
   c. Starting current
   d. Motor efficiency (ignore rotational and core losses)

7.10 A three-phase, 480 V, six-pole, Y-connected, 60 Hz, 10 kW, 1150 rpm induction motor is driving a constant-torque load of 60 Nm. The parameters of the motor are

$$R_1 = 0.4\ \Omega \qquad R_2 = 0.5\ \Omega \qquad X_{eq} = 4\ \Omega \qquad \frac{N_1}{N_2} = 2$$

Calculate the following:
   a. motor torque
   b. motor current
   c. starting torque
   d. starting current

A voltage is injected in the rotor circuit to reduce the motor speed by 40%.
   e. Calculate the magnitude of the injected voltage.
   f. Repeat (a) to (d) for the motor with injected voltage.
   g. Calculate the power delivered to the source of injected voltage.
   h. Determine the overall efficiency of the motor (ignore rotational and core losses).

7.11 A three-phase, six-pole, Y-connected, 60 Hz, 480 V induction motor is driving a 300 Nm constant-torque load. The motor has rotational losses of 1 kW.

The motor is driven by a slip energy recovery system. The triggering angle of the dc/ac converter is adjusted to 100°. Calculate the following:

a. Motor speed
b. Current in the dc link
c. Rotor rms current
d. Stator rms current
e. Power returned back to the source

7.12 A 480 V, four-pole, 60 Hz, Y-connected induction motor has a stator inductive reactance of 4 Ω and stator resistance of 0.2 Ω. The rotor inductive reactance and resistance referred to the stator are 4 Ω and 0.2 Ω, respectively. The magnetizing reactance is 150 Ω in the linear region and 50 Ω in the saturation region. The motor is driving a constant-torque load of 120 Nm and is driven by a CSI. The frequency of the CSI is adjustable. Calculate the frequency of the CSI for the needed speed without exceeding the ratings of the motor, and compute the starting torque.

7.13 A three-phase, 480 V, six-pole, 60 Hz induction motor is driving a constant-torque load of 80 Nm. The parameters of the motor are

$$R_1 = 0.5\ \Omega \qquad R_2 = 0.3\ \Omega \qquad X_{eq} = 4\ \Omega \qquad \frac{N_1}{N_2} = 2$$

a. Calculate the magnitude of the injected voltage that would reduce the motor speed to 800 rpm.
b. Calculate the power received by the injected voltage source.
c. Compute the starting current and the starting torque with the injected voltage.

7.14 A three-phase, four-pole, Y-connected, 480 V induction motor is driving a 400 Nm constant-torque load. The motor has the following parameters:

$$\frac{N_1}{N_2} = 1 \qquad P_{\text{rotational}} = 1\ \text{kW}$$

The motor is driven by a slip energy-recovery system. The triggering angle of the dc/ac converter is adjusted to 120°. Calculate the following:

a. Motor speed
b. Current in the dc link
c. Average voltage at the input of the dc/ac converter
d. Rotor rms current
e. Stator rms current
f. Power returned back to the source

7.15 A three-phase, 60 Hz, Y-connected, 480 V induction motor rotates at 3500 rpm at full load. The motor is driven by a slip energy-recovery system. Calculate the triggering angle for a motor speed of 2800 rpm. Assume the turns ratio is equal to 1.

7.16 A three-phase, 60 Hz, six-pole, Y-connected, 480 V induction motor has the following parameters:

$$R_1 = 0.2\ \Omega, \qquad R'_2 = 0.1\ \Omega, \qquad X_{eq} = 5\ \Omega$$

The load of the motor is a drilling machine. At 1150 rpm, the load torque is 150 Nm. The motor is driven by a constant v/f technique. When the frequency of the supply voltage is reduced to 50 Hz, calculate the following:

a. Motor speed
b. Maximum torque at 60 Hz and 50 Hz
c. Motor current at 50 Hz

8

Braking of Electric Motors

Braking is a generic term used to describe a set of operating conditions for electric drive systems. It includes rapid stopping of the electric motor, holding the motor shaft to a specific position, maintaining the speed to a desired value, or preventing the motor from overspeeding. All these aspects of braking are done electrically without any need for mechanical brakes. During the braking process, the energy can change its flow between the electric source and mechanical load. The mechanical load or the rotating mass can become the source of energy driving the machine as a generator, which pumps the energy back to the electrical supply. The utilization of this braking energy enhances overall system efficiency.

The complete operational cycle of an electric drive system is highly dependent on which braking method is used. The quickness and accuracy of braking techniques often determine the productivity and quality of the manufactured goods. A robot in an assembly line must be able to stop, hold its position, and reverse its motion with a high degree of accuracy. These functions can be achieved by electric braking.

Compared to the mechanical braking methods, electric braking is a highly efficient and low-maintenance technique. Nevertheless, braking can result in stressful electrical and mechanical transients. Therefore, the braking system must be designed to ensure effective and safe operation.

There are several forms of braking applicable to virtually all types of motors. Generally, we can group all braking methods into three types: regenerative, dynamic, and countercurrent braking. The next two chapters discuss the braking systems in more detail.

8.1 REGENERATIVE BRAKING

An electric motor is in regenerative braking when the load torque reverses its direction and causes the machine to run at a speed higher than its no-load speed but without changing the direction of rotation. An example of regenerative braking is given in Figure 8.1, where an electric motor is driving a trolley bus in the uphill and downhill directions. In the uphill direction, the gravity force can be resolved into two components: one perpendicular to the road surface F and the other parallel to the road surface F_l. The parallel force pulls the motor toward the bottom of the hill.

FIGURE 8.1
Example of regenerative braking

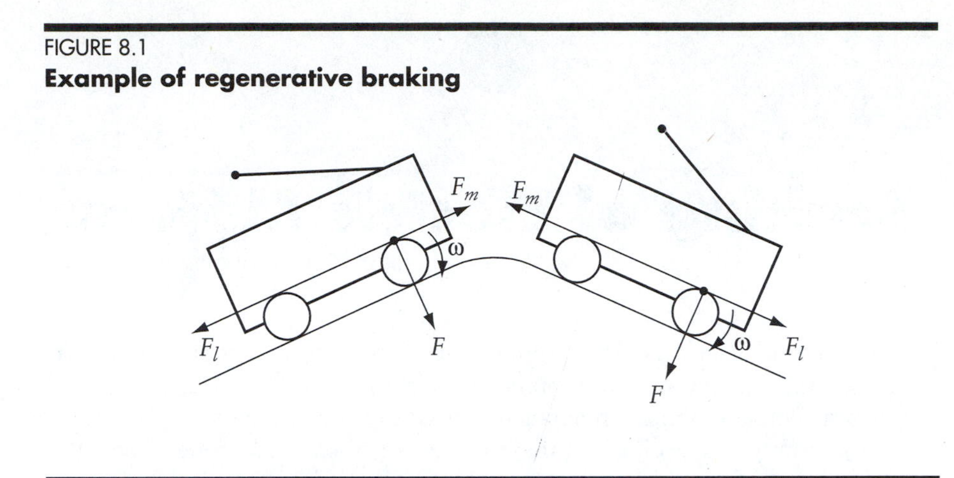

FIGURE 8.2
Regenerative braking in second quadrant

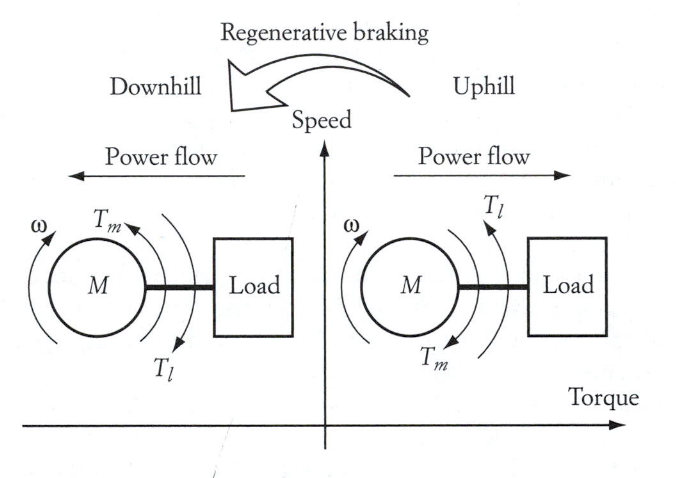

If we ignore the rotational losses, the motor must produce a force F_m opposite to F_l to move the bus in the uphill direction. This case is also depicted in Figure 8.2 in the first quadrant. Note that the motor torque and speed are in the same direction, and the load torque T_l is opposite to the motor torque T_m. The power flow is from the motor to the mechanical load.

Now assume that the same bus is traveling downhill. Since the gravitational force does not change its direction, the load torque pushes the motor toward the bottom of the hill. The direction of the motor torque is always opposite to the direction of the load torque, so the motor produces a torque in the reverse direction. Note that the rotation of the motor is still in the same direction on both sides of the hill. The downhill operation is shown in Figure 8.2 in the second quadrant. This is known as regenerative braking.

The energy exchange under regenerative braking is from the mechanical load to the electrical source. Hence, the load is driving the machine, and the machine is generating electric power that is returned back to the supply.

8.2 DYNAMIC BRAKING

When an electric motor spins, a kinetic energy is stored in its rotating mass. If the motor is disconnected from the power source, it continues to rotate for a period of time until the stored kinetic energy is totally dissipated in the form of rotational losses. The faster the dissipation of the kinetic energy, the more rapid is the braking.

With dynamic braking, the kinetic energy of the motor is transformed into electrical energy and dissipated in resistive elements. The rate of energy dissipation can be increased by the design of the braking resistance. A circuit for dynamic braking is shown in Figure 8.3. When the machine is connected to terminal *A*, it runs as a motor. While the motor is rotating, it acquires kinetic energy stored in its rotating mass. The current I_A flows into the machine. If the terminals of the motor are switched to position *B*, the energy stored in the rotating mass is dissipated in the braking resistance R_B. This is possible when the machine maintains its field. The braking current I_B flows out of the machine. The smaller the resistor is, the faster the energy is dissipated, and the faster the motor brakes.

FIGURE 8.3
Dynamic braking

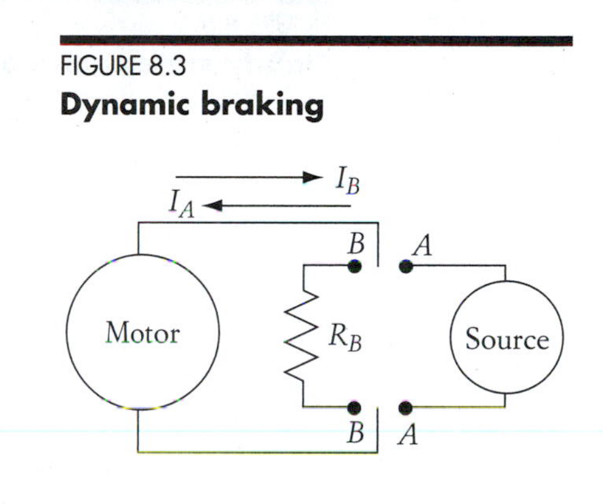

When the machine is operating in a dynamic braking mode, it acts as a generator. The speed of the machine does not change its direction of rotation during braking, but the machine torque reverses its direction (I_B is opposite to I_A). Thus, the motor is also in the second quadrant as depicted in Figure 8.2.

8.3 COUNTERCURRENT BRAKING

The direction of rotation of an electric motor is dependent on several variables. Among them are the phase sequence of the stator windings (for ac machines), and the polarities of the field or armature voltage (for dc machines).

For ac machines, the shaft of the machine rotates in the same direction as the magnetic field. If the phase sequence of the stator windings is reversed, the airgap field reverses its rotation, as shown in Figure 8.4. If we reverse the sequence while the motor is rotating, the rotor shaft decelerates until it stops; then it starts to accelerate in the reverse direction. If the power source is disconnected when the motor reaches standstill (zero speed), the motor is stopped by the countercurrent braking method. Countercurrent braking is also used to slow down the motor or

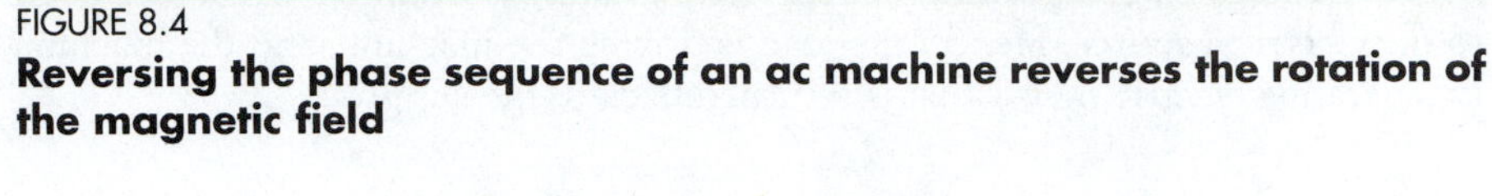

FIGURE 8.4
Reversing the phase sequence of an ac machine reverses the rotation of the magnetic field

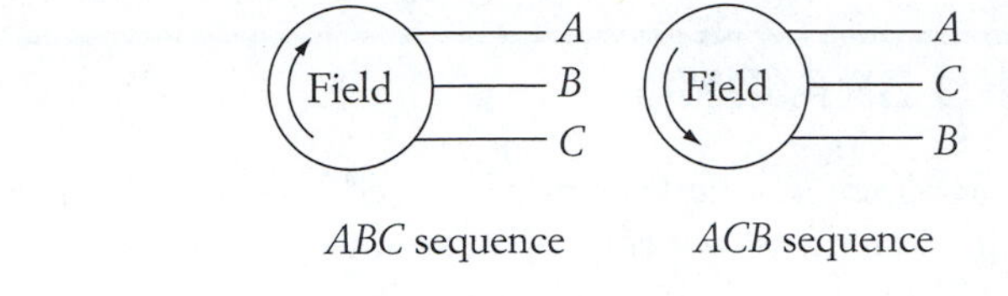

FIGURE 8.5
Steady-state operation of countercurrent braking

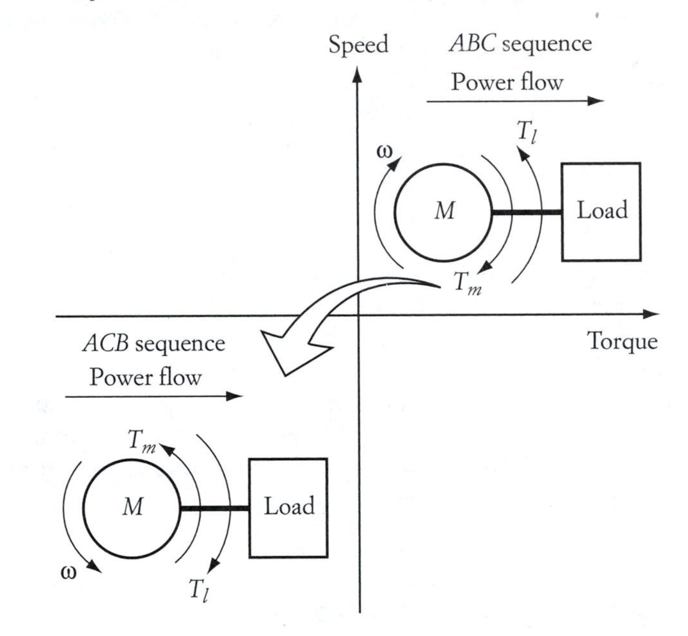

reverse its direction of rotation. Figure 8.5 shows two quadrants for the steady-state operation of two sequences (*ABC* and *ACB*). In each case, the machine is running as a motor.

Direct current machines can also utilize countercurrent braking. The braking occurs when the terminal voltage reverses its polarities, which eventually leads to the reversal of the motor's rotation. While the machine is decelerating, it operates temporarily in the second quadrant, where it acts as a generator. The machine settles in the third quadrant.

CHAPTER 8 PROBLEMS

8.1 An elevator consists of the cabin, motor, counterweight, cables, and pulleys. The elevator cabin is full and is moving downward. The mass of the cabin plus people is greater than the mass of the counterweight. Explain the operation of the motor in terms of energy transfer, and indicate the speed-torque quadrant for this motion.

8.2 Repeat Problem 8.1 for the elevator going upward.

8.3 Repeat Problems 8.1 and 8.2 where the mass of the counterweight is more than the mass of the cabin plus people.

8.4 If an electric car is moving downhill with its motor disconnected from the electrical source, the acceleration of the car increases according to Newton's laws of motion. Explain this motion in terms of energy.

8.5 Repeat Problem 8.4 where the motor is connected to the electrical source while moving downhill.

8.6 A windmill consists of a motor, rotating blades, and structure. Explain the operation of the system when wind speed rotates the shaft of the motor at a speed higher than its no-load speed. In which quadrant does the machine operate?

9

Braking of dc Motors

Braking of direct current machines can be done relatively easily by the three basic methods discussed in Chapter 8. Since dc machines have several field connections, braking circuits and methods can differ for different field connections. In this chapter, we shall discuss the braking of the separately excited, shunt, and series motors. Braking of the compound motor is very similar to that of the motors presented here.

In any of these braking methods, adequate safeguards should be implemented to limit the thermal and voltage levels to tolerable amounts. Braking tends to permit high transient currents to flow in the machine's windings that could be much higher than the starting currents. If excessive, the current during braking can result in permanent damage to the windings as well as to the power converter. In addition, mechanical stresses due to rapid stopping or starting of a motor, and the excessive torque that must be developed by the motor, can cause mechanical damage to the bearings, coupling, and the rotor itself. A drive system with repeated braking operations must have adequate mechanical coupling to withstand the sudden shear force. Large machines with braking operations are normally mounted on strong concrete slabs to prevent the movement of the stator frame.

9.1 REGENERATIVE BRAKING OF dc SHUNT MOTORS

Under given operating conditions, when the speed of the dc machine exceeds its no-load speed, the machine is in the regenerative braking mode. An example of this type of braking is given in Figure 8.1 for an electric bus going in the uphill and downhill directions. In the downward direction, the speed of the bus may exceed its no-load speed, and hence generate electric power that can be pumped back to the source.

As we saw earlier in Figure 5.4, the speed–torque characteristic of a dc motor (separately or shunt) is linear. The basic equations of the motor are repeated here.

$$V_t = E_a + R_a I_a = K\phi\omega + \frac{R_a}{K\phi} T_l \qquad (9.1)$$

$$\omega = \frac{V_t}{K\phi} - \frac{R_a}{(K\phi)^2} T_l \tag{9.2}$$

$$I_a = \frac{V_t - E_a}{R_a} = \frac{T_l}{K\phi} \tag{9.3}$$

Let us analyze these three equations under the regenerative braking condition for the bus in Figure 8.1. The electric bus travels uphill, then downhill. In the uphill direction, the dc machine acts as a motor represented by Equations (9.1) to (9.3). The load torque in this case is opposite to the direction of the bus motion, and the drive system is in the first quadrant as shown in Figure 8.2. The equivalent circuit of the system is shown in Figure 9.1. Under this condition, the back emf voltage E_a is less than the terminal voltage V_t due to the voltage drop across the armature resistance R_a.

Figure 9.2 shows the speed–torque characteristics of the dc machine. The figure is obtained by using Equation (9.2). The load torque in this figure is assumed to be bidirectional, which is the case for the electric bus we are discussing. In the first quadrant, the machine operates as a motor as described by the circuit in Figure 9.1. Let us assume that operating point 1 represents this case. When the bus reaches the peak of the hill, the load torque seen by the motor is zero, assuming that the frictional torque is ignored. This is because the gravitational torque at the top of the hill is perpendicular to the road surface and is not pulling the motor in either direction of motion. The subscript 2 is used here to represent the operation of the motor at the top of the hill, where the load torque seen by the motor is zero, and the motor speed is

$$\omega_2 = \frac{V_t}{K\phi} \tag{9.4}$$

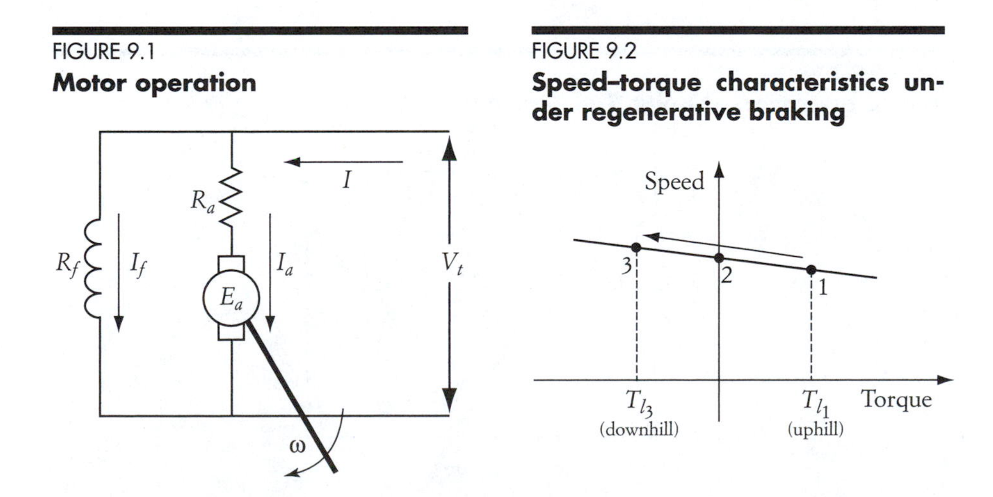

FIGURE 9.1
Motor operation

FIGURE 9.2
Speed–torque characteristics under regenerative braking

Because the load torque at the top of the hill is zero, the armature current must also be zero.

$$I_{a_2} = \frac{V_t - E_{a_2}}{R_a} = \frac{T_{l_2}}{K\phi} = 0 \tag{9.5}$$

Since the current is zero, the voltage drop across the armature resistance is also zero. Hence,

$$V_t = E_{a_2} \tag{9.6}$$

Equations (9.4) to (9.6) are represented by operating point 2 in Figure 9.2. This operating point is the no-load operating point of a dc machine.

Now assume that the electric bus is traveling in the downhill direction. Relative to the load torque at point 1, the load torque in the downhill motion is reversed and in the same direction as the speed. Note that the motor speed will not change its direction, and the motor moves to operating point 3 in Figure 9.2.

In the downhill operation, the load torque changes its direction while the field current remains in its original direction. The magnitude of the field current is constant because the terminal voltage is constant. Hence, the armature current at point 3 must reverse its direction, because

$$I_{a_3} = \frac{T_{l_3}}{K\phi}$$

This case is depicted in Figure 9.3. If the armature current is larger than the field current, the current $I = I_{a_3} - I_f$ flows into the source. Note that E_{a_3} does not change its direction since ω and $K\phi$ are in the same direction as at point 1. The motor is now generating electric power and delivering it to the source.

The speed and current of the machine at point 3 can be represented by the following equations:

FIGURE 9.3
Motor operation at point 3 of Figure 9.2

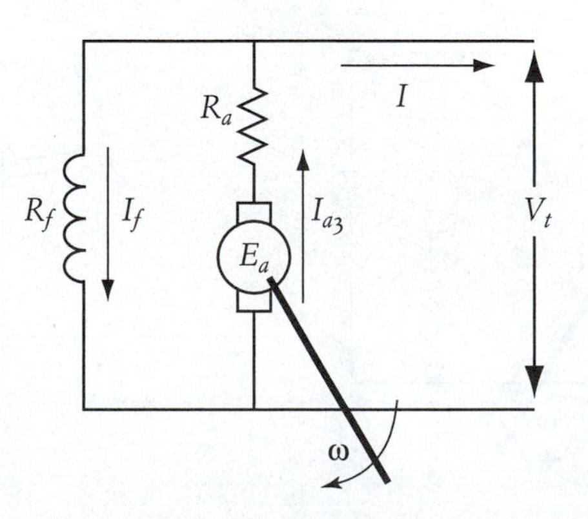

$$\omega_3 = \frac{V_t}{K\phi} - \frac{R_a}{(K\phi)^2} T_{l_3} \tag{9.7}$$

$$I_{a_3} = \frac{V_t - E_{a_3}}{R_a} = \frac{T_{l_3}}{K\phi} \tag{9.8}$$

Note that T_{l_3} is negative. If you wish, you can use the magnitude of T_{l_3} and introduce a negative sign in Equations (9.7) and (9.8) as follows:

$$\omega_3 = \frac{V_t}{K\phi} + \frac{R_a}{(K\phi)^2} T_{l_3} \tag{9.9}$$

$$I_{a_3} = \frac{V_t - E_{a_3}}{R_a} = \frac{-T_{l_3}}{K\phi} \tag{9.10}$$

Since I_{a_3} is negative, as seen in Equation (9.10), E_{a_3} must be larger than V_t. Thus, the motor is operating as a generator.

Figure 9.4 shows the motor speed versus E_a. The figure shows all three operating points discussed in this section. As you can see, the machine acts as a motor when $E_a < V_t$, and the machine is a generator when $E_a > V_t$.

A summary of the changes in the machine variables is given in Table 9.1. The arrows in the table represent directions. Arrows of operating point number 1 are considered to be the reference arrows. Examine the table against the cases just described.

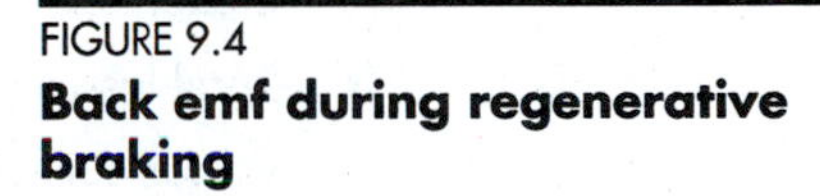

FIGURE 9.4
Back emf during regenerative braking

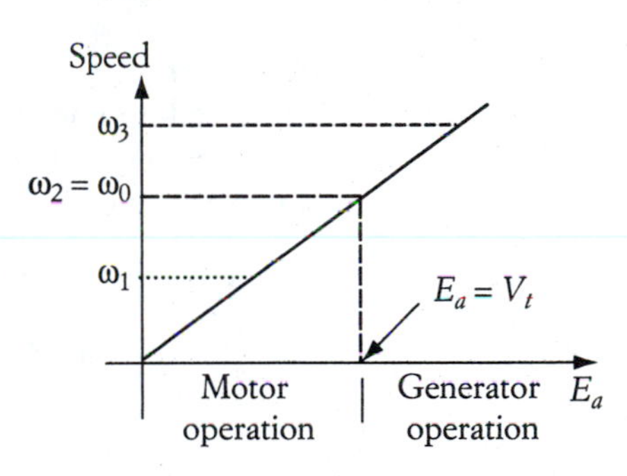

TABLE 9.1
Summary of regenerative braking

Operating Point	*Load Torque*	*Terminal Voltage*	*Armature Current*	*Speed*	*Field*	E_a	*Comments*
	→	→	→	→	→	→	
1	T_{l_1}			$\omega_1 < \omega_0$		$E_{a_1} < V_t$	Motor
		→		→	→		
2	0		0	$\omega_2 = \omega_0$		$E_{a_2} = V_t$	No load
	←	→	←	→	→	→	
3	T_{l_3}			$\omega_3 > \omega_0$		$E_{a_3} > V_t$	Generator

EXAMPLE 9.1

A 440 V, dc shunt motor has a rated armature current of 76 A at a speed of 1000 rpm. The armature resistance of the motor is 0.377 Ω, the field resistance is 110 Ω, and the rotational losses are 1 kW. The load of the motor is bidirectional. Calculate the following:

a. No-load speed of the motor
b. Motor speed, where the armature current is 60 A during regenerative braking
c. Developed torque during regenerative braking
d. E_a during regenerative braking
e. Power delivered by the source under normal motor operation
f. Terminal current under regenerative braking
g. Generated power during regenerative braking
h. Total losses under regenerative braking
i. Power delivered to the source under regenerative braking

SOLUTION

The speed–torque characteristic of this example is shown in Figure 9.5. Point 1 in the figure represents the motor operation at rated current and 1000 rpm. During motor operation,

$$E_a = V_t - R_a I_a = 440 - 0.377 \times 76 = 411.35 \text{ V}$$

The field constant $K\phi$, which remains unchanged during regenerative braking, can be computed as

$$K\phi = \frac{E_a}{\omega} = \frac{E_a}{2\pi \dfrac{n}{60}} = 3.93 \text{ V sec}$$

FIGURE 9.5
Speed–torque characteristic of the motor in Example 9.1

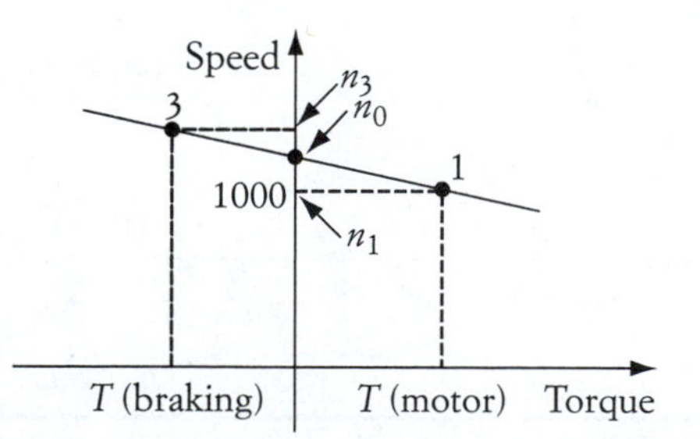

a. The no-load speed of the motor is equal to the terminal voltage divided by the field constant.

$$\omega_0 = \frac{V_t}{K\phi} = \frac{440}{3.93} = 111.96 \text{ rad/sec}$$

$$n_0 = 1069.1 \text{ rpm}$$

b. During regenerative braking (point 3), the speed of the motor can be computed by Equation (9.9).

$$\omega_3 = \frac{V_t}{K\phi} + \frac{R_a}{(K\phi)^2} T_{l_3} = \frac{V_t - R_a I_{a_3}}{K\phi} = \frac{440 + 0.377 \times 60}{3.93} = 117.72 \text{ rad/sec}$$

$$n_3 = 1124.1 \text{ rpm}$$

c. The developed torque at point 3 is

$$T_{l_3} = K\phi \, I_{a_3} = 3.93 \times 60 = 235.8 \text{ Nm}$$

d. The back emf at point 3 is

$$E_{a_3} = K\phi \, \omega_3 = 3.93 \times 117.72 = 462.64 \text{ V}$$

e. To calculate the terminal power, you must calculate the terminal current. When the motor operates at point 1, the terminal current is

$$I_1 = I_{a_1} + I_f = 76 + \frac{440}{110} = 80 \text{ A}$$

The total power delivered by the source P_s is

$$P_s = I_1 V_t = 80 \times 440 = 35.2 \text{ kW}$$

f. While the motor is in the regenerative region at point 3, the terminal current of the motor is

$$I_3 = I_{a_3} - I_f = 60 - \frac{440}{110} = 56 \text{ A}$$

g. The generated power P_g is

$$P_g = E_{a_3} I_{a_3} = 462.64 \times 60 = 27.76 \text{ kW}$$

h. The total losses P_{loss} are the sum of the losses in the armature resistance, the losses in the field resistance, and the rotational losses.

$$P_{\text{loss}} = R_a I_{a_3}^2 + \frac{V_t^2}{R_f} + \text{rotational losses} = 0.377 \times 60^2 + \frac{440^2}{110} + 1000 = 4.12 \text{ kW}$$

i. The power delivered to the source P_d is the generated power minus the losses.

$$P_d = 27.76 - 4.12 = 23.64 \text{ kW}$$

9.2 REGENERATIVE BRAKING OF dc SERIES MOTORS

Regenerative braking occurs when the motor speed exceeds the no-load speed (at zero torque). For a series motor, at zero torque, the no-load speed is theoretically infinity. Hence, one might conclude that the series motor could not operate under regenerative braking. Actually, the circuit of the series motor can be altered during regenerative braking to allow the machine to generate electric power that can be returned to the source. Consider the circuits in Figure 9.6. The circuit on the right side of the figure shows the normal motor operation of the series machine. The circuit on the left shows a configuration for regenerative braking. In this case, the field circuit is excited by a separate source. The voltage of the separate source must be low enough to prevent the field current from becoming excessive. This is because the field resistance of the series motor is small: The field coil is composed of a small number of turns with a large cross section.

Switching from the circuit on the right side to that on the left side is best done using solid-state switches. The switching should not allow the current in the field circuit to be interrupted. Uninterruptible field current reduces current transients and prevents the machine from overspeeding. A simple circuit for this operation is shown in Figure 9.7. In the figure, S_1 to S_3 are solid-state switches. During motor operation, S_1 is closed, and S_2 and S_3 are open as shown in Figure 9.7(a). When the machine operates under regenerative braking, S_1 is opened, and S_2 and S_3 are closed. This occurs in steps. In the first step, S_1 is opened and S_3 is closed as shown in Figure 9.7(b). In this case, the armature circuit is separated from the

FIGURE 9.6
Regenerative braking of dc series motor

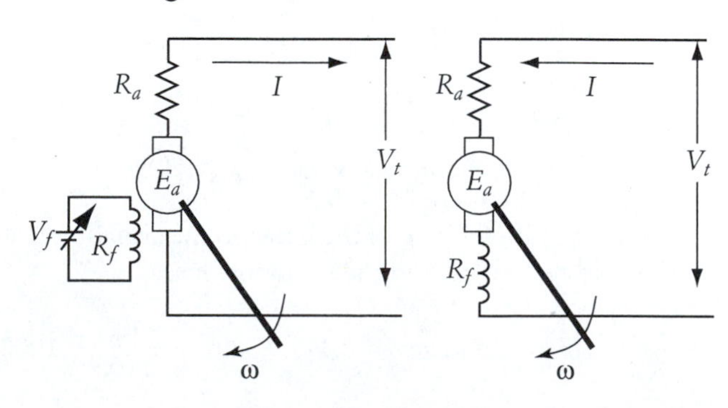

FIGURE 9.7
Regenerative braking circuit for series motor

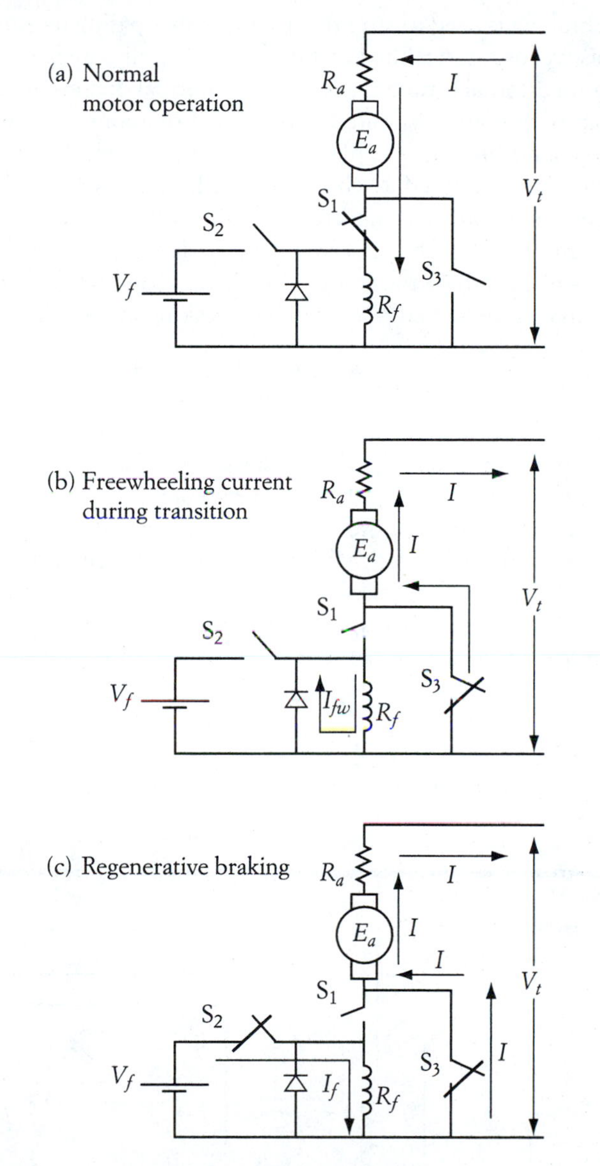

field winding. To prevent the collapse of the field current during the interval between opening S_1 and closing S_2, the freewheeling diode is used. The freewheeling current I_{fw} keeps the field current continuous. In the second step, S_2 is closed and the field current is provided by the separate source V_f as shown in Figure 9.7(c). The machine now operates under regenerative braking similar to that of the shunt machine.

9.3 DYNAMIC BRAKING OF dc SHUNT MOTORS

Dynamic braking is used to stop the motor by dissipating its stored kinetic energy into a resistive load. Once the kinetic energy is totally dissipated, the motor stops rotating if no external torque is exerted. The normal operation of the dc shunt motor is depicted in Figure 9.1 and described by Equations (9.1) to (9.3).

The dynamic braking is explained here using Figure 9.8. Assume that the machine is running at a speed ω when dynamic braking is applied. The terminals of the armature circuit are disconnected from the power source and connected across a braking resistance R_b. The field circuit is also disconnected from the armature circuit but is still excited by the source. Under this condition, the back emf E_a is the voltage source of the armature circuit. The braking current in this case is

$$I_b = -\frac{E_a}{R_a + R_b} = -\frac{K\phi\omega}{R_a + R_b} \tag{9.11}$$

The negative sign in Equation (9.11) indicates that the braking current is in the reverse direction of the armature current of Figure 9.1. The power dissipated during the dynamic braking is composed of two major components. The first is mechanical losses (or rotational losses), which are due to friction and windage losses. The second component is electrical losses P_b in the armature and braking resistances. The electrical loss is mainly responsible for dissipating the motor's kinetic energy. The larger are the electric losses, the shorter is the braking time. These losses can be calculated as

$$P_b = \frac{E_a^2}{R_a + R_b} = \frac{(K\phi\omega)^2}{R_a + R_b} \tag{9.12}$$

FIGURE 9.8
Dynamic braking

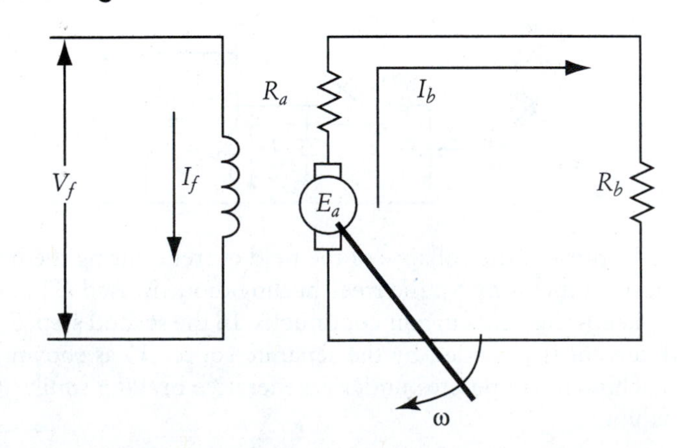

Equation (9.12) indicates that more electric power is dissipated if the braking resistance R_b is small and the field ϕ is strong.

Equation (9.11) is represented by the graph in Figure 9.9. The first quadrant of the graph is for normal motor operation, where the motor is operating at point A. Since the speed direction is unchanged and the current direction is reversed, the motor during dynamic braking is in the second quadrant.

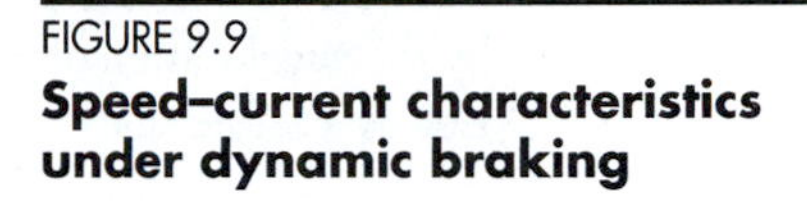

FIGURE 9.9
Speed–current characteristics under dynamic braking

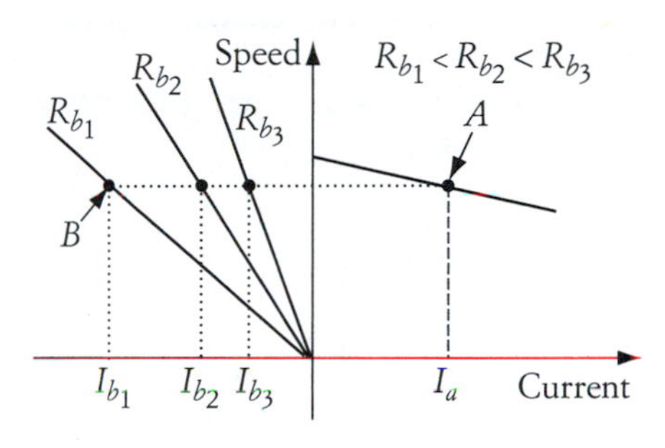

During dynamic braking, the speed–current characteristics are all straight lines with negative slopes that intercept at the point of origin. The figure shows the characteristics for three different values of braking resistance, where $(R_{b_1} < R_{b_2} < R_{b_3})$. The smaller the braking resistance is, the larger is the braking current, and the higher is the rate by which the kinetic energy is dissipating. This situation results in faster braking.

The circuit for dynamic braking is shown in Figure 9.10. When the switch is in position *A,* the motor operates at point *A* in Figure 9.9. The armature current I_a flows from the source to the motor. By connecting the switch in Figure 9.9 to terminal *B,* the voltage source is disconnected and the resistance R_b is inserted across the motor terminals. The operating point is shown in Figure 9.9 and labeled *B* for $R_b = R_{b_1}$. If the switching from *A* to *B* is done quickly enough, one can assume that the motor speed during the switching interval is unchanged. After switching to *B,* the operating point of the motor moves horizontally to point *B.* At this point the armature current is I_{b_1}, which flows in the opposite direction to I_a. This current is flowing from the machine to the

FIGURE 9.10
Circuit for dynamic braking

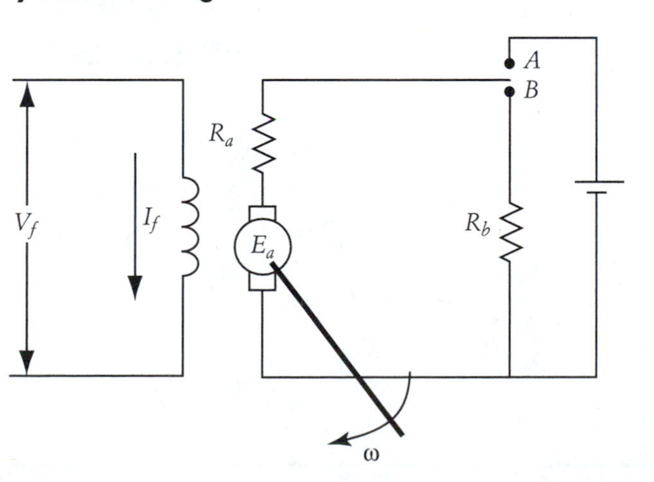

braking resistance. Depending on the loading condition of the motor, the operating point may not stay at point *B*. For example, if the load torque is frictional, the speed slows down until the motor stops. In this case, the operating point moves from *B* to the origin along the speed–current characteristics.

Another interesting case is shown in Figure 9.11, where the load torque is assumed constant and gravitational. The load force is equal to the load mass multiplied by the acceleration of gravity. Such a torque is constant regardless of the motor speed. Let us assume that the original operating point is *A* under normal motor operation. Just after dynamic braking is applied, the motor operating point moves to *B*. The final destination of the operating point is when the motor torque meets the load torque, which occurs only in the fourth quadrant at point *C*. Hence, the operating point of the motor moves from *B* to the point of origin, then continues to point *C*. The motor stops momentarily when the operating point reaches the origin. If the load is disconnected or a mechanical brake is applied at the origin, the motor stops. Otherwise, the motor speed reverses its direction until the machine reaches point *C*. The operation at point *C* is a typical generator operation in which the motor is driven mechanically by a unidirectional torque. The motor under this condition delivers electric power to the electrical load resistance R_b.

Table 9.2 summarizes the general operation of the dynamic braking in the case just discussed for a gravitational load.

FIGURE 9.11
Dynamic braking of gravitational torque load

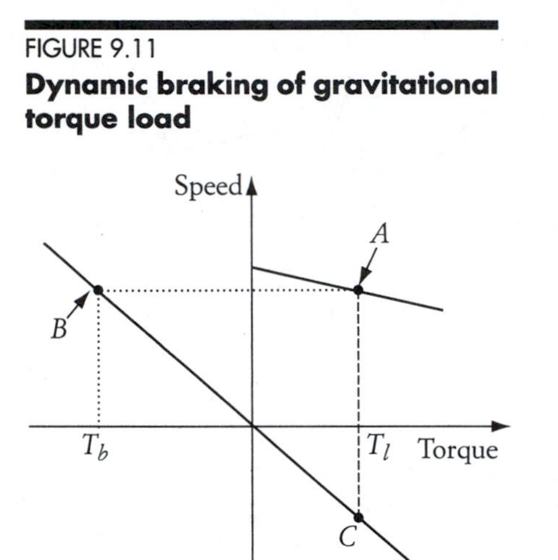

TABLE 9.2
Summary of dynamic braking

Operating Point	*Motor Torque*	*Terminal Voltage*	*Armature Current*	*Speed*	*Field*	E_a	*Comments*
	→	→	→	→	→	→	
A	T_l			ω_A		$E_{a_A} < V_t$	Motor
	←		←	→	→	→	
B	T_b	0		$\omega_B = \omega_A$		$E_{a_B} = E_{a_A}$	Generator
					→		
Origin	0	0	0	0		0	No load
	→		→	←	→	←	
C	T_l	0		ω_C		E_{a_C}	Generator

EXAMPLE 9.2

For the dc motor given in Example 9.1, assume that the load torque is gravitational. The current of the motor is 40 A at the steady-state condition. A dynamic braking technique employing a braking resistance of 2 Ω is used. Calculate the speed at the new steady-state operating point.

SOLUTION

With dynamic braking, the terminal voltage of the motor is zero. The braking current at point *C* in Figure 9.11 is equal to the steady-state current at point *A* because the load is gravitational. Hence,

$$I_b = -\frac{K\phi\omega}{R_a + R_b}$$

$$40 = -\frac{3.93\ \omega}{(0.377 + 2)}$$

$$n = -231 \text{ rpm}$$

9.4 DYNAMIC BRAKING OF dc SERIES MOTORS

Dynamic braking requires a strong magnetic field to convert the mechanical energy into electrical energy, and to allow the kinetic energy to be dissipated at a higher rate. In series machines, the field is proportional to the armature current. At the beginning of dynamic braking, the field is strong, but gradually weakens because of the reduction in armature current, which may prolong the braking time. To brake the motor faster, the series field can be separated from the armature circuit and excited by a different voltage source, as shown in the schematic of Figure 9.12. Keep in mind that the voltage applied to the separated field circuit must be reduced to

FIGURE 9.12
Dynamic braking of series motor

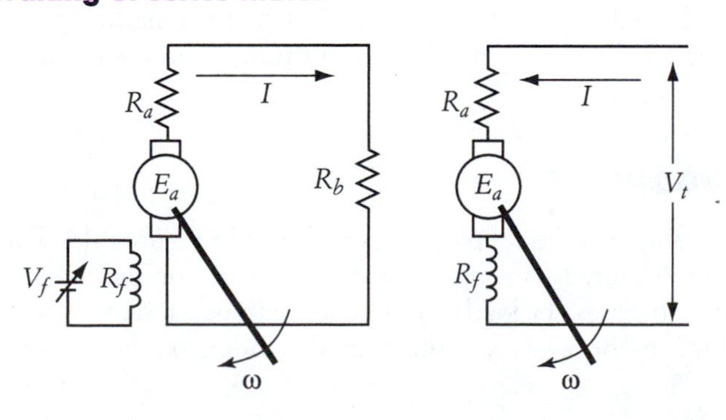

prevent the field current from exceeding its limits. Also, the braking resistance should be selected to limit the braking current in the armature circuit. Under this form of braking, and by using the circuit described in Figure 9.12, the motor behaves as a separately excited motor.

FIGURE 9.13
Dynamic braking characteristics of series motor

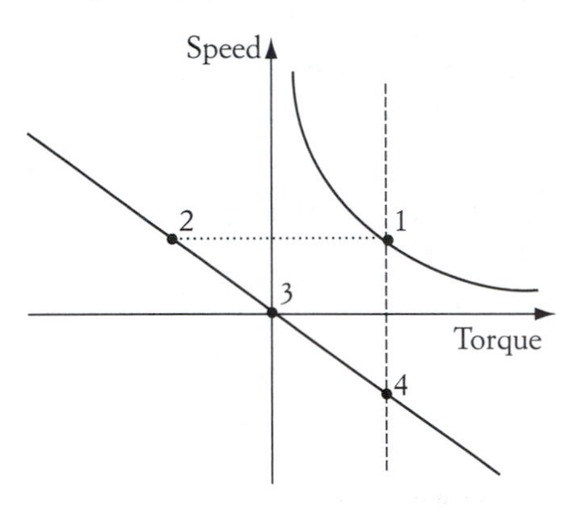

The characteristics of the motor are shown in Figure 9.13. The original operating point of the motor is at 1. When dynamic braking is applied and the field circuit is separately excited, the motor moves to point 2. Note that the motor characteristic is linear under dynamic braking because the field is constant. The motor speed starts to slow down until it is fully stopped at point 3. If the load torque is unidirectional and the motor is not disconnected at 3, the motor operating point moves to point 4, where the load torque and the machine torque are equal. At point 4, the machine is a generator with input mechanical power from the load, and the output electric power is dissipated in the braking resistance. The motor equation at points 2 and 4 are given here:

$$\omega_2 = \omega_1 = -\frac{R_a + R_b}{K\phi_2} I_{a_2}$$

$$\omega_4 = -\frac{R_a + R_b}{K\phi_4} I_{a_4}$$

9.5 COUNTERCURRENT BRAKING OF dc SHUNT MOTORS

Countercurrent braking of the dc shunt motor is done by two methods known as plugging and terminal voltage reversal (TVR). The plugging method is suitable for the gravitational-type load where the motor stops, reverses its direction of rotation, or operates under holding conditions. The TVR is also a method that can stop the motor rather rapidly or reverse its rotation. It cannot hold the motor at zero speed if the load is gravitational.

9.5.1 PLUGGING

Consider the example of the simple elevators shown in Figure 9.14. The elevator on the left side of the figure is moving upward, and the one on the right side is moving downward. In both cases, the load force is unidirectional, and so is the motor torque. The speed of the motor reverses its direction depending on the motion of the elevator cabin.

FIGURE 9.14
Bidirectional speed

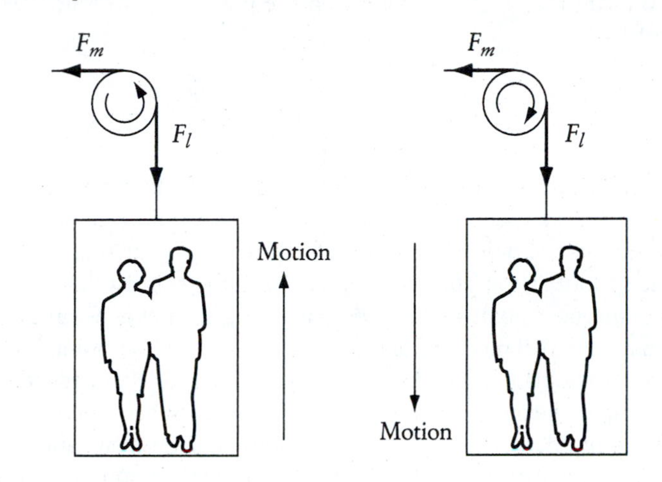

FIGURE 9.15
Plugging of dc, separately excited motor

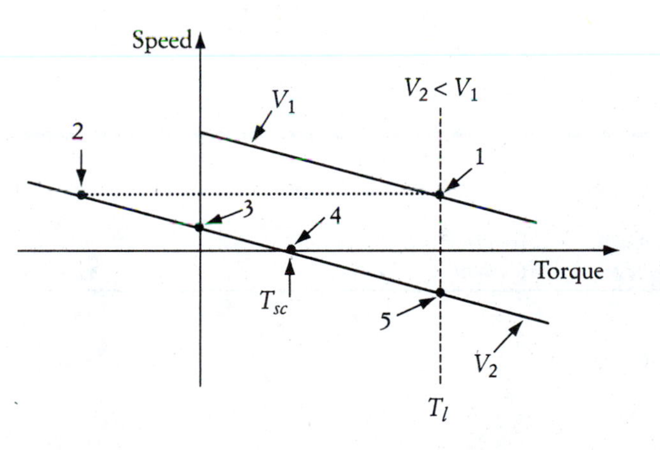

Figure 9.15 shows the characteristics of the motor operating under plugging. Let operating point 1 represent the upward motion of the elevator. The load torque is T_l, and the voltage applied to the armature of the motor is V_1. One way to achieve the downward operation is to reduce the armature voltage to V_2. The characteristics of the motor are shown in Figure 9.15 as parallel lines. If the change in the armature voltage is done quickly, the operating point of the motor moves rapidly from 1 to 2 without any change in the speed due to system inertia. The motor does not settle at point 2 since the load torque and the motor characteristic do not meet at point 2. Hence, the motor operating point moves from 2 to

5. While traveling from 2 to 5, the motor passes through points 3 and 4. Note that while the operating point is moving from 2 to 3, the motor is in the regenerative braking region. At point 3, the motor speed is equal to the no-load speed for the given value of V_2.

$$\omega_3 = \frac{V_2}{K\phi} \tag{9.13}$$

The operating point does not settle at point 3, but continues toward 4. At point 4, the motor stops and its speed is zero. If the braking technique is designed to stop the motor, the motor should be disconnected electrically at this point, and mechanical brakes should be applied to keep the motor at standstill.

The motor torque at point 4 is labeled T_{sc} in the figure. Let us assume that the motor does not have external mechanical brakes to hold its shaft at point 4. In this case, the motor moves to point 5, where the load torque and motor characteristics intersect.

In this example, the motor operates in three quadrants: first, second, and fourth. In the first quadrant, the machine operates as a motor. In the second and fourth quadrants, the machine is a generator. The operation in the second quadrant is explained earlier, in Section 9.1. In the fourth quadrant, the operation of the motor can be described by a set of equations for points 4 and 5. At point 4, the speed equation is

$$\omega_4 = \frac{V_2}{K\phi} - \frac{R_a}{(K\phi)^2} T_{sc} = 0 \tag{9.14}$$

TABLE 9.3
Plugging

Operating Point	*Motor Torque*	*Terminal Voltage*	*Armature Current*	*Speed*	*Field*	E_a	*Comments*
	→	→	→	→	→	→	
1	T_l	V_1		ω_1			Motor
	←	→	←	→	→	→	
2		$V_2 < V_1$		$\omega_2 = \omega_1$			Generator
		→		→	→	→	
3	0	$V_2 < V_1$	0	ω_3			No load
	→	→	→		→		
4	T_{sc}	$V_2 < V_1$		0		0	Holding
	→	→	→	←	→	←	
5	T_l	$V_2 < V_1$		ω_5			Generator

Hence,

$$T_{sc} = \frac{K\phi\, V_2}{R_a} \tag{9.15}$$

Since the torque at point 5 is larger than the torque at 4, the speed of the motor is negative at 5.

$$\omega_5 = \frac{V_2}{K\phi} - \frac{R_a}{(K\phi)^2} T_l < 0 \tag{9.16}$$

Table 9.3 summarizes the general operation of plugging for the example discussed in this section.

EXAMPLE 9.3

A dc motor has an armature resistance of 0.5 Ω, and $K\phi$ of 3 V sec. The motor is driven by an SCR, full-wave, ac/dc converter. The input to the converter is an ac source of 277 V. The motor is used as a prime mover of a forklift. In the upward direction, the mechanical load is 100 Nm, and the triggering angle of the converter is 20°. In the downward direction, the load torque is 200 Nm. Calculate the triggering angle required to keep the downward speed equal in magnitude to the upward speed. Assume that the motor current is always continuous.

SOLUTION

The operation of the system is depicted in Figure 9.16. Operating point 1 is for the upward motion and 2 is for the downward motion. The nonlinear shape of the characteristics is due to the presence of the converter, as explained in Section 6.1.4.3. Since the downward torque is higher than the upward torque, the voltage in the downward direction must be less than that in the upward direction to maintain the same magnitude of speed.

The equation for operating point 1 is

$$V_{eq_1} = E_{a_1} + R_a I_{a_1} = K\phi\, \omega_1 + R_a \frac{T_{\text{up}}}{K\phi}$$

where V_{eq_1} is given in Equation (6.22).

$$V_{eq_1} = \frac{2\, V_{\max}}{\pi} \cos \alpha_1 = \frac{2\sqrt{2} \times 277}{\pi} \cos(20) = 234.35 \text{ V}$$

The motor speed at point 1 is

$$\omega_1 = \frac{234.35}{3} - \frac{0.5 \times 100}{(3)^2} = 72.56 \text{ rad/sec}$$

$$n_1 = 693 \text{ rpm}$$

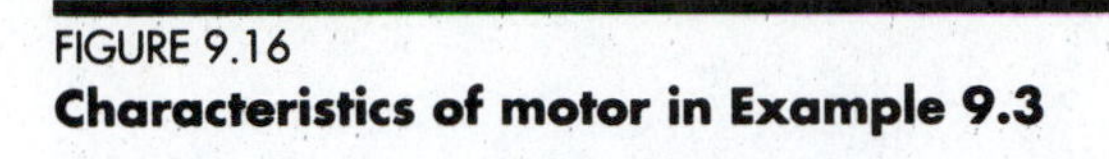

FIGURE 9.16
Characteristics of motor in Example 9.3

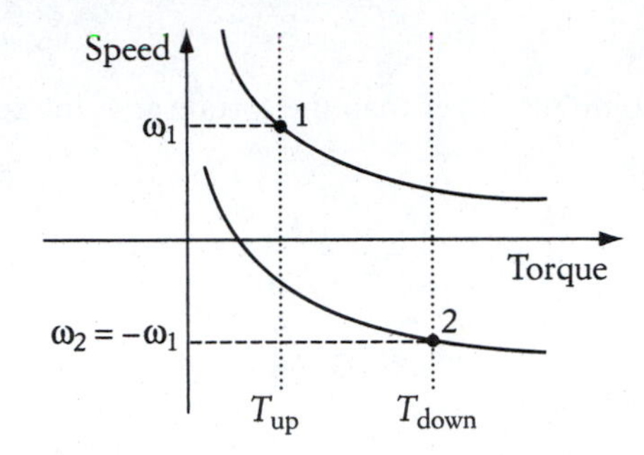

For operating point 2,

$$V_{eq_2} = \frac{2\,V_{\max}}{\pi}\cos\alpha_2 = E_{a2} + R_a I_{a2} = -K\phi\,\omega_1 + R_a\frac{T_{\text{down}}}{K\phi}$$

$$\frac{2\sqrt{2}\times 277}{\pi}\cos(\alpha_2) = -3\times 72.56 + \frac{0.5\times 200}{3}$$

$$\alpha_2 = 137.7°$$

EXAMPLE 9.4

For the motor in Example 9.3, the operator during the upward motion changes the triggering angle to keep the motor at holding position. Calculate the triggering angle.

SOLUTION

Point 1 in Figure 9.17 represents the normal operation of the motor. The holding operation is represented by point 3. The holding position is reached when the load torque is equal to the motor torque and the speed of the motor is zero. In this case, $E_{a_3} = 0$, and

$$V_{eq_3} = R_a\,I_{a3} = \frac{R_a\,T_{\text{up}}}{K\phi}$$

$$V_{eq_3} = \frac{2\,V_{\max}}{\pi}\cos\alpha_3 = \frac{R_a\,T_{\text{up}}}{K\phi}$$

$$\frac{2\sqrt{2}\times 277}{\pi}\cos\alpha_3 = \frac{0.5\times 100}{3}$$

$$\alpha_3 = 86.2°$$

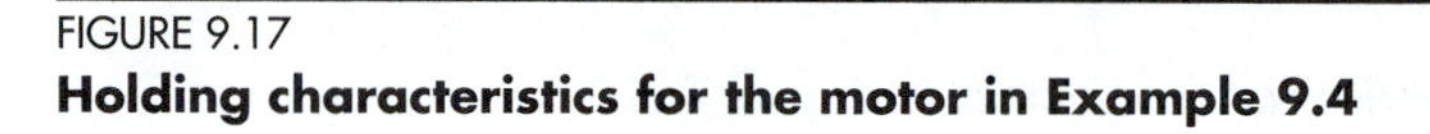

FIGURE 9.17
Holding characteristics for the motor in Example 9.4

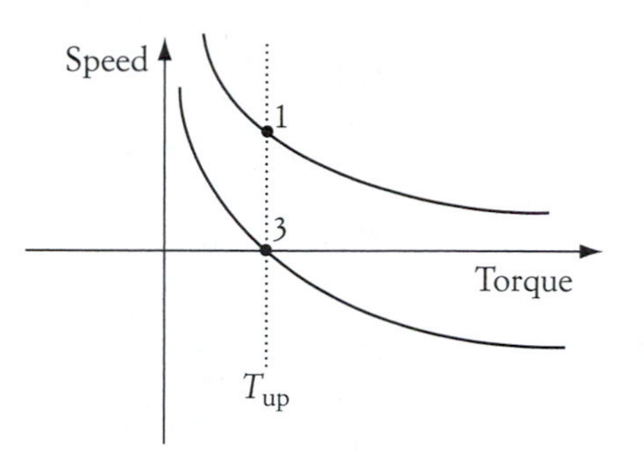

9.5.2 BRAKING BY TERMINAL VOLTAGE REVERSAL (TVR)

The TVR braking scheme is a method based on reversing the motor terminal voltage. In doing so, the motor speed stops abruptly, then reverses its rotation—but only if the field current of a dc motor is not reversed. Figure 9.18 shows a schematic of a TVR braking circuit. Examine the circuit of Figure 9.18(a). Note the direction of the armature current and the polarities of the armature voltage. During the steady-state operation, the polarities of E_a are the same as the polarities of the terminal voltage.

Now examine Figure 9.18(b). In this circuit, the polarities of the terminal voltage are reversed. Also, the polarities of E_a are reversed, and the current I_a flows in the opposite direction. Keep in mind that the field current is kept in its original direction. In the steady state, since E_a reverses its polarities while ϕ is unchanged, the speed of the motor reverses its direction.

The characteristics of the motor with different polarities are shown in Figure 9.19. If the motor in Figure 9.18(a) operates in the first quadrant, the motor in Figure 9.18(b) operates in the third quadrant.

Now assume that the motor is originally running at steady state and is connected as shown in Figure 9.18(a). Also, assume that the load torque is bidirectional. The motor in this case operates at point 1, shown in Figure 9.20. When the motor's terminal voltage is reversed, the motor eventually operates at point 4 in the steady state. If the reversal of the voltage is sudden, the operating point moves first from point 1 to point 2. The speed of the motor does not change during this time due to the inertia of the system. The motor does not settle at point 2 because the load torque and the motor torque are not equal. The motor continues to move until it reaches point 4, which is the new steady-state point.

FIGURE 9.18
Armature circuit for TVR braking

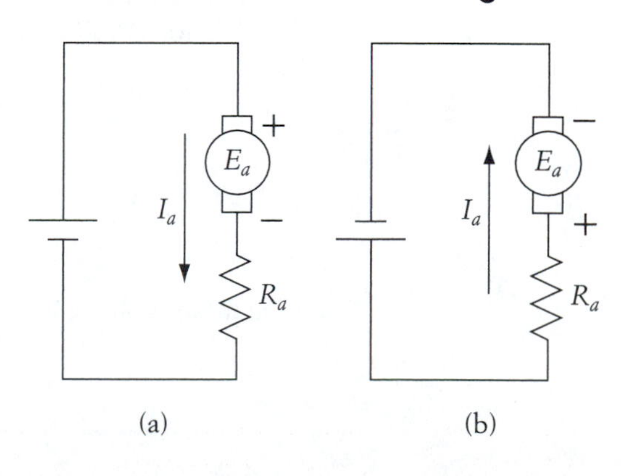

While traveling, the motor passes through point 3, where the speed of the motor is zero. If the objective of the TVR is to stop the motor, the

FIGURE 9.19
Motor characteristics for the circuits in Figure 9.18

FIGURE 9.20
TVR braking

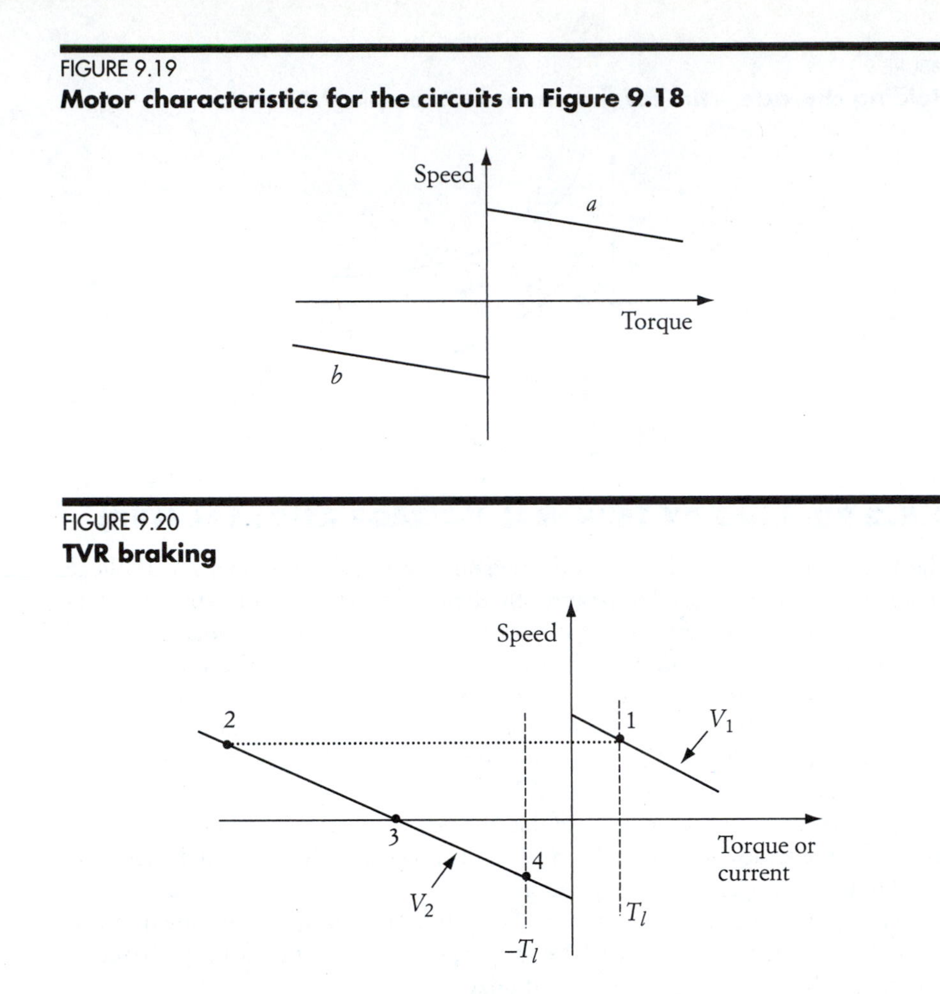

motor should be disconnected electrically at this point and mechanical brakes should be applied. While traveling from 2 to 3, the motor operates in the plugging mode. To realize this, turn Figure 9.20 upside down.

The equations describing operating point 1 are

$$V_1 = K\phi\,\omega_1 + R_a \frac{T_l}{K\phi} \tag{9.17}$$

$$I_1 = \frac{V_1 - E_{a_1}}{R_a} \tag{9.18}$$

When the motor voltage is reversed, the equations at point 2 are

$$V_2 = -V_1 = K\phi\,\omega_1 + R_a \frac{T_2}{K\phi} \tag{9.19}$$

$$I_2 = \frac{-V_1 - E_{a_1}}{R_a} \tag{9.20}$$

Keep in mind that T_2 is a negative value. Note that while $E_{a_1} = E_{a_2}$, I_2 is much larger than I_1 (almost double the starting current) and is flowing in the opposite direction to I_1. This high magnitude of I_2 can have a damaging effect and must be controlled.

At point 3, the speed is zero and so is E_{a_3}. The equations of the system at point 3 are

$$-V_1 = R_a \frac{T_3}{K\phi} \tag{9.21}$$

$$I_3 = \frac{-V_1}{R_a} \tag{9.22}$$

The current I_3 is less than the current in Equation (9.20). However, it is still very large and equal to the starting current of the motor without starters. Remember that the starting current without starter could be damaging to the motor.

The motor operating point at 4 is described by

$$-V_1 = K\phi\,\omega_4 + R_a \frac{-T_l}{K\phi} \tag{9.23}$$

$$I_4 = \frac{-V_1 - E_{a_4}}{R_a} \tag{9.24}$$

$$E_{a_4} = K\phi\,\omega_4$$

where ω_4 and E_{a_4} are negative. Table 9.4 summarizes the general operation of TVR. It is assumed in this table that the load torque is bidirectional.

The current I_4 is equivalent in magnitude to the normal operating current. Figure 9.21 is a conceptual graph that shows the armature current of the motor at all four operating points. All currents are per unit with I_1 the base value. As explained earlier, the current at operating point 2 is excessively large and can cause permanent damage to the motor; it must be kept within safe limits. One way to do this is to simultaneously insert a resistance (braking resistance) in the armature circuit when the armature voltage is reversed. The schematic of such a circuit is shown in Figure 9.22.

Assume that the contacts of switch S are in the A position while the motor operates at point 1. When the

FIGURE 9.21
TVR braking current for selected operating points

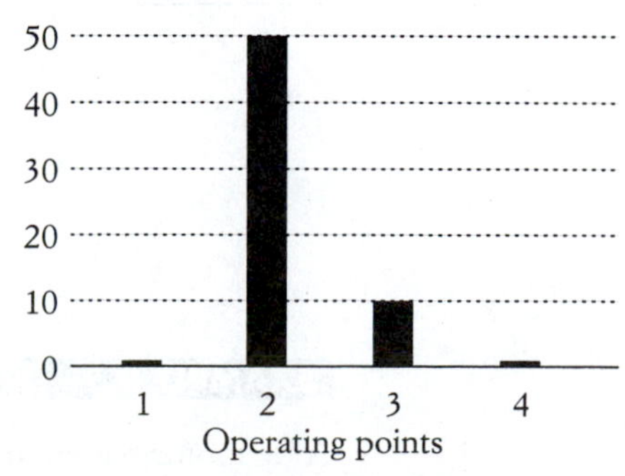

TABLE 9.4
Terminal voltage reversal

Operating Point	*Motor Torque*	*Terminal Voltage*	*Armature Current*	*Speed*	*Field*	E_a	*Comments*
	→	→	→	→	→	→	
1				ω_1			Motor
	←	←	←	→	→	→	
2				$\omega_2 = \omega_1$			Generator
	←	←	←		→		
3				0		0	Holding
	←	←	←	←	→	←	
4				ω_4			Motor

FIGURE 9.22
Reduction of armature current during TVR braking

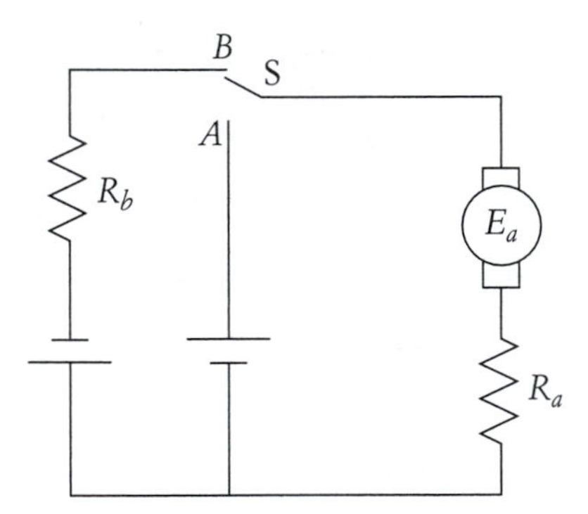

switch moves to position *B*, the terminal voltage of the motor is reversed and a braking resistance R_b is inserted. Figure 9.23 shows three characteristics that describe the operation of the circuit in Figure 9.22. The first is for normal operation when the switch contacts are in position *A*. The second, which has operating point 3, is for the case when a braking resistance is inserted and the switch contacts are in position *B*. The third is for the case when the switch contacts are in position *B* but no braking resistance is inserted. Note that point 2 in this figure is the same as point 2 in Figure 9.20.

When the braking resistance is applied, the slope of the motor characteristic is steeper, and the current at point 3 is smaller than at point 2. The equations of the motor at point 3 are

$$-V_1 = K\phi\,\omega_1 + (R_a + R_b)\frac{T_3}{K\phi} \tag{9.25}$$

$$I_3 = \frac{-V_1 - E_{a_1}}{R_a + R_b} \tag{9.26}$$

EXAMPLE 9.5

A dc motor has an armature resistance of 1 Ω, and $K\phi = 3$ V sec. When the motor's terminal voltage is adjusted to 320 V, the motor speed is 1000 rpm. A TVR

FIGURE 9.23
Effect of braking resistance during TVR braking

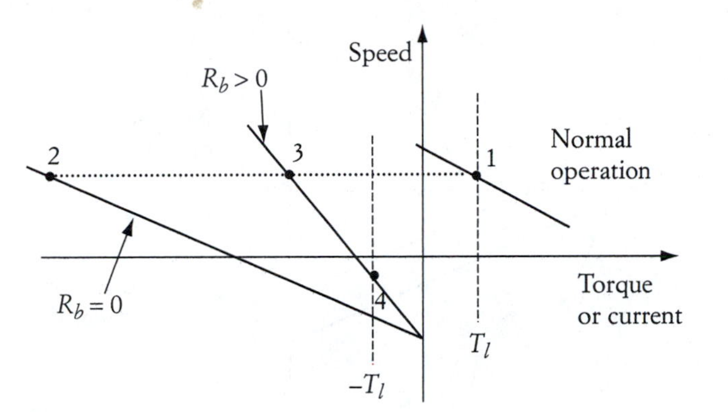

braking is applied. Calculate the value of the braking resistance that would reduce the maximum braking current to twice the rated current.

SOLUTION

First, let us find normal operating current.

$$I_a = \frac{V_1 - E_{a_1}}{R_a} = 6\ \text{A}$$

The maximum braking current occurs at point 3 in Figure 9.23. The braking current is twice the rated current.

$$I_3 = I_b = 2\,I_a = 12\ \text{A}$$

At point 3,

$$I_b = \frac{-V_1 - E_{a_1}}{R_a + R_b} = \frac{-V_1 - K\phi\,\omega_1}{R_a + R_b} = \frac{-320 - 3\,\frac{2\pi}{60}\,1000}{1 + R_b} = 12\ \text{A}$$

$$R_b = 51.8\ \Omega$$

Another common method for current reduction during TVR braking is to reduce the terminal voltage. Figure 9.24 shows the basic concept. Assume that the normal operating point is 1. When the terminal voltage of the motor is reversed and the voltage magnitude is unchanged, the operating point moves rapidly to point 2, which results in an excessive braking current. However, when the voltage is reversed and reduced, the operating point is 3. The braking current at 3 is smaller than at point 2. If the voltage remains unchanged, the new steady-state operating point is 4.

FIGURE 9.24
Effect of voltage reduction during TVR braking

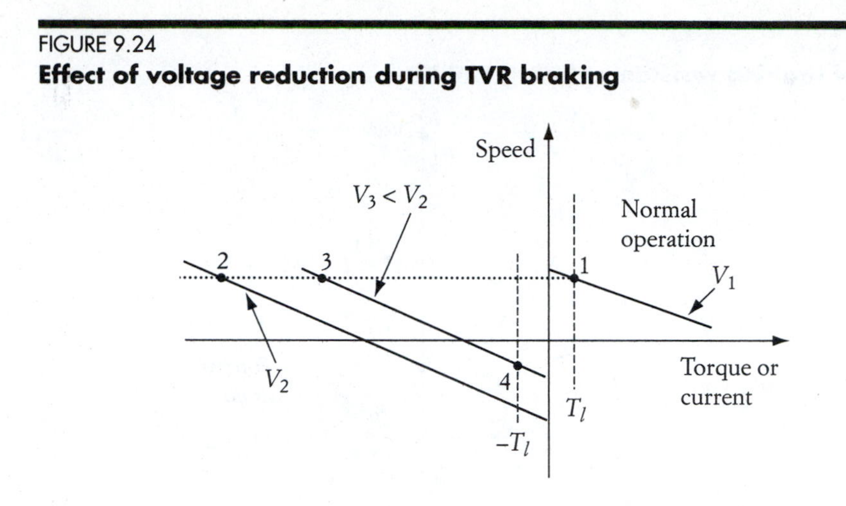

FIGURE 9.25
TVR braking circuit with dc source

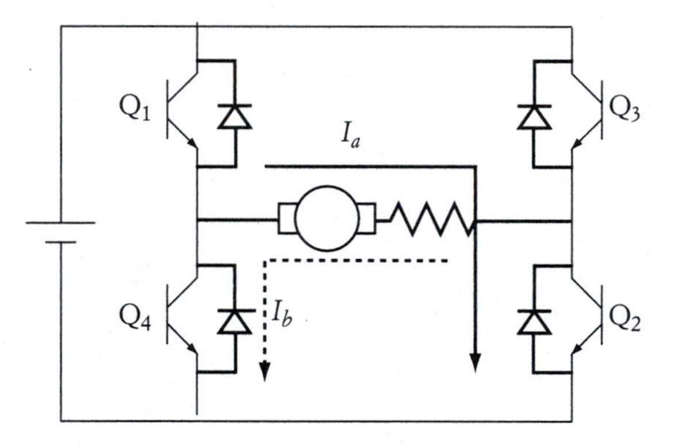

Figure 9.25 shows one schematic that can be used to achieve this type of braking. When transistors Q_1 and Q_2 are triggered, the terminal voltage of the motor has the positive polarity at the emitter side of Q_1. If Q_3 and Q_4 are triggered, and Q_1 and Q_2 are turned off, the voltage across the motor has its polarities reversed. The diodes in the figure are for freewheeling. To reduce the braking current due to the TVR, transistors Q_3 and Q_4 must be pulse-width modulated to reduce the voltage across the motor terminals.

For an ac source, a TVR braking circuit can be constructed from solid-state switches such as SCRs. Such a circuit is shown in Figure 9.26. The SCRs labeled 1 and 2 are used in normal operation. SCRs 1 are triggered when terminal *A* is positive with respect to *B*. SCRs 2 are triggered when terminal *B* is positive with respect to *A*. This makes terminal *C* of the motor positive regardless of the polarity of the source volt-

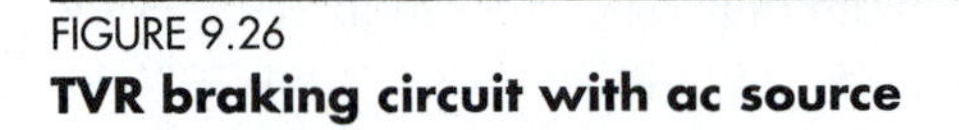
FIGURE 9.26
TVR braking circuit with ac source

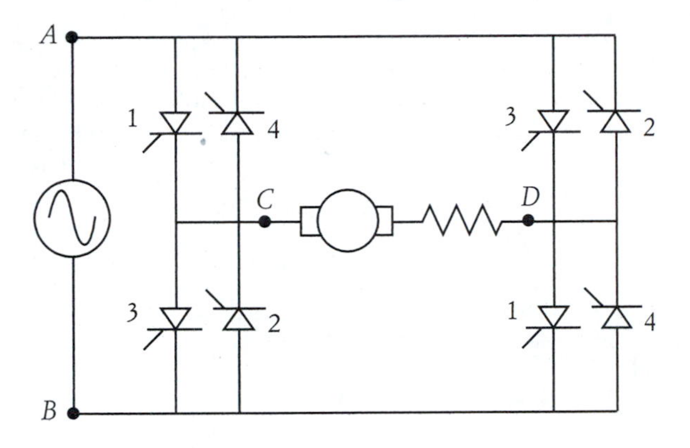

age. There is no need here for freewheeling diodes because the SCRs are naturally commutated, and the circuit is a full-wave bridge.

To do the TVR braking, SCRs 3 are turned on when terminal A is positive with respect to B. Also, SCRs 4 are triggered when B is positive with respect to A. This switching sequence reverses the polarities of the motor (D is positive with respect to C). Keep in mind that the triggering angle of SCRs 3 and 4 must be adjusted to reduce the voltage across the motor during the TVR braking.

The motor characteristics for the TVR braking circuit of Figure 9.26 are shown in Figure 9.27. Here we are assuming that the load torque is bidirectional. The original motor operating point is point 1. After SCRs 1 and 2 are commutated and SCRs 3 and 4 are triggered, the motor moves to point 2. The motor characteristic is mainly in the third quadrant, but because of the reduction in the terminal voltage, the characteristics tend to extend to the second quadrant. The motor settles at an operating point where the load and motor characteristics meet.

FIGURE 9.27
TVR braking characteristics using the circuit in Figure 9.26

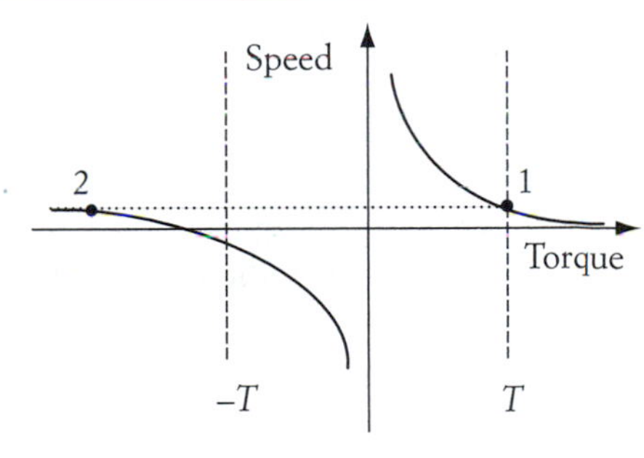

EXAMPLE 9.6

Consider the circuit in Figure 9.26. Assume that the dc, separately excited motor has an armature resistance of 1 Ω and a field constant $K\phi = 3$ V sec. The ac voltage source of the circuit is 480 V. Assume that the load torque is bidirectional and equal to 120 Nm in either direction of rotation. The triggering angle of SCRs 1 and 2 is 30° and the current is continuous. A TVR braking is used. Calculate the triggering angle of SCRs 3 and 4 to reduce the maximum braking current to three times the current before the TVR braking is applied. Also assume that the current is continuous during normal and braking operations.

SOLUTION

Let point 1 in Figure 9.27 represent the operating point at which SCRs 1 and 2 are triggered. At this point, the motor speed can be obtained from Equation (6.23).

$$\frac{2\,V_{max}}{\pi}\cos\alpha_1 = K\phi\,\omega_1 + R_a I_{ave\,1} \tag{9.27}$$

$$I_{ave\,1} = \frac{T_1}{K\phi} = \frac{120}{3} = 40 \text{ A}$$

Substituting $I_{ave\,1}$ in Equation (9.27) yields

$$\frac{2\sqrt{2} \times 480}{\pi}\cos(30) = 3\,\omega_1 + 40$$

$$\omega_1 = 111.418 \text{ rad/sec}$$

$$n_1 = 1{,}063.96 \text{ rpm}$$

At point 2, when SCRs 3 and 4 are triggered, the motor speed is unchanged momentarily. The terminal voltage and the motor current are reversed. The speed equation in this case can be written as follows:

$$V_b = K\phi\omega + R_a I_b$$

where V_b is the braking voltage and I_b is the braking current at point 2. The braking voltage and current are negative in magnitude as compared to the voltage and current at point 1. Equation (9.27) can then be modified to reflect these changes as follows:

$$-\frac{2\,V_{max}}{\pi}\cos\alpha_2 = K\phi\omega_1 + R_a I_b \tag{9.28}$$

For $I_b = -3\,I_{ave\,1}$,

$$-\frac{2\sqrt{2} \times 480}{\pi}\cos(\alpha_2) = 3 \times 111.418 - 120$$

$$\alpha_2 = 119.72°$$

EXAMPLE 9.7

Assume that it is necessary to keep the motor at holding after TVR braking is applied. Calculate the triggering angle of SCRs 3 and 4.

FIGURE 9.28
Holding by TVR braking

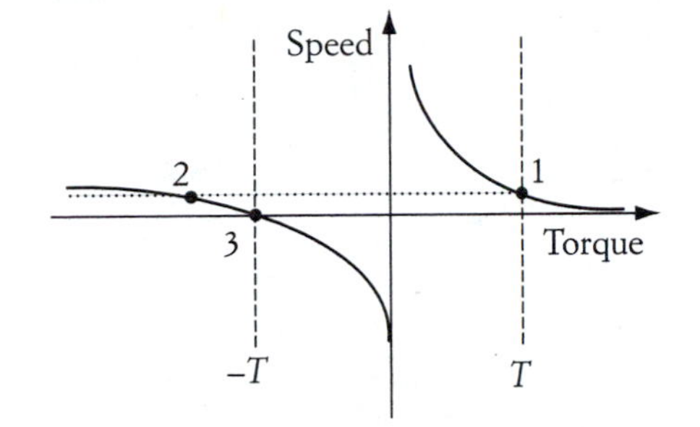

SOLUTION
To keep the motor at holding, the motor torque must be equal to the load torque at the point where the speed is zero. This is shown in Figure 9.28 as point 3. In this case, Equation (9.27) can be modified so that the speed is equal to zero, the load current is equal to the normal current in magnitude but opposite in sign, and the terminal voltage is negative.

$$-\frac{2\,V_{max}}{\pi}\cos(\alpha_3) = K\phi\,\omega_3 + R_a I_{ave\,3} \tag{9.29}$$

$$-\frac{2\sqrt{2}\,480}{\pi}\cos(\alpha_3) = -40$$

$$\alpha_3 = 84.69°$$

When TVR braking is used with a gravitational load or any other unidirectional load torque, the new steady-state operating point is in the fourth quadrant. Examine Figure 9.29. Assume that the original operating point before TVR is point 1. If the terminal voltage of the motor is reversed, the operating point moves to point 2 without substantial change in speed. Since the motor torque and the load torque at point 2 are not equal, the operating point continues moving to 3. At point 3, the motor is momentarily at zero speed. The motor continues moving to point 4 due to the mismatch between the load and motor torques. At point 4, the motor torque is equal to zero, and the speed of the motor is equivalent to the no-load speed expressed by

$$\omega_4 = \frac{V_2}{R_a}$$

where V_2 is the terminal voltage after TVR is applied, which is a negative value. The motor does not settle at point 4, but continues moving to the new steady-state point 5, where the load and motor torques are equal. At point 5, the motor speed exceeds

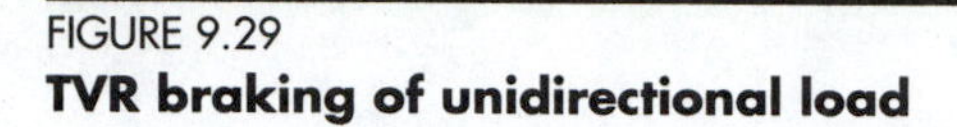

FIGURE 9.29
TVR braking of unidirectional load

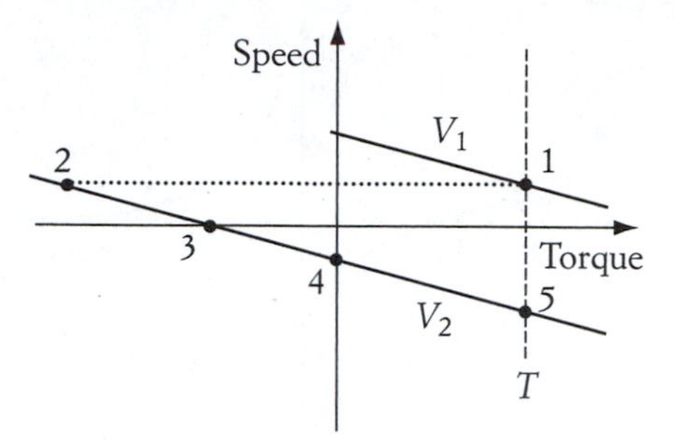

the no-load speed; this corresponds to regenerative braking when the motor operates at a reversed voltage. The motor equations at point 5 are

$$V_2 = -V_1 = K\phi\,\omega_5 + R_a \frac{T_5}{K\phi}$$

$$I_5 = \frac{-V_1 - K\phi\,\omega_5}{R_a} \tag{9.30}$$

Table 9.5 shows the machine variables for the five operating points. Note that during this TVR braking, the motor passes through all four quadrants.

EXAMPLE 9.8

A dc, separately excited motor has an armature resistance of 0.5 Ω and a field constant $K\phi = 3$ V sec. The dc voltage source of the circuit is 200 V. The motor is driving a forklift whose torque is 180 Nm. A TVR braking is applied by switching the terminal voltage of the motor to a 30 V reversed-polarity dc supply. Calculate the new steady-state speed and the armature current at the new speed.

SOLUTION

Since the load torque is constant, the new operating point is in the fourth quadrant at point 5, as shown in Figure 9.29.

$$\omega_5 = \frac{V_2}{K\phi} - \frac{R_a}{(K\phi)^2} T_l$$

$$\omega_5 = \frac{-30}{3} - \frac{0.5}{9} 180 = -20 \text{ rad/sec}$$

$$n_5 = -190.98 \text{ rpm}$$

The current at point 5 is

$$I_5 = \frac{V_2 - K\phi\,\omega_5}{R_a} = \frac{-30 - 3(-20)}{0.5} = 60 \text{ A}$$

9.6 COUNTERCURRENT BRAKING OF dc SERIES MOTORS

The two basic methods for countercurrent braking can also be applied to the series motor. The reversal of E_a can be achieved by simply adding resistance to the armature circuit, as shown in Figure 9.30. The left side of the figure shows the normal operation, and the right side shows the countercurrent braking by reversing E_a. The equation of the series motor under this type of braking is

$$E_a = V_t - (R_a + R_b)I_b \qquad (9.31)$$

where I_b is the steady-state armature current after braking, which can be computed by Equation (5.20).

$$I_b = \sqrt{\frac{T_l}{KC}}$$

TABLE 9.5

TVR braking for the case in Figure 9.29

Operating Point	*Motor Torque*	*Terminal Voltage*	*Armature Current*	*Speed*	*Field*	E_a	*Comments*
	$\rightarrow$	$\rightarrow$	$\rightarrow$	$\rightarrow$	$\rightarrow$	$\rightarrow$	
1		V_1		ω_1			Motor
	$\leftarrow$	$\leftarrow$	$\leftarrow$	$\rightarrow$	$\rightarrow$	$\rightarrow$	
2		$V_2 = -V_1$		$\omega_2 = \omega_1$			Generator
	$\leftarrow$	$\leftarrow$	$\leftarrow$		$\rightarrow$		
3		V_2		0		0	Holding
		$\leftarrow$		$\leftarrow$	$\rightarrow$	$\leftarrow$	
4	0	V_2	0	ω_4			No load
	$\rightarrow$	$\leftarrow$	$\rightarrow$	$\leftarrow$	$\rightarrow$	$\leftarrow$	
5		V_2		$\omega_5 > \omega_4$			Generator

If the terminal voltage and load torque are constant, E_a in Equation (9.31) can become negative when the braking resistance R_b is large enough. The characteristics of this type of braking are given in Figure 9.31. Let point 1 be the original operating point and point 2 be the new steady-state operating point under countercurrent braking. At point 2, the motor operation is described by Equation (9.31).

TVR braking can also be implemented for the series motor. The idea of the TVR braking is to reverse the source voltage while maintaining the field in its original direction. For a series motor, however, the reversal of the supply voltage leads to the reversal of the armature current, which in turn reverses the field. To prevent the field from being reversed, the circuit in Figure 9.32 can be implemented. In normal operation, the switch S is at position *A*, and transistors Q_1 and Q_2 are closed. Diode D_1 is not conducting and diode D_2 is conducting. D_2 shorts the braking re-

FIGURE 9.30

Countercurrent braking by reversing E_a for series motor

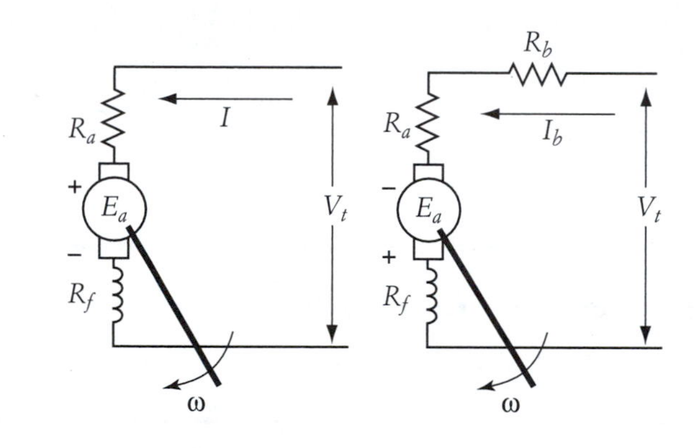

FIGURE 9.31

Characteristics of countercurrent braking of series motor by reversing E_a

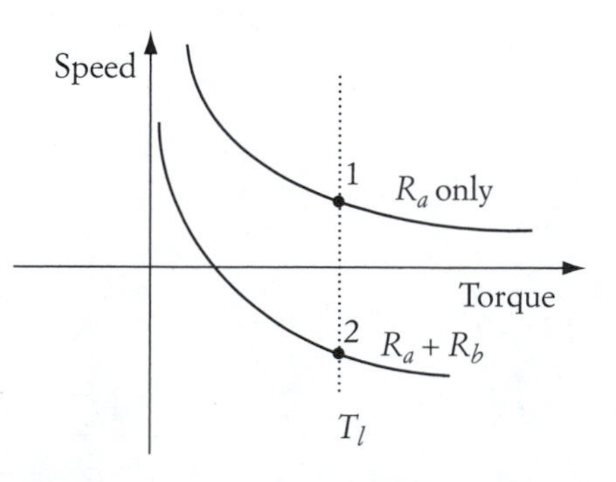

sistance during normal motor operation. To apply TVR braking, the switch S changes its position to *B*. At the same time, Q_1 and Q_2 are turned off, and transistors Q_3 and Q_4 are turned on, which allows the field current to remain in its original direction, and the armature current to reverse its flow. Diode D_1 does not allow the current in the field circuit to be interrupted while the transistors are switching. Diode D_2 is now in its reverse conduction and allows the braking current to flow in the braking resistance.

The characteristics of the circuit in Figure 9.32 are shown in Figure 9.33. Assume that the original operating point of the motor is point 1. When the TVR is implemented as we have described, the operating point moves to 2. Note that the original characteristic is in the first quadrant only, while the TVR characteristic is in the third and second. This is because of the presence of the braking resistance during the TVR braking.

After the motor reaches point 2, it moves to point 3, which is at zero speed. The motor should be disconnected electrically at this point. If the load torque is unidirectional, the motor cannot produce a torque that meets the load demand. There is no operating point at which the motor torque matches the load torque.

FIGURE 9.32
TVR braking circuit for series motor

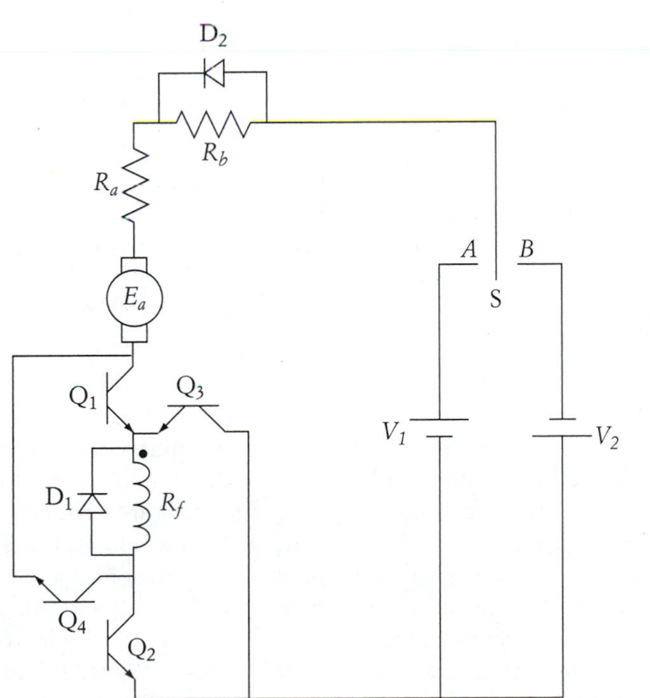

FIGURE 9.33
Characteristics of TVR braking for series motor

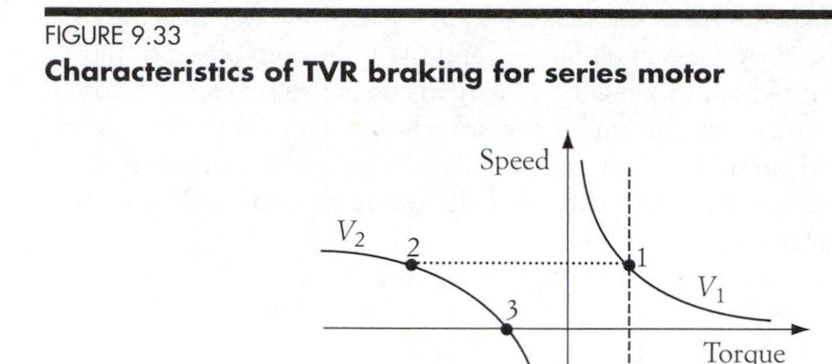

CHAPTER 9 PROBLEMS

9.1 A 23 hp, dc shunt motor is running at 1000 rpm when the armature voltage is 600 V. The armature resistance of the motor is 0.8 Ω, and the field resistance is 900 Ω. The efficiency of the motor at full load is 92%.

a. Calculate the value of the resistance of the dynamic braking at full load. The armature current should be limited to 150% of the rated value.

b. If the motor operates in regenerative braking at a speed of 1100 rpm, calculate the line current.

c. Design a circuit for countercurrent braking (by reversing the supply voltage) so that the line current will not exceed 120% of the rated value.

9.2 A 16 kW, 220 V, dc series motor has its armature current equal to 86 A when the motor speed is 600 rpm. The armature resistance of the motor is 0.198 Ω, and the field resistance is 0.1 Ω. The motor voltage can be adjusted by means of solid-state devices. The motor is to be used to lower a load by countercurrent braking (reversing E_a) at an armature current of double the rated value, and at a speed of 10 rpm. Determine the terminal voltage required for this operation.

9.3 A 220 V, dc series motor has an armature current of 86 A, armature resistance of 0.2 Ω, and a field resistance of 0.1 Ω. The terminal voltage of the motor can be adjusted by means of solid-state devices. Ignore the field saturation.

a. When the motor runs at 600 rpm, the armature current is 20 A. A 0.1 Ω resistance is added in shunt to the field windings. If the load torque remains constant, calculate the speed of the motor.

b. The resistance in shunt of the field windings is removed. The motor is used to lower the same mechanical torque described in (a) by a countercurrent braking method. The desired speed in the countercurrent braking is 10 rpm. The braking current is twice the rated value. Calculate the armature voltage of the motor.

9.4 A dc, separately excited motor is driving an elevator. At a load torque of 300 Nm, the motor speed is 100 rpm. The motor terminal voltage is 210 V, and the field constant $K\phi$ is 3 V sec.

a. If the terminal voltage is reversed and reduced to 90 V, calculate the steady-state speed of the motor.
b. Calculate the terminal voltage required to block (stop) the rotor.
c. Sketch the speed–torque characteristics and show all operating conditions.

9.5 A dc, separately excited motor is used in a forklift. The motor is driven by a full-wave, ac/dc converter. The voltage on the ac side is 120 V (rms). The field constant of the motor $K\phi = 3$ V sec. The armature current is continuous under loading conditions. At a load torque of 300 Nm and triggering angle of 40°, the motor speed is 100 rpm.

a. Calculate the triggering angle required to block (stop) the motor.
b. Calculate the average voltage across the motor terminals if the triggering angle increases to 120°.
c. Calculate the new speed of the motor at the condition described in part (b).
d. Sketch the speed–torque characteristics and show all the operating conditions.

9.6 A 150 V, dc, separately excited motor is used to lift material. The motor has an armature resistance of 1 Ω, and $K\phi = 3$ V sec. The load torque is 9 Nm. The armature voltage of the motor is controlled by a single-phase, full-wave, ac/dc solid-state converter employing SCRs. The voltage on the ac side of the converter is 120 V (rms). The triggering angle of the SCRs is 30° when the motor is lifting up the load. Calculate the motor speed. Also calculate the triggering angle required to lower the same load at the same speed.

9.7 A 250 V, 500 rpm, dc, separately excited motor is driving a constant-load torque. The motor has a field constant $K\phi = 4$ V sec and an armature resistance of 1 Ω. Ignore the frictional losses.

a. Calculate the full-load torque.
b. Determine the terminal voltage required to block (stop) the motor.
c. Determine the terminal voltage needed to rotate the motor at 100 rpm in the reverse direction.
d. Calculate the motor speed while the motor is operating at the conditions of part (c), and the polarities of the terminal voltage are reversed.
e. Sketch the speed–torque characteristics and show all the operating conditions.

9.8 A 240 V, 70 rad/sec, dc, separately excited motor has a field constant $K\phi = 3$ V sec and an armature resistance of 1 Ω.

a. Calculate the dynamic braking resistance that will not allow the braking current to be larger than twice the rated value.

b. If the motor drives a constant-torque load of 60 Nm, determine the new steady-state speed and the current after the dynamic braking is applied.

9.9 A dc, separately excited motor is driven by a full-wave, SCR, ac/dc converter. The voltage on the ac side is 120 V (rms). The field constant of the motor $K\phi = 3$ V sec. At a load torque of 300 Nm and triggering angle of 40°, the motor speed is 100 rpm, and the armature current is continuous.

a. The polarities of the terminal voltage of the motor (armature voltage) are reversed, and the triggering angle is simultaneously changed. Assume that the load torque is also reversed and is equal to 300 Nm. Calculate the triggering angle that stops the motor and keeps it at the holding position. Assume that the current is still continuous.

b. Sketch the speed–torque characteristics and indicate the steady-state operating points before and after polarity reversal.

9.10 A dc, separately excited motor has an armature resistance of 1 Ω and field constant $K\phi = 3$ V sec. The motor is driven by a full-wave, ac/dc, SCR circuit. The triggering angle of the SCRs is 30°. The voltage on the ac side is 277 V (rms). The motor load is bidirectional and equals 120 Nm at steady state, regardless of the direction of rotation. A countercurrent braking is implemented by reversing the polarities of the motor terminals. The triggering angle of the SCRs is adjusted so that the maximum braking current is twice the steady-state value. Calculate the new steady-state speed.

10

Braking of Induction Motors

Electrical braking is used to maintain the speed of the motor at a certain range without overspeeding, to stop the motor, or to hold the motor at a specific rotor position. If no form of braking is applied, the motor stops only when the kinetic energy stored in its rotating mass is dissipated due to friction and windage losses. For larger motors with good bearing systems, this could take a long time. If faster braking is needed, the kinetic energy must be dissipated much faster through other means, as described in Chapters 8 and 9.

The basic types of braking discussed in Chapter 9 for dc motors are also applicable to induction motors. The principles are the same for regenerative, dynamic, or countercurrent braking, but the implementation is different due to the difference in topology and the principles of rotation. Although more involved, the braking of the induction motor is less stressful than that of dc motors in terms of transients.

10.1 REGENERATIVE BRAKING

Regenerative braking occurs when the motor speed exceeds the synchronous speed. This may happen when the load torque drives the electric motor beyond its synchronous speed. In this case, the load is the source of energy and the induction machine is converting the mechanical power into electrical power, which is delivered back to the electrical system.

Wind generators using induction machines are good examples of regenerative braking. These wind-generating machines are very popular and widely used in wind farms. The photo in Figure 10.1 shows a wind farm located at Tehachapi near Los Angeles, California. Induction machines are popular in wind applications because they are ideally suited for variable-power-profile applications. Unlike the synchronous or dc machines, induction machines become automatically synchronized with the external power system.

The basic components of a wind generator system are shown in Figure 10.2. There are several design variations for the wind machine, but the horizontal design in the figure is common. It consists of a housing box mounted on top of a

FIGURE 10.1
Photo of a wind farm located near Los Angeles, California

FIGURE 10.2
Wind-generating system

Horizontal design

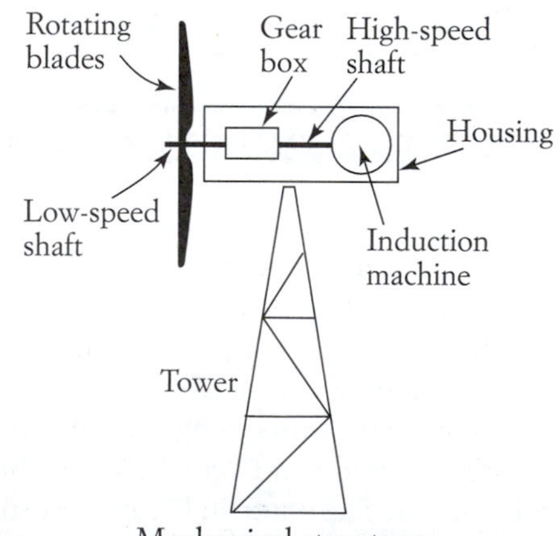

Mechanical structure

tower, which contains the induction machine, a gearbox, and the rotating blades. The high-speed shaft of the gearbox is connected to the induction machine, and the low-speed shaft is connected to the rotating blades. At moderate wind speed (called cutoff speed), the gearbox is designed to rotate the high-speed shaft near the synchronous speed of the induction machine. If the wind speed increases beyond the cutoff speed, the induction machine rotates at a speed higher than its synchronous speed. This is the regenerative braking operation. The housing box can swivel at the top of the tower to point the blades at the direction of maximum wind effects. When the wind speed becomes excessive, the blades can be locked to prevent any mechanical damage to the system.

Regenerative braking can be explained using Figure 10.3 and the torque equation (5.57) given in Chapter 5.

$$T_d = \frac{V^2 R'_2}{s\omega_s\left[\left(R_1 + \frac{R'_2}{s}\right)^2 + X_{eq}^2\right]} \tag{10.1}$$

The voltage V is a line-to-line quantity and T_d is the developed torque for the three phases. Figure 10.3 is a graph representing Equation (10.1) but is expanded to include negative slip, which yields an electric torque in the second quadrant. The negative slip occurs when the speed of the machine exceeds the synchronous speed.

$$s = \frac{n_s - n}{n_s}$$

Since the torque of the machine is negative during regenerative braking, but the direction of rotation of the machine is the same as that in the first quadrant, the flow of power is reversed. The mechanical power is the source of energy and is converted to electrical power by the machine. This electrical power is delivered to the electrical system, and the machine is acting as generator.

The induction machine of the wind system is designed to operate at regenerative braking only (second quadrant). When the wind speed is low so that the rotor speed is near or below the synchronous speed, the blades are locked and the motor is disconnected from the electrical supply to prevent the machine from running as a motor.

When used in wind applications, induction machines demand a significant amount of reactive power from the utility system, mainly because they do not have their own field circuit. When used in the regenerative mode, the induction machine consumes reactive power from the system while delivering real power. In some cases, the amount of reactive power consumed exceeds the amount of real power generated. This reactive power is also dependent on the speed of the machine, which can be explained by using the equivalent circuit of Figure 5.26(b). The inductive reactive power Q is consumed in the magnetizing inductive reactance X_m and the equivalent winding reactance X_{eq}.

FIGURE 10.3
Regenerative braking of induction machines

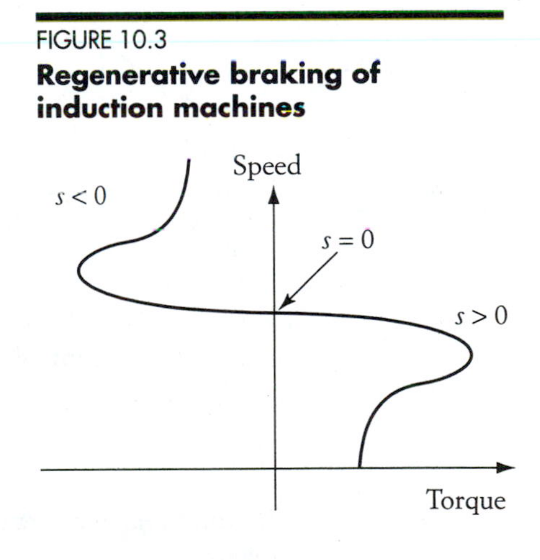

$$Q = \frac{V^2}{X_m} + 3(I'_2)^2 X_{eq}$$

$$Q = V^2 \left[\frac{1}{X_m} + \frac{X_{eq}}{\left(R_1 + \frac{R'_2}{s}\right)^2 + X_{eq}^2} \right] \tag{10.2}$$

where V is the line-to-line voltage. The reactive power in Equation (10.2) is plotted in Figure 10.4. Q is at its minimum value at synchronous speed, but is almost linearly proportional to speed when $n > n_s$. In a wind farm when a large number of induction machines are producing energy, the reactive power demand is excessive and could lead to poor voltage regulations, especially when the machines are installed in remote areas. Another problem arises from the fact that the wind's speed is continually changing, which changes the speeds of all induction machines in the farm. This may result in a cyclic variation in reactive power demand, which leads to voltage flickers for nearby customers. One way to alleviate this problem is to install a reactive power controller at the wind farm site, as shown in Figure 10.5. The controller must be fast and adaptive to compensate for the cyclic variations of the reactive power of the induction machine. Such a controller is usually made out of several capacitor banks switched by solid-state devices. The cyclic inductive reactive power of the induction machine Q is sensed and the controller produces a capacitive reactive power Q_c, which is equal in magnitude to Q.

FIGURE 10.4
Reactive power of induction machine in regenerative braking

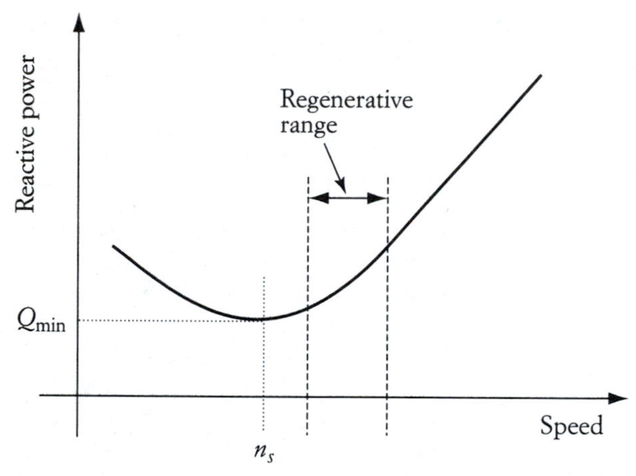

$$Q_c = -Q$$

The total reactive power seen by the system Q_s is the sum of these two reactive powers.

$$Q_s = Q_c + Q$$

With ideal compensation, the system delivers no reactive power to the induction machine. This describes an ideal wind farm operation.

In more general drive systems, the induction machine may operate in the first or second quadrant (as motor or generator). An example is shown in Figure 10.6. In this example, the load torque is considered constant but reversible. The reference

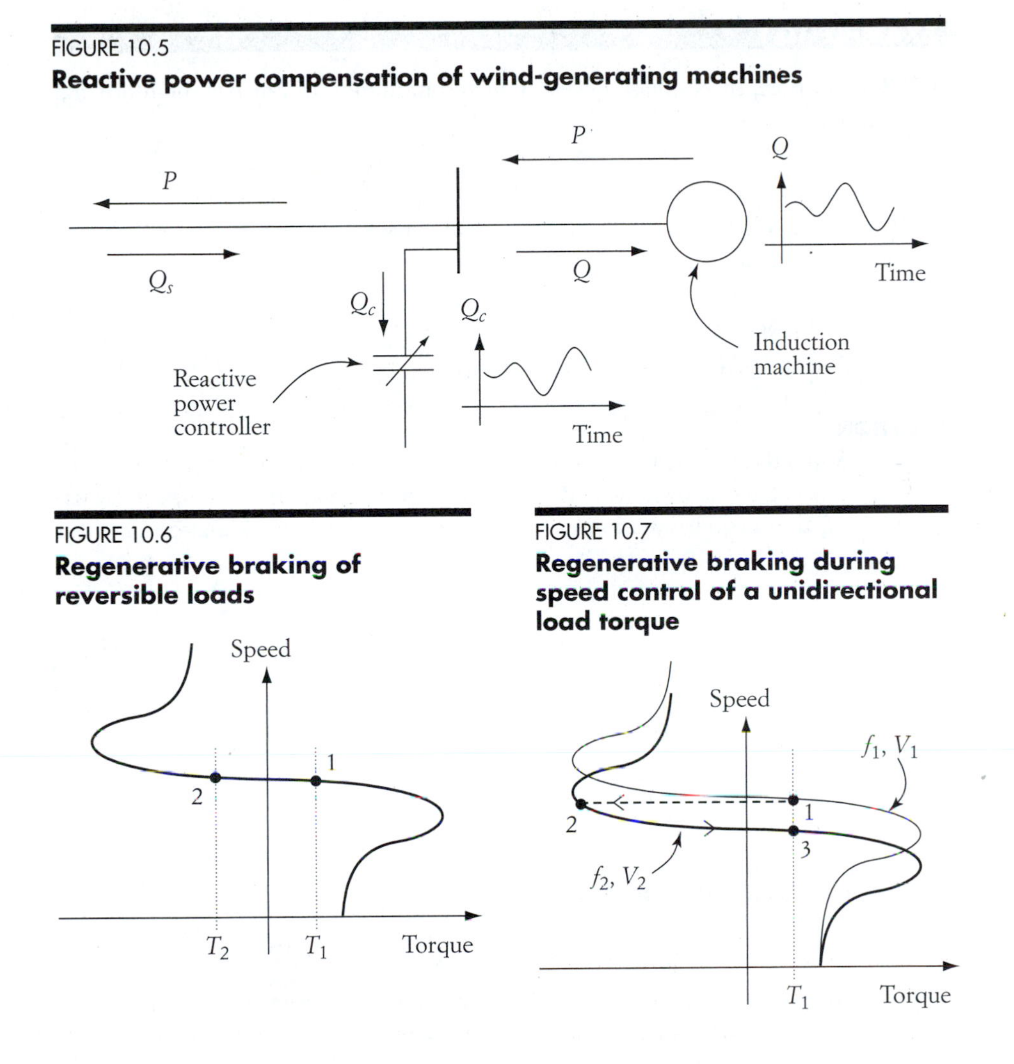

FIGURE 10.5
Reactive power compensation of wind-generating machines

FIGURE 10.6
Regenerative braking of reversible loads

FIGURE 10.7
Regenerative braking during speed control of a unidirectional load torque

operating point 1 represents a motor operation where the motor speed is less than the synchronous speed. When the load torque changes its direction from T_1 to T_2, the motor operates in the second quadrant and the speed of the motor exceeds its synchronous speed. Keep in mind that the motor still rotates in its original direction.

Another example of regenerative braking is shown in Figure 10.7. The figure shows two characteristics for two different values of v/f control. The v/f control is used to regulate the speed of the motor, as discussed in Chapter 7. The load torque is assumed to be constant and unidirectional, and the original operating point is 1. If v/f control is applied to reduce the speed of the motor, the operating point moves to 2, and eventually settles at point 3 in the first quadrant. However, during the transition from point 2 to point 3, the motor operates in the second quadrant under regenerative braking.

EXAMPLE 10.1

A 208 V, six-pole, three-phase, wye-connected induction motor has the following parameters:

$$R_1 = 0.6\ \Omega \qquad R'_2 = 0.4\ \Omega \qquad X_{eq} = 5\ \Omega$$

The motor is loaded by a 30 Nm bidirectional constant torque. If the load torque is reversed, calculate the following:

a. Motor speed
b. Power delivered to the electrical supply

SOLUTION

a. When the torque is reversed, the motor operates in the second quadrant at point 2 as shown in Figure 10.6. We can calculate the new speed by using the torque Equation (10.1) or the small-slip approximation given in Equation (5.60).

$$T_d = \frac{V^2 s}{\omega_s R'_2}$$

$$-30 = \frac{208^2 s}{2\pi \frac{1200}{60} 0.4}$$

$$s = -0.035$$

The regenerative speed n is

$$n = n_s(1 - s) = 1200(1 + 0.035) = 1242 \text{ rpm}$$

b. To calculate the electrical power, we need to subtract the winding losses (electrical losses) from the developed power since the machine is operating as a generator.

$$P_d = T_d \omega = (30) 2\pi \frac{1242}{60} = 3.902 \text{ kW}$$

To calculate the winding losses, we need to compute the current at point 2.

$$P_d = 3 (I'_2)^2 \frac{R'_2}{s} (1 - s)$$

$$-3902 = 3 (I'_2)^2 \frac{0.4}{-0.035} (1 + 0.035)$$

$$I'_2 = -10.5 \text{ A}$$

The negative sign indicates that the current is in the reverse direction with respect to the current at point 1.

The winding losses P_{loss} are

$$P_{loss} = 3(I'_2)^2(R_1 + R'_2) = 330 \text{ W}$$

The power delivered to the electrical source P_{ds} is

$$P_{ds} = 3902 - 330 = 3.572 \text{ kW}$$

10.2 DYNAMIC BRAKING

Dynamic braking of electric motors occurs when the energy stored in the rotating mass is dissipated in an electrical resistance. This requires the motor to operate as a generator to convert this mechanical energy into electrical.

For dc machines, dynamic braking requires a stationary magnetic field in the airgap, which does not exist in induction motors. However, we can create a temporary stationary field for the induction machine by applying a dc voltage to the stator terminals. When the rotor windings pass through this stationary field, voltages and currents are induced in the rotor windings. The rotor current produces losses in the rotor resistance. Since the rotor is spinning solely because of its stored kinetic energy, the rotor losses reduce the overall kinetic energy of the motor, thus assisting the motor to stop.

The stationary field for dynamic braking can be created using the circuit in Figure 10.8, which is the same as the power electronic circuit given in Figure 3.29. Instead of switching the transistor sequentially, here we close only three transistors for the duration of the dynamic braking. Note that no two transistors on the same leg can be closed. Now, let us assume that S_1, S_5, and S_6 are closed. As seen in Figure 10.9(a), the terminals of phases a and c are positive potentials and that for phase b is negative. The current in the stator winding is dc and will produce a stationary field in the airgap.

FIGURE 10.8
Six-pulse drive circuit

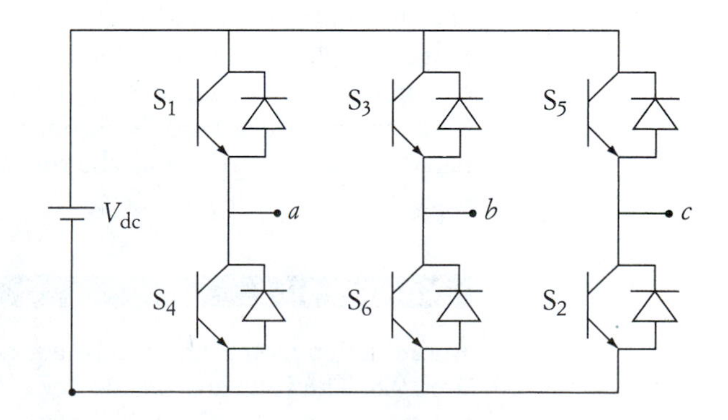

Keep in mind that the resistance of the stator windings is usually very small and the inductive reactance has no impact on dc currents. Therefore, the current in the stator windings could be excessive unless the terminal voltage during braking

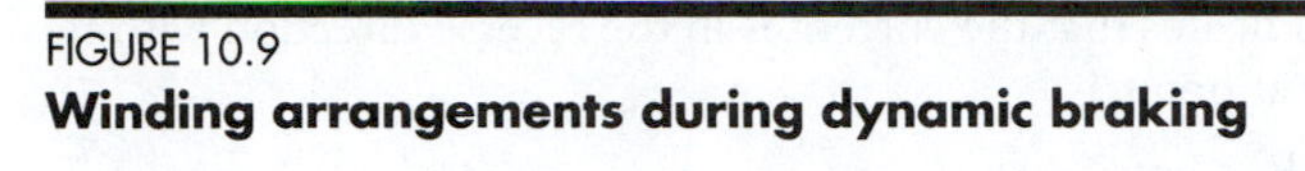

FIGURE 10.9
Winding arrangements during dynamic braking

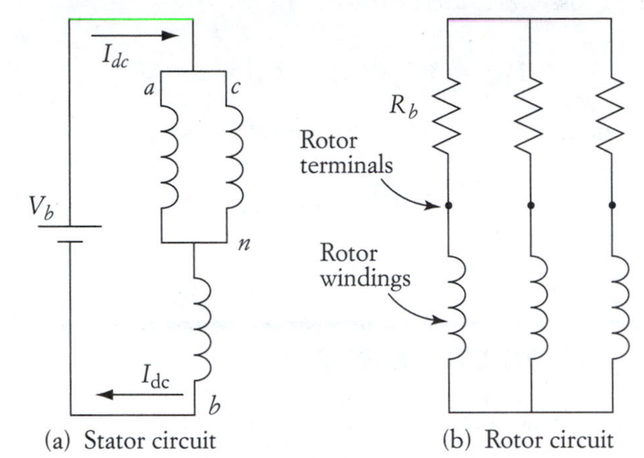

(a) Stator circuit (b) Rotor circuit

is small enough. To reduce the voltage across the stator windings, PWM or FWM techniques can be used, as explained in Chapter 3.

To calculate the maximum braking voltage, let us examine the stator circuit in Figure 10.9(a). Assume that the stator windings have only resistive elements. The total dc current in this circuit can be calculated by

$$I_{dc} = \frac{V_b}{1.5\, R_1} \leq I_b \tag{10.3}$$

where V_b is the reduced voltage applied to the stator windings during dynamic braking. R_1 is the resistance of single-phase windings. I_b is the upper limit of the stator current during braking. Depending on the size of the motor and the braking time, I_b could be selected as high as three times the rated current. Remember that the shorter the braking time is, the higher is the braking current. A larger braking current results in a stronger stationary field in the airgap, which induces larger current and higher losses in the rotor circuit. If the motor is a slip-ring type, we can insert an external resistance R_b in the rotor circuit, as shown in Figure 10.9(b). The function of this resistance is to control the rate at which the kinetic energy is dissipated. It is also used to limit the current in the rotor circuit.

EXAMPLE 10.2

An induction motor is driven by a six-step converter similar to that shown in Figure 10.8. The voltage at the dc link, V_{dc}, is 200 V. At normal full-load operation, the motor current is 25 A. The stator resistance is 0.5 Ω. The FWM technique is used during the dynamic braking. Calculate the duty ratio of the FWM.

SOLUTION

To calculate the braking voltage V_b of the stator, let us assume that the braking current is three times the rated value.

$$I_b = \frac{V_b}{1.5\, R_1}$$

$$75 = \frac{V_b}{1.5(0.5)}$$

$$V_b = 56.25 \text{ V}$$

The FWM is shown in Figure 3.35. The duty ratio d is the ratio of the closing subintervals to the conduction period. The equation of the duty ratio is given in Equation (3.74).

$$V_b = \sqrt{\frac{2d}{3}}\, V_{dc}$$

Then the duty ratio d is

$$d = 1.5\left(\frac{V_b}{V_{dc}}\right)^2 = 0.119$$

10.3 COUNTERCURRENT BRAKING

Countercurrent braking can be accomplished by reversing the direction of the field. As shown in Figure 10.10, the direction of rotation of the field in the airgap depends on the sequential connection of the stator windings. When the three-phase stator windings are connected in the *ABC* sequence, the airgap field rotates in one direction at synchronous speed. When the sequence of the stator windings is changed, say, to *ACB,* the field reverses its rotation. The change in sequence can be implemented by using the circuit of Figure 10.8. If you swap the triggering of S_1 with S_5, and S_4 with S_2, the sequence of the applied voltage is reversed to *ACB*.

FIGURE 10.10
Effect of voltage sequence on direction of rotation of airgap field

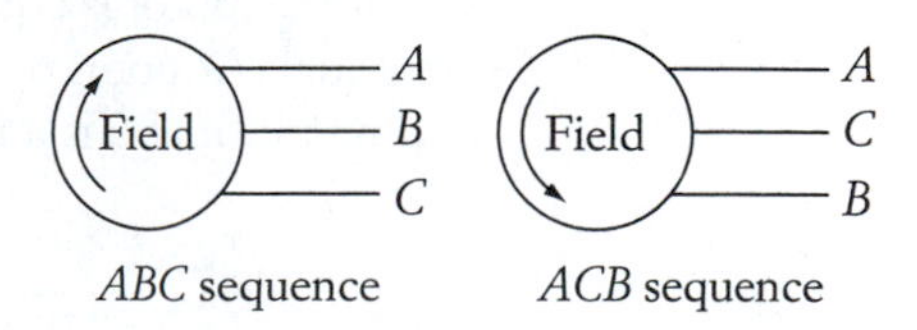

During steady state, assume that the operation of the motor in the *ABC* sequence is in the first quadrant. Then the motor in the reverse sequence operates in the third quadrant. We can utilize this feature to perform countercurrent braking, as shown in Figure 10.11. The figure shows the motor characteristics due to both sequences. Assume that the operating point of the motor in the *ABC* sequence is point 1. When the sequence of the stator voltage is reversed, the eventual steady-state operating

FIGURE 10.11
Induction motor characteristics during countercurrent braking

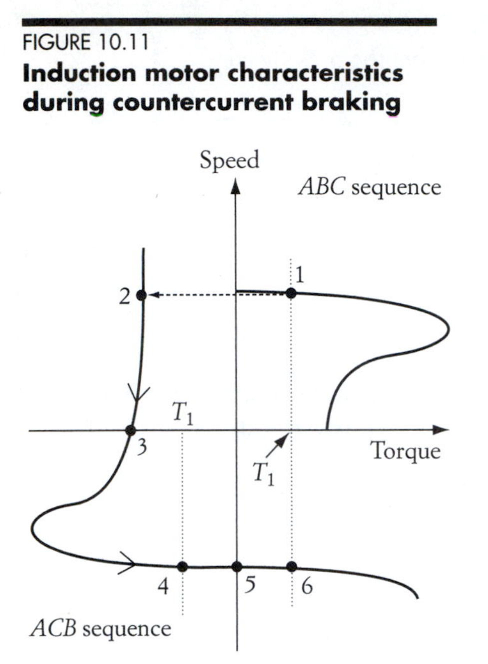

point is located on the *ACB* characteristics where the load torque and the developed torque of the motor are equal. Before reaching this final operating point, however, the motor initially moves from point 1 to point 2 without any appreciable change in speed due to system inertia. Point 2 is not a steady-state point because of the inequality of the motor and load torques. The motor continues moving to point 3. At 3 the motor stops. If the motor is not disconnected at 3, it moves toward point 4. Point 4 is the new steady-state point if the load torque is bidirectional. However, if the motor torque is unidirectional, the motor continues moving toward the new steady-state point. It will reach point 5, at which the motor speed is equal to the synchronous speed. Then the operating point moves toward point 6. At 6, the load and developed torques are equal, and the new steady-state operating point is achieved. Note that point 6 represents regenerative braking for the *ACB* sequence.

The analysis of the induction motor at any of these six points is dependent on the computation of the slip. The slip at point 1 is less than one, but is very small since the induction motor rotates at nearly synchronous speed. At point 2, the motor speed is still the same as the speed at point 1, but the airgap field rotates at synchronous speed in the reverse direction. Thus, the slip at point 2 is

$$s_2 = \frac{-n_s - n}{-n_s} = \frac{n_s + n}{n_s} > 1$$

Slip s_2 is almost equal to two. At point 3, the speed of the motor is zero, but the field is still rotating in the new direction, so the slip is equal to one. At point 4, the motor reverses its rotation and is now in the same direction as the airgap field. The slip in this case is

$$s_4 = \frac{-n_s - (-n)}{-n_s} = \frac{-n_s + n}{-n_s} = \frac{n_s - n}{n_s}$$

Assuming symmetry, the slip at point 4 is equal to that at point 1. At point 5, the slip of the motor is equal to zero, since the motor speed and the synchronous speed are equal. At point 6, the motor speed exceeds the synchronous speed, so the operating point at 6 is a regenerative braking ($n > n_s$). The motor slip at 6 is

$$s_6 = \frac{-n_s - (-n)}{-n_s} = \frac{n_s - n}{n_s} < 0$$

Table 10.1 shows the essential parameters and equations at each operating point. Using either the small-slip or the large-slip approximations, we can derive the

TABLE 10.1
Summary of countercurrent braking

Operating Point	*Motor Speed*	*Field Speed*	*Slip*	*Approximation Method*	*Motor Torque*	*Motor Current*
1	→	→ n_s	→ $0 < s_1 < 1$	Small slip	→ $\frac{V^2 s_1}{\omega_s R'_2}$	→ $\frac{V s_1}{\sqrt{3} R'_2}$
2	→	←	→ $s_2 > 1$	Large slip	← $\frac{V^2 R'_2}{s_2(-\omega_s)X_{eq}^2}$	→ $\frac{V}{\sqrt{3} X_{eq}}$
3	0	←	→ $s_3 = 1$	Large slip	← $\frac{V^2 R'_2}{s_3(-\omega_s)X_{eq}^2}$	→ $\frac{V}{\sqrt{3} X_{eq}}$
4	←	←	→ $0 < s_4 < 1$	Small slip	← $\frac{V^2 s_4}{(-\omega_s)R'_2}$	→ $\frac{V s_4}{\sqrt{3} R'_2}$
5	←	←	$s_5 = 0$	Small slip	0	0
6	←	←	← $s_6 < 0$	Small slip	→ $\frac{V^2 s_6}{(-\omega_s)R'_2}$	← $\frac{V s_6}{\sqrt{3} R'_2}$

torque and current equations. The operating point at 1 is considered to be the reference point. The arrows indicate changes in the direction of rotation, direction of torque, or direction of current with respect to the reference operating point 1.

EXAMPLE 10.3

A three-phase, wye-connected induction motor is loaded by a constant-torque load and is rotating at 1150 rpm. A PWM power converter is used to drive the motor. The line-to-line rms value of the fundamental component of voltage is 300 V, and its frequency is 80 Hz. The rotor resistance of the motor is 0.5 Ω, and the equivalent inductive reactance is 3 Ω at 80 Hz. A countercurrent braking is performed without any change in the frequency or voltage. Calculate the slip, current, and torque at all operating points.

SOLUTION

Using the equations in Table 10.1, the slip, current, and torque can be directly computed. However, before we attempt to solve this problem, we must first find the synchronous speed. Note that the full-load speed is 1150 rpm. The synchronous speed must be slightly higher than this. For 80 Hz, the machine must have eight poles so that the synchronous speed is 1200 rpm.

$$\omega_s = 2\pi \frac{n_s}{60} = 125.7 \text{ rad/sec}$$

The results are tabulated in Table 10.2. The computation is straightforward substitution in the formulas of Table 10.1.

The slip at point 6 can be computed using the torque equation at 6, where the developed torque T_6 is equal to the torque at point 1.

$$T_6 = \frac{V^2 s_6}{(-\omega_s)R'_2} = T_1$$

T_1 is computed in Table 10.2.

$$s_6 = \frac{(60)0.5(-125.7)}{300^2} = -0.042$$

$$n_6 = -1200(1 + 0.042) = -1250 \text{ rpm}$$

Note that unlike the current in the dc motor, the current of the induction motor does not surge to high values when countercurrent braking is implemented. This is because the inductance of the stator windings limits the magnitude of the braking current.

CHAPTER 10 PROBLEMS

10.1 A 60 Hz, 480 V, three-phase, wye-connected, six-pole induction machine has the following parameters:

$$R_1 = 0.3\ \Omega \qquad R'_2 = 0.1\ \Omega \qquad X_{eq} = 10\ \Omega$$

The machine operates in the regenerative braking mode as a wind energy generator. A gearbox of 5:1 speed ratio is between the machine and the turbine blades, with the high-speed side connected to the induction machine. If the wind speed exerts 200 Nm on the shaft of the blades, calculate the speed of the induction machine.

10.2 For Problem 10.1, calculate the mechanical power input to the induction machine and the power delivered to the electrical system. Assume that the rotational losses are 200 W.

TABLE 10.2
Results of Example 10.3

Operating Point	*Motor Speed*	*Field Speed*	*Slip*	*Motor Torque*	*Motor Current*
1	1150	1200	$\frac{1200 - 1150}{1200} = 0.042$	$\frac{300^2\,(0.042)}{125.7 \times 0.5} = 60\text{ Nm}$	$\frac{\frac{300}{\sqrt{3}}\,0.042}{0.5} = 14.55\text{ A}$
2	1150	−1200	$\frac{-1200 - 1150}{-1200} = 1.96$	$\frac{300^2 \times 0.5}{1.96(-125.7)3^2} = -20\text{ Nm}$	$\frac{\frac{300}{\sqrt{3}}}{3^2} = 19.12\text{ A}$
3	0	−1200	$s_3 = 1$	$\frac{300^2 \times 0.5}{(-125.7)3^2} = -40\text{ Nm}$	$\frac{\frac{300}{\sqrt{3}}}{3^2} = 19.12\text{ A}$
4	−1150	−1200	$\frac{-1200 + 1150}{-1200} = 0.042$	$\frac{300^2\,(0.042)}{(-125.7)\,0.5} = -60\text{ Nm}$	$\frac{\frac{300}{\sqrt{3}}\,0.042}{0.5} = 14.55\text{ A}$
5	−1200	−1200	$s_5 = 0$	0	0
6	1250	−1200	$s_6 = -0.042$	$\frac{300^2(-0.042)}{(-125.7)0.5} = 60\text{ Nm}$	$\frac{\frac{300}{\sqrt{3}}\,(-0.042)}{0.5} = -14.55\text{ A}$

10.3 A 60 Hz, 480 V, three-phase, wye-connected induction machine has the following parameters:

$$R_1 = 0.2\ \Omega \qquad R'_2 = 0.2\ \Omega \qquad X_{eq} = 15\ \Omega$$

At full-load torque, the motor speed is 580 rpm. A FWM converter drives the motor. Calculate the duty ratio that limits the dynamic braking current to twice the rated value.

10.4 Repeat Problem 10.3 for the case when only transistors S_1 and S_2 in Figure 10.8 are closed during the dynamic braking.

10.5 A 60 Hz, four-pole, 480 V, three-phase, wye-connected induction machine has the following parameters:

$$R_1 = 0.2\ \Omega \qquad R'_2 = 0.3\ \Omega \qquad X_{eq} = 12\ \Omega$$

The motor is loaded by a unidirectional constant-torque load of 40 NM. A countercurrent braking is performed on the motor. Calculate the following:

a. Motor current before the countercurrent braking
b. Motor current just after the voltage sequence is reversed
c. Current at the moment when the motor stops
d. Motor torque at the condition of part (c)
e. Motor current at the new steady-state point
f. Motor torque at the new steady-state point
g. Motor speed at the new steady-state point

10.6 Repeat Problem 10.5 for the case when the motor frequency is reduced to 50 Hz and the voltage sequence is reversed. Assume that the converter is a fixed v/f controller.

11

Dynamics of Electric Drive Systems

In modern high performance drives (HPD), such as robotics, guided manipulation, and supervised actuation, controlling the motor's final speed is not the only goal. A multirobot system performing complementing functions must have the end effectors (moving terminals) proceed about the space of operation according to preselected time-tagged trajectories. To achieve this, every motor in the robot arm must follow a specific track so that the aggregated motion of all motors keeps the end effector on its trajectory at all times. This must be achieved even when the system loads, inertia, and parameters are varying. To realize this level of performance, the traveling time of the motor in a HPD system must be controlled during starting, braking, and change of speed. Traveling time is defined here as the time required to change the motor speed (or rotor position) from one steady-state operating point to another. The traveling time is determined by the mechanical parameters of the system, such as inertia and load torque, and by electrical quantities, such as motor voltage and developed torque.

The general torque expression of an electromechanical system under dynamic conditions is

$$T_d - T_l = T_i \tag{11.1}$$

where T_d is the developed torque of the motor, T_l is the load torque, and T_i is the inertia torque of the entire rotating mass (motor, load, and mechanical interface). The inertia torque is only present during acceleration and is expressed by

$$T_i = J\frac{d\omega}{dt} \tag{11.2}$$

where J is the moment of inertia, and $d\omega/dt$ is the angular acceleration. In linear motion, the inertia force F_i is computed as

$$F_i = m\frac{dv}{dt}$$

where m is the mass, and dv/dt is the linear acceleration.

11.1 MOMENT OF INERTIA

The moment of inertia of a rotating object is defined by

$$J = \frac{GD^2}{4g} = m\frac{D^2}{4} \tag{11.3}$$

where G is the weight of the rotating mass in Newtons, g is the gravity acceleration, D is the diameter of gyration, and m is the mass. The diameter of gyration is the diameter of the rotating path of the object at which the mass of the body is concentrating. The moment of inertia of some commonly used objects in electric drives is given in Table 11.1.

Depending on the system configuration and design, the moment of inertia can vary during motion. However, there are several applications for which the moment of inertia is constant, such as hoists and forklifts. A simple hoist system is shown in Figure 11.1. In the figure, the motor motion is angular while the motion of the load is translation (linear). The load torque is the load force multiplied by the radius of the rotating wheel. The load force is the load mass times the acceleration of gravity. The inertia of the load can be computed using the kinetic energy for both motions (angular and translation).

$$KE = \frac{1}{2}J\omega^2 = \frac{1}{2}mv^2$$

where ω is the angular velocity of the rotating wheel, v is the linear vertical velocity of the load, and m is the load mass. The load inertia in this case can be computed as

$$J = m\left(\frac{v}{\omega}\right)^2$$

Since the ratio v/ω is constant and equal to the radius of the wheel, the inertia of the load J is constant.

TABLE 11.1
Moment of inertia for selected rotating mass

Object	*Moment of inertia*
Thin-walled hollow cylinder with diameter D	$m\frac{D^2}{4}$
Hollow cylinder with inner diameter D_1 and outer diameter D_2	$\frac{1}{2}m\frac{D_1^2 + D_2^2}{4}$
Solid cylinder with diameter D	$\frac{1}{2}m\frac{D^2}{4}$

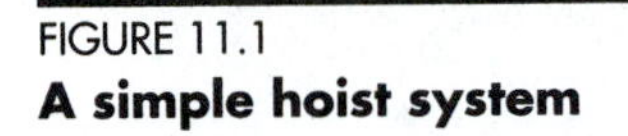

FIGURE 11.1
A simple hoist system

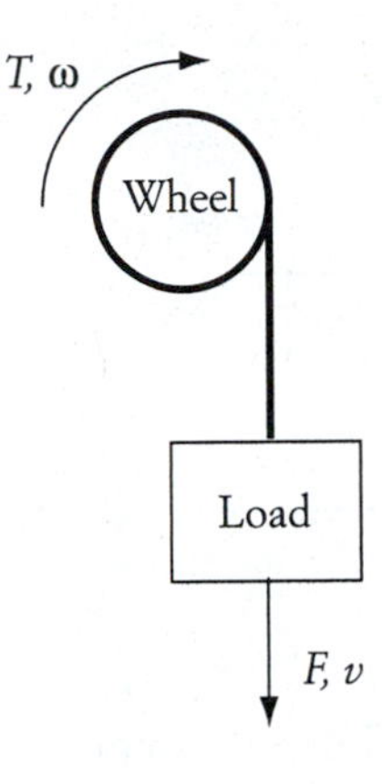

FIGURE 11.2
A system with linear and rotating motion

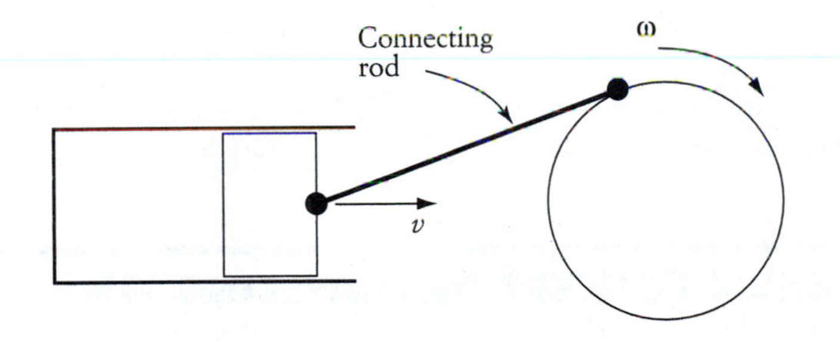

A variable moment of inertia can occur when the system configuration leads to asymmetrical motions. Figure 11.2 shows a common drive system for a rotating motor that drives a piston-type load. The motor wheel is attached to the piston via a connecting rod. The motion of the piston is linear, and its velocity changes depending on the position of the motor wheel. Figure 11.3 shows the velocity vectors of this system. The linear velocity of the piston v is in the direction shown in the figure during half of the rotating cycle; then it reverses its direction. The kinetic energy of the system can be calculated using the linear and rotating speed, as shown in Equation (11.4).

$$KE = \frac{1}{2} m v^2 = \frac{1}{2} J \omega^2 \tag{11.4}$$

FIGURE 11.3
Velocity vectors of the system in Figure 11.2

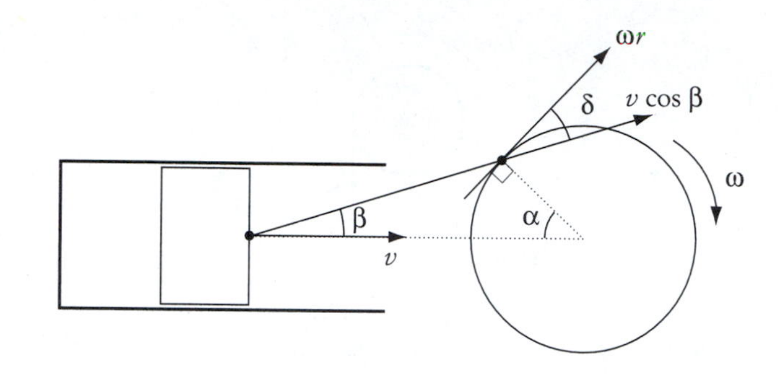

At the point where the connecting rod is mounted on the rotating wheel, the velocity vectors are governed by the following equation:

$$v \cos\beta = \omega r \cos\delta = \omega r \sin(\alpha + \beta) \tag{11.5}$$

Substituting (11.5) in (11.4) yields

$$J = mr^2\left(\frac{\sin(\alpha + \beta)}{\cos\beta}\right)^2$$

In this system, the inertia is dependent on β, which is a function of the wheel position, so the system inertia varies during the motor rotation.

11.2 BASIC CONCEPT OF TRAVELING TIME

Let us examine the traveling time of an electric drive system by using the characteristics shown in Figure 11.4. In the figure, an induction motor characteristic is used. The motor torque is T_d, and the load torque is T_l. The inertia torque T_i, at any speed ω, is the difference between the motor torque and the load torque at the given speed. The inertia torque compensates for the total system inertia.

$$T_i = J\frac{d\omega}{dt} = T_d - T_l$$

For given motor and load torques, the higher the inertia torque, the higher the motor acceleration. Using the previous equation, when the motor speed changes from ω_1 to ω_2, the traveling time can be computed by

$$t = \int_{\omega_1}^{\omega_2} \frac{J}{T_d - T_l}\, d\omega \tag{11.6}$$

FIGURE 11.4
Developed inertia torque for induction motor

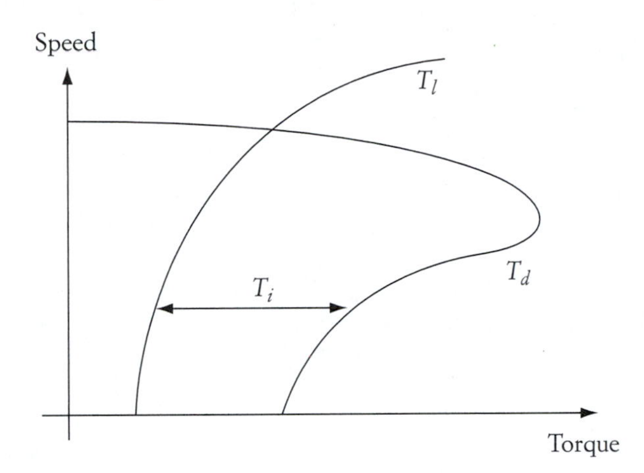

Equation (11.6) shows that the traveling time between two speeds can be reduced if at least one of the following conditions is met:

1. The inertia torque increases.
2. The moment of inertia decreases.

The first condition can be met by adjusting the motor characteristics. For example, an increase in the induction motor voltage increases T_i, as shown in Figure 11.5. The motor characteristic labeled T_1 is for a motor voltage V_1. The inertia torque at an arbitrary speed is T_{i_1}. If the motor voltage increases to V_2, the motor characteristic labeled T_2 stretches, and the inertia torque increases to T_{i_2}.

The second condition can be met by using gears or belt systems. These interface systems alter the load inertia seen by the motor and are the subject of the next section.

11.3 GEARS AND BELTS

A simple gear system is shown in Figure 11.6. It consists of a motor, a mechanical load, and a gear. The gear ratio is defined as the ratio of its diameters or speeds.

$$\frac{n_1}{n_2} = \frac{d_2}{d_1}$$

where n is the shaft speed and d is the gear diameter. Now let us assume that the load inertia is J_l, and the motor inertia is J_m. Also assume that the gear inertia is J_{d_1}

FIGURE 11.5
Change in inertia torque for induction motor due to changes in voltage

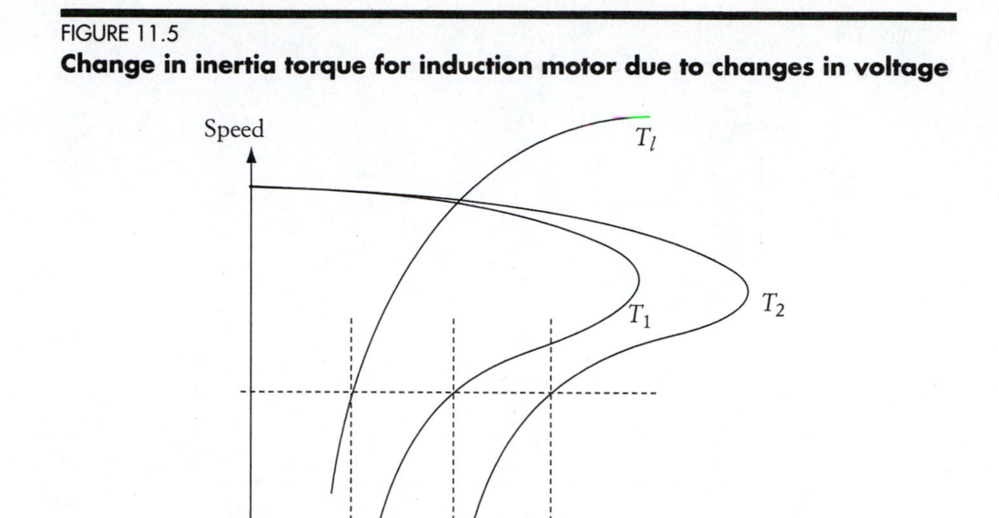

FIGURE 11.6
A simple gear system

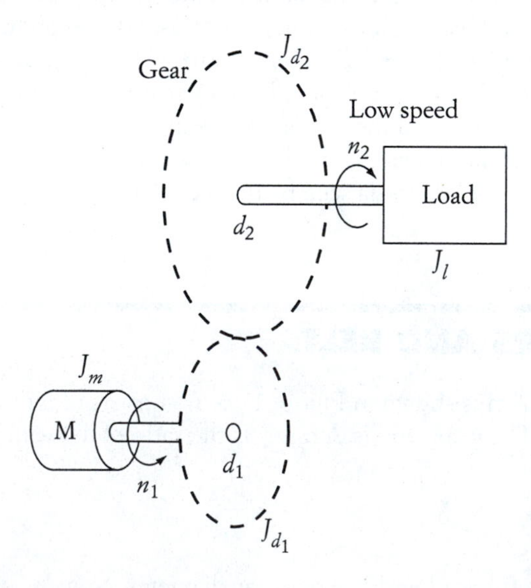

for one gear wheel and J_{d_2} for the other. If a gear is not present, the inertia seen by the motor is $J_l + J_m$. With the gear in place, the moment of inertia seen by the motor can be computed using the kinetic energy (KE) for each component of the system. The total kinetic energy seen by the motor is

$$\text{Total } KE = KE \text{ of motor} + KE \text{ of gear} + KE \text{ of load} \tag{11.7}$$

$$\text{Total } KE = \frac{1}{2} J_{eq}\, \omega_1^2$$

where J_{eq} is the equivalent moment of inertia seen by the motor. Equation (11.7) can be written in more detail as

$$\frac{1}{2} J_{eq}\, \omega_1^2 = \frac{1}{2} J_m\, \omega_1^2 + \frac{1}{2} J_{d_1} \omega_1^2 + \frac{1}{2} J_{d_2}\, \omega_2^2 + \frac{1}{2} J_l\, \omega_2^2$$

Hence,

$$\begin{aligned} J_{eq} &= J_m + J_{d_1} + J_{d_2}\left(\frac{n_2}{n_1}\right)^2 + J_l\left(\frac{n_2}{n_1}\right)^2 \\ J_{eq} &= J_m + J_{d_1} + (J_{d_2} + J_l)\left(\frac{d_1}{d_2}\right)^2 \end{aligned} \tag{11.8}$$

Equation (11.8) shows that the gear ratio can change the inertia seen by the motor. Accordingly, the load torque seen by the motor changes. This can be verified by computing the load power at either side of the gear, assuming that the gear is lossless.

$$T_l\, \omega_2 = T_{eq}\, \omega_1$$

where T_{eq} is the equivalent torque of the load seen at the motor side of the gear. Thus, the load torque seen by the motor is

$$T_{eq} = T_l\left(\frac{n_2}{n_1}\right) = T_l\left(\frac{d_1}{d_2}\right) \tag{11.9}$$

If the gear system is designed so that d_1/d_2 is less than 1, the load inertia seen by the motor is reduced, and the load torque seen by the motor is also reduced. Consequently, the traveling time is reduced. (Examine Equation 11.6.)

A belt system can have an effect identical to that of a gear system. Figure 11.7 shows a simple system consisting of load, motor, and two wheels connected by a belt. The equations just described for the gear system can be directly applied to the belt system.

FIGURE 11.7
A simple belt drive system

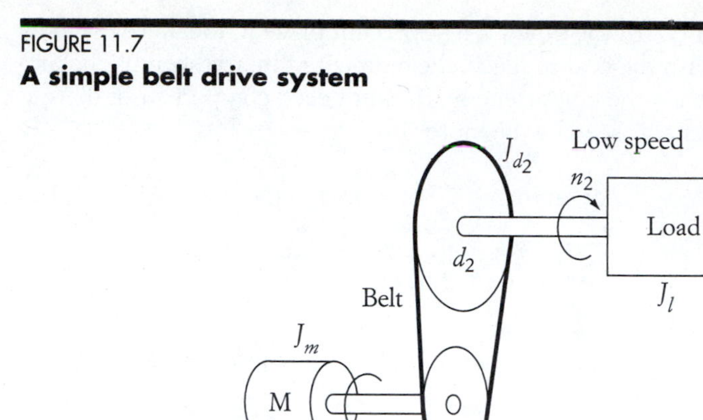

11.4 TRAVELING TIME OF dc MOTORS

The traveling time of any electric motor can be computed using the electrical model of the motor and the torque equation of the rotating mass. For dc machines, the electrical equation is

$$V_t = E_a + i_a R_a + L_a \frac{di_a}{dt} = K\phi\omega + i_a R_a + L_a \frac{di_a}{dt} \tag{11.10}$$

where V_t is the terminal voltage of the motor, i_a is the armature current, R_a is the armature resistance, $K\phi$ is the field constant, and L_a is the inductance of the armature windings. The torque equation of the motor can be written as

$$K\phi\, i_a = T_l + J\frac{d\omega}{dt} \tag{11.11}$$

where T_l is the load torque. For a constant-load torque and constant field current, the first derivative of the current in Equation (11.11) is

$$\frac{di_a}{dt} = \frac{J}{K\phi}\frac{d^2\omega}{dt^2} \tag{11.12}$$

Substituting (11.11) and (11.12) into (11.10) yields

$$\frac{d^2\omega}{dt^2} + \frac{R_a}{L_a}\frac{d\omega}{dt} + \frac{(K\phi)^2}{JL_a}\omega = \frac{(K\phi)^2}{JL_a}\omega_f \tag{11.13}$$

where ω_f is the final steady-state speed defined by

$$\omega_f = \frac{V_t}{K\phi} - \frac{R_a}{(K\phi)^2}T_l$$

Solving the differential equation (11.13) yields

$$\omega = \omega_f\left[1 - \frac{e^{-\xi\omega_n t}}{\sqrt{1-\xi^2}}\sin(\omega_n\sqrt{1-\xi^2}t + \theta)\right] + \omega(0)\left[\frac{2\,\xi e^{-\xi\omega_n t}}{\sqrt{1-\xi^2}}\sin(\omega_n\sqrt{1-\xi^2}t)\right] \quad (11.14)$$

where

$$\omega_n = \frac{K\phi}{\sqrt{JL_a}}$$

$$\xi = \frac{R_a}{2\,K\phi}\sqrt{\frac{J}{L_a}}$$

$$\theta = \cos^{-1}\xi$$

$$\omega(0) = \omega(t = 0)$$

EXAMPLE 11.1

A dc shunt motor has the following parameters:

$$K\phi = 3.0 \text{ V sec} \qquad R_a = 1\,\Omega \qquad L_a = 10 \text{ mH}$$

The rated voltage of the motor is 600 V. The voltage is reduced to 150 V at starting. The load connected to the motor is a constant torque of 20 Nm. The total moment of inertia of the entire drive system is 6 Nm sec^2. Calculate and plot the motor speed versus time. Also calculate the motor speed after 5 seconds.

FIGURE 11.8
Speed of dc shunt motor during starting

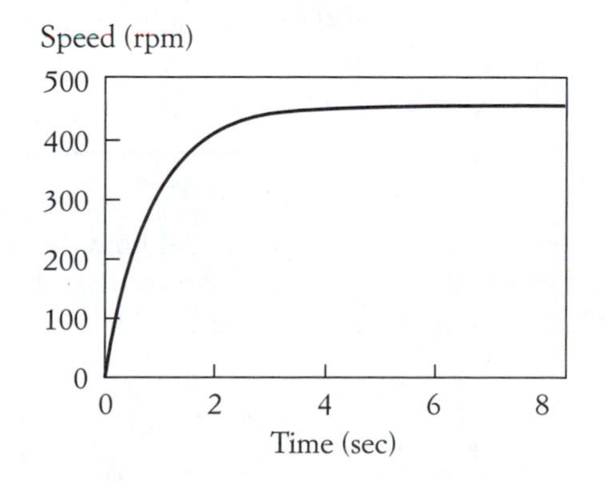

SOLUTION

The solution of this problem is by direct substitution in Equation (11.14). The initial speed in this case is zero since the motor is starting from rest. The plot in Figure 11.8 is the result obtained by a simulation software. After about 5 seconds, the motor reaches the steady-state speed of

$$\omega_f = \frac{V_t}{K\phi} - \frac{R_a}{(K\phi)^2} T_l = \frac{150}{3} - \frac{20}{9} = 47.78 \text{ rad/sec}$$

$$n_f = 456 \text{ rpm}$$

EXAMPLE 11.2

Ignore the armature reactance of the motor and repeat Example 11.1.

SOLUTION

An approximation of Equation (11.13) can be made by ignoring the armature inductance. In this case the equation is modified to

$$\frac{d\omega}{dt} + \frac{(K\phi)^2}{JR_a}\omega = \frac{(K\phi)^2}{JR_a}\omega_f \tag{11.15}$$

This is a first-order differential equation, and its solution is

$$\omega = \omega_f (1 - e^{-t/\tau}) + \omega(0)e^{-t/\tau} \tag{11.16}$$

where

$$\tau = \frac{JR_a}{(K\phi)^2}$$

Equation (11.16) is used to simulate the speed of the dc motor. The result is shown in Figure 11.9, which is almost identical to the result obtained in Figure 11.8. We can then conclude that the starting time of the dc motor is more dependent on the system inertia than on the armature reactance.

By using Equation (11.16), we can understand how the starting time of a dc motor is controlled. Rewrite the equation and assume that the initial condition of the speed is zero. (The motor is starting from rest.)

$$\frac{\omega}{\omega_f} = 1 - e^{-t/\tau}$$

FIGURE 11.9
Traveling time of dc motor during starting when armature reactance is ignored

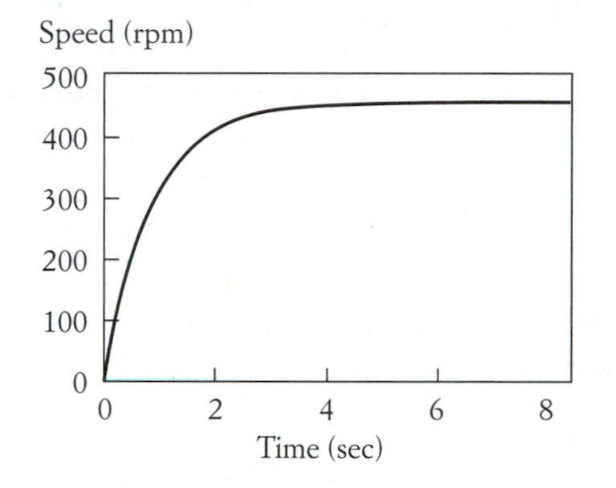

Let us assume that the motor reaches its new steady-state operating point when the speed of the motor is about 95% of the final speed.

$$0.95 = 1 - e^{-t_{st}/\tau}$$

Hence,

$$t_{st} = 3\tau = \frac{3\,JR_a}{(K\phi)^2} \tag{11.17}$$

Equation (11.17) shows that the starting time of the dc machine is dependent on the system inertia, armature resistance, and field current. It may look surprising that the motor voltage has no effect on the starting time. This is true—the motor voltage controls the magnitude of the final speed, but not the starting time. To reduce the starting time, the system inertia must be reduced either by using a gear or belt system or by increasing the field current.

EXAMPLE 11.3

A dc shunt machine is used in high-performance operation. The starting time of the motor must be limited to 2 sec. The motor has a moment of inertia equal to 1 Nm sec^2. The load moment of inertia is 5 Nm sec^2. The field constant of the motor $K\phi$ is 3 V sec, and the armature resistance is 2 Ω. Show how we can achieve the desired starting time.

SOLUTION

The starting time of the motor based on the given data is

$$t_{st} = 3\tau = \frac{3\,JR_a}{(K\phi)^2} = \frac{3(1+5)2}{9} = 4 \text{ sec}$$

which is higher than the desired starting time. Using a gear system can reduce the system inertia. The equivalent moment of inertia J_{eq} must then be equal to

$$J_{eq} = \frac{t_{st}(K\phi)^2}{3\,R_a} = \frac{2(9)}{6} = 3 \text{ Nm sec}^2$$

If we ignore the moment of inertia of the gear system, the ratio of the gear can be computed using Equation (11.8).

$$J_{eq} = J_m + J_l\left(\frac{n_2}{n_1}\right)^2$$

$$3 = 1 + 5\left(\frac{n_2}{n_1}\right)^2$$

Then the gear ratio must be

$$\frac{n_2}{n} = \sqrt{\frac{2}{5}}$$

EXAMPLE 11.4

For the dc motor in Example 11.1, the motor operates at a steady-state speed when the terminal voltage is at 500 V. To increase the motor speed, the terminal voltage increases to 600 V, while the field remains constant. Calculate the time required to change the motor speed.

SOLUTION

The first step is to calculate the initial and final speeds. The initial speed is

$$\omega(0) = \frac{V_t}{K\phi} - \frac{R_a}{(K\phi)^2}T_d = \frac{500}{3} - \frac{20}{9} = 164.44 \text{ rad/sec}$$

and the final speed is

$$\omega_f = \frac{600}{3} - \frac{20}{9} = 197.78 \text{ rad/sec}$$

We can use Equation (11.16) to compute the traveling time, but first let us calculate τ.

$$\tau = \frac{JR_a}{(K\phi)^2} = 0.67 \text{ sec}$$

The traveling time can be computed by assuming that the motor reaches the new steady-state operating point when the motor speed is 95% of the final value.

$$\omega = \omega_f(1 - e^{-t/\tau}) + \omega(0)e^{-t/\tau}$$

$$0.95\ \omega_f = \omega_f(1 - e^{-t/\tau}) + \omega(0)e^{-t/\tau}$$

Then the traveling time is

$$t = -\tau \ln\left[\frac{0.05\ \omega_f}{\omega_f - \omega(0)}\right] = 0.81 \text{ sec}$$

EXAMPLE 11.5

For the dc motor in Example 11.1, the motor operates at a steady-state speed when the terminal voltage is 500 V. Assume that the load torque is constant. Calculate the terminal voltage that stops the motor and keeps it at holding. Also calculate the traveling time during braking.

SOLUTION

To stop the motor, ω_f is set to zero.

$$\omega_f = \frac{V_b}{3} - \frac{20}{9} = 0$$

$$V_b = 6.66 \text{ V}$$

The dynamic braking time can be computed by assuming that the motor reaches the holding state when its speed is about 5% of the initial speed.

$$\omega = \omega_f(1 - e^{-t/\tau}) + \omega(0)e^{-t/\tau}$$

$$0.05\ \omega(0) = \omega(0)e^{-t/\tau}$$

$$t = 3\tau = 2 \text{ sec}$$

11.5 TRAVELING TIME OF INDUCTION MOTORS

The torque and speed of the induction motor are related by a nonlinear function, which makes the solution of Equation (11.6) more involved. To simplify the calculations, let us consider the three basic equations for the induction motor given in Chapter 5 and repeated here:

$$T_d = \frac{P_d}{\omega} = \frac{V^2 R'_2}{s\omega_s\left[\left(R_1 + \frac{R'_2}{s}\right)^2 + X_{eq}^2\right]} \tag{11.18}$$

$$s_{\max} = \frac{R'_2}{\sqrt{R_1^2 + X_{eq}^2}} \tag{11.19}$$

$$T_{\max} = \frac{V^2}{2\omega_s\left[R_1 + \sqrt{R_1^2 + X_{eq}^2}\,\right]} \tag{11.20}$$

Let us find the ratio of the developed torque to the maximum torque.

$$\frac{T_d}{T_{\max}} = \frac{2\,R'_2\left(R_1 + \sqrt{R_1^2 + X_{eq}^2}\right)}{s\left[\left(R_1 + \frac{R'_2}{s}\right)^2 + X_{eq}^2\right]} \tag{11.21}$$

Now insert the expression $s_{\max}$ of Equation (11.19) into (11.21).

$$\frac{T_d}{T_{\max}} = \frac{2\,R'_2\left(R_1 + \frac{R'_2}{s_{\max}}\right)}{s\left[\left(\frac{R'_2}{s_{\max}}\right)^2 + \left(\frac{R'_2}{s}\right)^2 + \frac{2\,R_1 R'_2}{s}\right]} \tag{11.22}$$

It is reasonable to make the following assumption:

$$R_1 \cong R'_2$$

Then Equation (11.22) can be approximated by

$$T_d = \frac{2K\,T_{\max}}{\frac{s}{s_{\max}} + \frac{s_{\max}}{s} + 2\,s_{\max}} \tag{11.23}$$

where $K = s_{\max} + 1$. The inertia torque equation used in the computation of the traveling time can now be rewritten as

$$T_i = J_{eq}\frac{d\omega}{dt} = T_d - T_l = \frac{2\,T_{\max}\,K}{\frac{s}{s_{\max}} + \frac{s_{\max}}{s} + 2\,s_{\max}} - T_l \tag{11.24}$$

Consider the derivative of s with respect to time,

$$\frac{ds}{dt} = \frac{d}{dt}\left(\frac{\omega_s - \omega}{\omega_s}\right) = \frac{-1}{\omega_s}\frac{d\omega}{dt} \tag{11.25}$$

Substituting Equation (11.25) into (11.24) yields

$$-\tau\frac{ds}{dt} = \frac{2K - S_R T_R}{S_R} \tag{11.26}$$

where

$$\tau = \frac{J_{eq}\,\omega_s}{T_{\max}}$$

$$S_R = \frac{s}{s_{\max}} + \frac{s_{\max}}{s} + 2\,s_{\max}$$

$$T_R = \frac{T_l}{T_{\max}}$$

τ is known as the system time constant. To compute the traveling time, we must solve the differential equation given in (11.26),

$$t = \int_{s_1}^{s_2} \frac{-\tau\,S_R}{2K - S_R\,T_R}\,ds \tag{11.27}$$

where s_1 is the initial slip and s_2 is the final slip.

11.5.1 UNLOADED INDUCTION MOTOR

The traveling time of an unloaded induction machine ($T_l = 0$) can be computed by a simple closed-form solution, since T_R is zero. In Equation (11.27), if the motor voltage, frequency, and rotor resistance are maintained constant, τ and $S_{\max}$ are constant quantities.

$$t = \frac{\tau}{2K}\int_{s_1}^{s_2}(-S_R)\,ds = \frac{\tau}{2K}\int_{s_2}^{s_1}\left[\frac{s}{s_{\max}} + \frac{s_{\max}}{s} + 2\,s_{\max}\right]ds$$

$$t = \frac{\tau}{2K}\left[\frac{s_1^2 - s_2^2}{2\,s_{\max}} + s_{\max}\ln\frac{s_1}{s_2} + 2\,s_{\max}\,(s_1 - s_2)\right] \tag{11.28}$$

EXAMPLE 11.6

A 480 V, three-phase induction motor has a rated speed at full load of 1120 rpm, stator resistance of 1 Ω, rotor resistance referred to stator of 1 Ω, and equivalent winding reactance of 5 Ω. The inertia of the motor is 4 Nm sec^2. Compute the starting time of the motor at no load and at full voltage and frequency.

SOLUTION

Since the rated speed of the induction motor at steady state is very close to the synchronous speed, this motor must be a six-pole type at 60 Hz with a synchronous speed of 1200 rpm. To compute the starting time of the motor, we must first compute T_{max} and s_{max}:

$$T_{max} = \frac{V^2}{2\omega_s(R_1 + \sqrt{R_1^2 + X_{eq}^2})} = \frac{480^2}{2\left(2\pi \frac{1200}{60}\right)(1 + \sqrt{1 + 25})} = 150 \text{ Nm}$$

$$\tau = \frac{J_{eq}\,\omega_s}{T_{max}} = \frac{4\left(2\pi \frac{1200}{60}\right)}{150} = 3.36 \text{ sec}$$

$$s_{max} = \frac{R'_2}{\sqrt{R_1^2 + X_{eq}^2}} = \frac{1}{\sqrt{1 + 25}} = 0.196$$

The slip at starting is equal to 1. A good approximation is to assume that the final slip at no load is about 2%. The starting time in this case is

$$t_{st} = \frac{\tau}{2K}\left[\frac{1 - s_2^2}{2\,s_{max}} + s_{max}\ln\frac{1}{s_2} + 2\,s_{max}\,(1 - s_2)\right]$$

$$t_{st} \approx \frac{\tau}{K}\left(\frac{0.25}{s_{max}} + 1.95\,s_{max} + s_{max}\right) = 5.2 \text{ sec}$$

EXAMPLE 11.7

Assuming that the motor in Example 11.6 is unloaded and a starting resistance R_{add} of 1 Ω is inserted in the rotor circuit, compute the starting time of the induction machine.

SOLUTION

The starting resistance increases the magnitude of s_{max} but not T_{max}:

$$s_{max} = \frac{R'_2 + R'_{add}}{\sqrt{R_1^2 + X_{eq}^2}} = \frac{2}{\sqrt{1 + 25}} = 0.392$$

Although it is slightly higher, we can still assume that the slip at steady state is 2%.

$$t_{st} \approx \frac{\tau}{K}\left[\frac{0.25}{0.392} + 1.95(0.392) + 0.392\right] = 4.34 \text{ sec}$$

Note that the starting time is shorter when the starting resistance is added to the rotor of the induction motor. This is due to the higher starting torque and the lower final speed.

EXAMPLE 11.8

A countercurrent braking is applied to the induction motor in Example 11.6. Compute the magnitude of the motor voltage that limits the traveling time (time taken to stop the motor) for countercurrent braking to 15 sec.

SOLUTION

When the voltage sequence of the motor is reversed and the voltage magnitude is reduced, the motor operates in the second and third quadrants, as shown in Figure 11.10. If we assume that the reversal of sequence and the change in voltage are done simultaneously, the motor initial operating point moves from point 1 to point 2. The initial slip at point 2 is almost equal to 2 since the field reversed its rotation, but the motor rotation is still in the same direction as it was before braking. Although s_{max} remained unchanged, the magnitude of the maximum torque is reduced due to the reduction of the terminal voltage. The braking time t_{br} is then

$$t_{br} = \frac{\tau}{2}\left[\frac{s_1^2 - s_2^2}{2\, s_{max}} + s_{max} \ln \frac{s_1}{s_2} + 2\, s_{max}(s_1 - s_2)\right]$$

$$t_{br} = \frac{\tau}{2}\left[\frac{4 - 1}{2(0.196)} + 0.136 + 2(0.196)(2 - 1)\right] = 3.42\,\tau$$

$$\tau = \frac{15}{3.42} = 4.39 \text{ sec}$$

We can now use τ to compute T_{max}

$$T_{max} = \frac{J\omega_s}{\tau} = \frac{4\left(2\pi \frac{1200}{60}\right)}{4.39} = 114.5 \text{ Nm}$$

Since the maximum torque is proportional to the square of the voltage,

$$\frac{114.5}{150} = \left(\frac{V_{br}}{480}\right)^2$$

where V_{br} is the motor voltage during braking.

$$V_{br} = 419 \text{ V}$$

FIGURE 11.10
Countercurrent braking

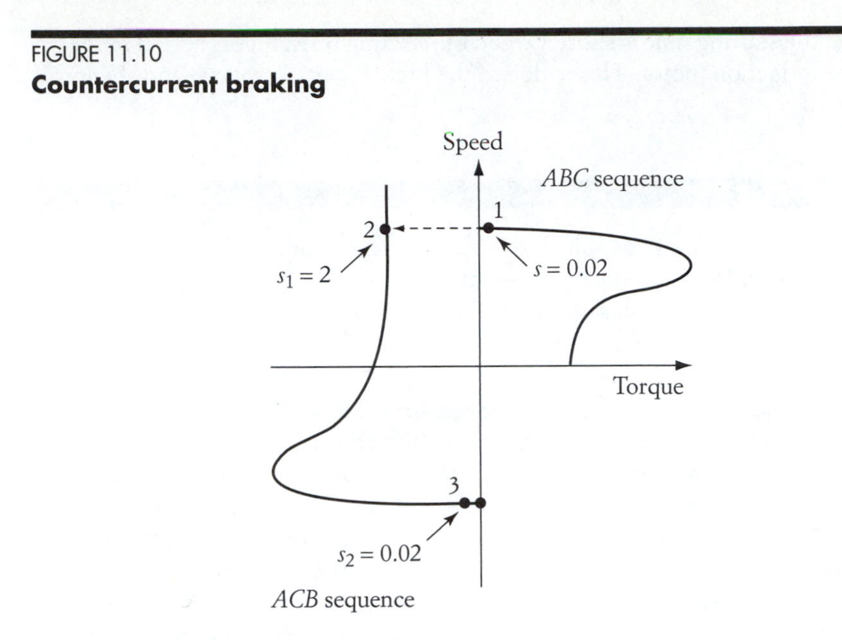

EXAMPLE 11.9

A countercurrent braking is applied to stop the induction motor in Example 11.6 by reversing the terminal voltage and simultaneously inserting a resistance in the rotor circuit. Compute the value of the braking resistance that minimizes the braking time.

SOLUTION

When the voltage sequence of the motor is reversed and a resistance is added to the rotor circuit, the motor operates in the second and third quadrants, as shown in Figure 11.11. If we assume that the reversal of sequence and insertion of rotor resistance are done at the same time, the motor initial operating point is point 2, and the final operating point is point 3. The slip at point 2 is almost equal to 2, as explained in the previous example. The braking time t_{br} is then

$$t_{br} = \frac{\tau}{2K}\left[\frac{s_1^2 - s_2^2}{2\, s_{max}} + s_{max}\,\ln\frac{s_1}{s_2} + 2\, s_{max}(s_1 - s_2)\right]$$

To minimize the braking time, the derivative of this equation must be set equal to zero.

$$\frac{dt_{br}}{ds_{max}} = \frac{\tau}{2K}\left[\frac{s_2^2 - s_1^2}{2\, s_{max}^2} + \ln\frac{s_1}{s_2} + 2(s_1 - s_2)\right] = 0$$

or

$$\frac{-3}{2s_{max}^2} + 2.693 = 0$$

$$s_{max} = 0.75$$

FIGURE 11.11
Countercurrent braking with resistance added to rotor circuit

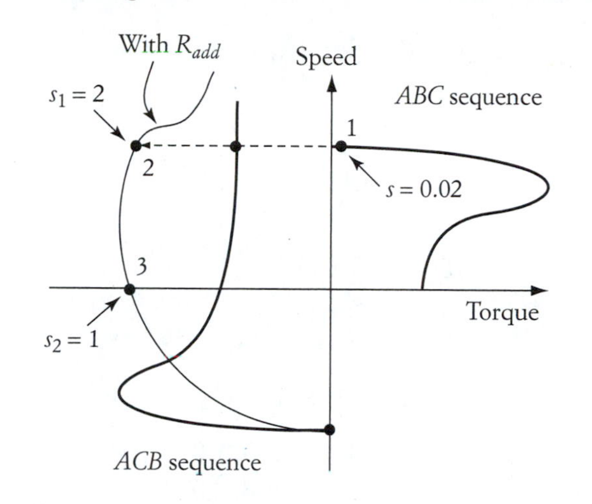

The value of the added resistance is computed using Equation (5.61).

$$s_{max} = \frac{R'_2 + R'_{add}}{\sqrt{R_1^2 + X_{eq}^2}}$$

$$0.75 = \frac{1 + R'_{add}}{\sqrt{26}}$$

$$R_{add} = 2.8 \, \Omega$$

In this case, $t_{br} = 4.5$ sec.

11.5.2 LOADED INDUCTION MOTOR

The solution of Equation (11.27) is more involved when the load torque is present. If we assume that the load torque is independent of speed, the solution of Equation (11.27) takes the following form:

$$t = \int_{s_1}^{s_2} \frac{\tau S_R}{S_R \, T_R - 2K} \, ds = \int_{s_1}^{s_2} \frac{\tau\left(\frac{s}{s_{max}} + \frac{s_{max}}{s} + 2 \, s_{max}\right)}{T_R\left(\frac{s}{s_{max}} + \frac{s_{max}}{s} + 2 \, s_{max}\right) - 2K} \, ds$$

$$t = \frac{\tau}{T_R} \int_{s_1}^{s_2} \frac{(s^2 + s_{max}^2 + 2 \, s_{max}^2 \, s)}{(s^2 + s_{max}^2 + 2s_{max}^2 \, s) - \frac{2K \, s_{max} \, s}{T_R}} \, ds$$

The solution for the starting time of the loaded induction motor in the previous equation is

$$t_{st} = \frac{\tau}{T_R}[1 - (0.5\,A - s_{max}^2)\,(m\,AD + \log(mB))] \tag{11.29}$$

where

$$A = 2\left(s_{max}^2 - \frac{K\,s_{max}}{T_R}\right)$$

$$D = \frac{-2}{\sqrt{Q}}\tanh^{-1}\frac{m(2 + A)}{\sqrt{Q}}$$

$$Q = A^2 - 4\,s_{max}^2$$

$$B = 1 + A + s_{max}^2$$

$$m = \begin{cases} 1 & \text{if } B \geq 0 \\ 0 & \text{if } B < 0 \end{cases}$$

EXAMPLE 11.10

Compute the starting time of the induction machine in Example 11.6, assuming that the load torque is constant and equal to 60 Nm.

SOLUTION

The first step is to compute the constant T_R.

$$T_R = \frac{T_l}{T_{max}} = 0.4$$

$$A = 2\left(s_{max}^2 - \frac{K\,s_{max}}{T_R}\right) = 2\left(0.196^2 - \frac{1.196 \times 0.196}{0.4}\right) = -1.1$$

$$Q = A^2 - 4\,s_{max}^2 = 1.056$$

$$B = 1 + A + s_{max}^2 = -0.0616$$

$$D = \frac{-2}{\sqrt{Q}}\tanh^{-1}\frac{\mathrm{m}(2 + A)}{\sqrt{Q}} = 2.642$$

$$t_{st} = \frac{\tau}{T_R}[1 - (0.5\,A - s_{max}^2)\,(m\,AD + \log\,(mB))]$$

$$= \frac{3.36}{0.4}[1 - (-0.55 - 0.196^2)\,(1.1 \times 2.642 + \log\,(0.0616)] = 16.8 \text{ sec}$$

Note that the starting time with load is more than twice the no-load case computed in Example 11.6.

11.6 TRAVELING TIME OF SYNCHRONOUS MOTORS

The expression of a rotating mass is given in Equation (11.1). The expression is expanded as shown in Equation (11.30) to include system damping D.

$$J_{eq}\frac{d\omega}{dt} + D\omega = T_d - T_l \tag{11.30}$$

where J_{eq} is the equivalent inertia of the system including the load, the motor, and any gear or belt system. D is the system damping due to friction, windage, or damper windings that might be present in the rotor circuit. T_d is the developed torque of the motor, and T_l is the load torque. The developed torque of the synchronous machine is given by Equation (5.73). The equation is repeated here:

$$T_d = \frac{P}{\omega_s} = \frac{3}{\omega_s}\frac{V_t E_f}{X_s}\sin\delta = K\sin\delta \tag{11.31}$$

where V_t and E_f are phase quantities, and δ is known as the torque angle or power angle. Substituting the torque equation in (11.31) into (11.30) yields

$$J_{eq}\frac{d\omega}{dt} + D\omega = K\sin\delta - T_l \tag{11.32}$$

The relationship between δ and ω can be explained by examining Figure 11.12. In the figure, the synchronous motor is connected to a fixed-frequency system. The frequency of the system is producing a field in the airgap rotating at the synchronous speed ω_s. Meanwhile, the motor shaft is rotating at an angular speed ω. Both speeds are equal at the steady state, but during starting, braking, or speed control, the speed of the shaft differs from the synchronous speed.

FIGURE 11.12
Synchronous motor connected to a fixed-frequency system

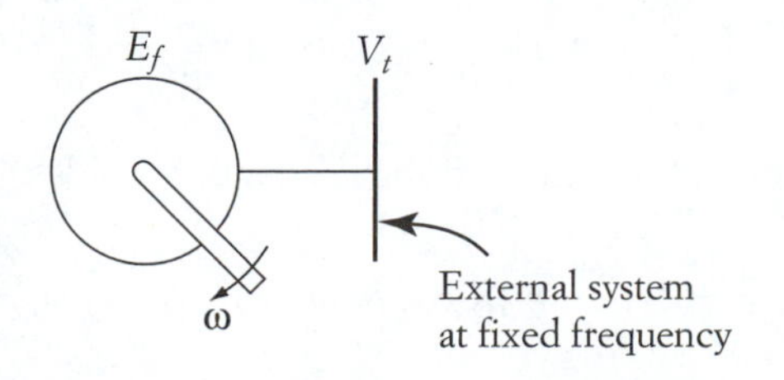

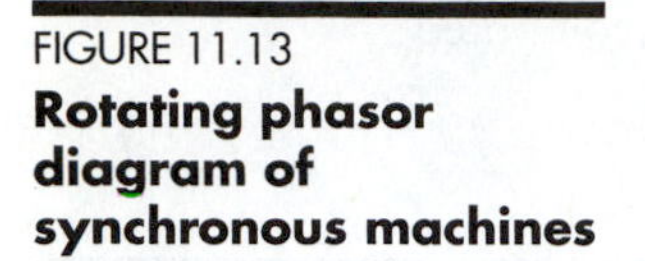

FIGURE 11.13
Rotating phasor diagram of synchronous machines

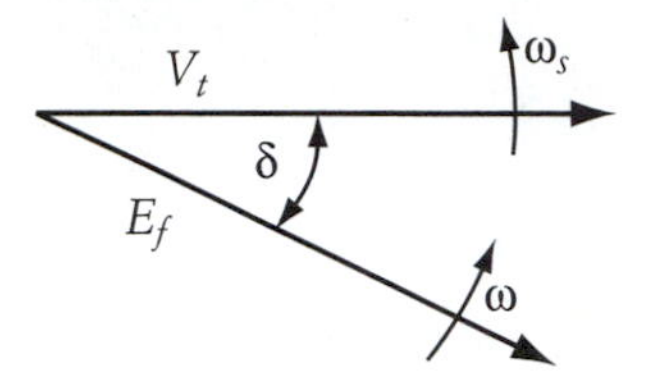

The phasor diagram of Figure 5.39 is modified in Figure 11.13 to show the effect of the speeds on δ. Keep in mind that the vector in any phasor diagram rotates at its own frequency. The frequency is seldom shown in phasor diagrams because all variables are assumed to have the same frequency during steady state, which is not the case here. The stator voltage, which is connected to the constant-frequency utility system, rotates at the frequency that corresponds to ω_s. The rotor equivalent field voltage E_f is also rotating, but at the frequency of the rotor circuit ω. Hence, δ is a function of the difference between the two speeds.

$$\delta = f(\omega_s - \omega)$$

or, in an explicit form,

$$\frac{d\delta}{dt} = (\omega_s - \omega) \tag{11.33}$$

Hence,

$$\delta = \omega_s t - \int \omega \, dt \tag{11.34}$$

Equations (11.32) and (11.34) form a nonlinear model for the traveling time of the synchronous motor. If we also assume that the load torque is speed dependent, we can rewrite Equation (11.32) as

$$J_{eq} \frac{d\omega}{dt} + D\omega = K \sin\left(\omega_s t - \int \omega \, dt\right) - T_l(\omega) \tag{11.35}$$

Even if we assume that the synchronous speed, load torque, and all voltages are constant, Equation (11.35) cannot be solved easily without using numerical methods.

For small changes in speed, we can approximate the solution of Equation (11.32) by assuming that the variation in δ is small. If this approximation is valid, we can arrive at an approximate solution in a few steps.

Let us find the second derivative of Equation (11.32).

$$J_{eq} \frac{d^2\omega}{dt^2} + D \frac{d\omega}{dt} = (K \cos \delta) \frac{d\delta}{dt} - \frac{dT_l}{dt}$$

Substituting equation (11.33) into the above equation yields

$$J_{eq} \frac{d^2\omega}{dt^2} + D \frac{d\omega}{dt} + (K \cos \delta)\, \omega = (K \cos \delta)\, \omega_s - \frac{dT_l}{dt}$$

Rewrite the equation as

$$\frac{d^2\omega}{dt^2} + \frac{D}{J_{eq}}\frac{d\omega}{dt} + \frac{K\cos\delta}{J_{eq}}\omega = \frac{\omega_s K\cos\delta}{J_{eq}} - \frac{1}{J_{eq}}\frac{dT_l}{dt} \qquad (11.36)$$

If the motor operates at a relatively small value of δ, a little variation in δ does not affect the value of (cos δ). Also, if we assume that the load torque is constant, $dT_1/dt = 0$.

$$\frac{d^2\omega}{dt^2} + \frac{D}{J_{eq}}\frac{d\omega}{dt} + \frac{K\cos\delta}{J_{eq}}\omega = \frac{\omega_s K\cos\delta}{J_{eq}}$$

This equation can be written in the popular second-order form used often by control engineers:

$$\frac{d^2\omega}{dt^2} + 2\,\xi\omega_n\frac{d\omega}{dt} + \omega_n^2\,\omega = C$$

where

$$\omega_n = \sqrt{\frac{K\cos\delta}{J_{eq}}}$$

$$\xi = \frac{D}{2\sqrt{J_{eq}K\cos\delta}}$$

$$C = \frac{\omega_s K\cos\delta}{J_{eq}} = \omega_s\,\omega_n^2$$

The solution is in the form

$$\omega = \omega_f\left[1 - \frac{e^{-\xi\omega_n t}}{\sqrt{1-\xi^2}}\sin\left(\omega_n\sqrt{1-\xi^2}\,t + \theta\right)\right] + \omega_i\left[\frac{e^{-\xi\omega_n t}}{\sqrt{1-\xi^2}}\sin\left(\omega_n\sqrt{1-\xi^2}\,t + \theta\right)\right] \qquad (11.37)$$

where

$$\theta = \cos^{-1}\xi$$

$$\omega_i = \omega(t=0); \quad \text{initial speed}$$

$$\omega_f = \omega(t=t_f); \quad \text{final speed}$$

CHAPTER 11 PROBLEMS

11.1 The system in Figure 11.2 consists of a wheel driven by a dc motor. The wheel is driving a piston pump. The system mass is 10 kg and the wheel radius is

1 meter. The length of the piston rod is 3 meters. Compute the maximum and minimum values of the system's moment of inertia.

11.2 A dc, separately excited motor is rated at 10 hp and 300 V. The rotor resistance of the motor is 1 Ω, with an inductance of 12 mH. The field constant ($K\phi$) of the motor is 4 V sec. The inertia of the entire system is 10 Nm sec^2, and the load torque is constant and equal to 50 Nm. The motor rotates at 1500 rpm. If the voltage of the motor is reduced to 200 V, compute the new steady-state speed. Also, plot the speed of the motor as a function of time during the transition.

11.3 A 500 V, dc shunt motor has the following parameters:

$$K\phi = 3.0 \text{ V sec} \qquad R_a = 1\ \Omega \qquad L_a = 10 \text{ mH}$$

The load connected to the motor is a constant torque of 20 Nm. The total moment of inertia of the entire drive system is 4 Nm sec^2. The voltage of the motor at starting is 150 V.

a. Calculate and plot the motor speed versus time.
b. Calculate the motor speed after 5 seconds.
c. Calculate the starting time.
d. The motor reaches the steady-state speed and the terminal voltage is at full value. To decrease the motor speed, the terminal voltage is reduced by 10% while the field remains constant. Calculate the traveling time.

11.4 A 480 V, 60 Hz, three-phase induction motor has a rated speed at full load of 1720 rpm, stator resistance of 1 Ω, rotor resistance of 1 Ω, and equivalent winding reactance of 4 Ω. Compute the starting time of the motor at no load and at full voltage and frequency. The inertia of the motor is 3 Nm sec^2.

11.5 A 20 hp, 480 V, 60 Hz, three-phase induction motor has a rated speed at full load of 3500 rpm, stator resistance of 1 Ω, rotor resistance of 1 Ω, and equivalent winding reactance of 5 Ω. Compute the starting time of the motor at full load and at full voltage and frequency. Assume that the load torque is constant. The inertia of the motor is 3 Nm sec^2.

11.6 Compute the countercurrent traveling time of the induction machine in Problem 11.4, assuming that the motor is running at no load.

11.7 A four-pole induction motor with the following parameters is running at no load:

$$T_{max} = 30 \text{ Nm} \qquad R_1 = 0.2\ \Omega \qquad R'_2 = 0.5\ \Omega$$

$$X_{eq} = 1.5\ \Omega \qquad J_{eq} = 1.4 \text{ Nm sec}^2$$

a. At starting, a resistance equal to the rotor resistance is added to the rotor circuit. Calculate the starting time.
b. At countercurrent braking, a resistance equal to half the rotor resistance is inserted in the rotor circuit. Compute the braking time.

11.8 A 480 V, six-pole, 60 Hz, three-phase synchronous motor has a synchronous reactance of 5 Ω and equivalent field voltage of 520 V. At steady state, the load torque is 300 Nm; assume the load torque to be constant. The equivalent inertia of the entire system is 4 Nm sec^2. The damping of the system D is 0.5 Nm sec. If the frequency of the motor supply decreased to 50 Hz, compute the settling time by using a numerical solution for Equation (11.35).

11.9 The motor in Problem 11.8 is driving a small-load torque of 10 Nm. Repeat the problem using Equation (11.37).

Index